Digital Physics and Gravitation;

Probabilistic, Relativistic, Kinetic, Entropic

Entropic

With Integer Physical Constants

S. M. Zoledziowski

&

Paul E. Platt

Digital Physics and Gravitation; Probabilistic, Relativistic, Kinetic, Entropic With Integer Physical Constants

Stanislaw (Stan) ZOLEDZIOWSKI is a retired academic from U.K.; the above unsolicited research was done in the period 2002-2013 at the author's residence in Alberta, T7X-1H2, Canada.

iUniverse books may be ordered through booksellers or by contacting:

iUniverse LLC
1663 Liberty Drive
Bloomington, IN 47403
www.iuniverse.com
1-800-Authors (1-800-288-4677)

ISBN: 978-1-4759-2190-8 (sc)

Printed in the United States of America

iUniverse rev. date: 04/25/2014

Preface

To prospective purchasers

For optimum benefit the readers should

  a) **Have a high interest in the source of our everyday reality**
  b) Understand the meaning of majority of common physical constants
  c) Be able to follow mathematical equations involving power indexes
  d) Be able to use simple science calculator to check calculations themselves
  e) Be able to find WEB pages on the computer for further explanations.

The book is aimed at people deeply interested in the fundamental physics and in gravitation willing on their own to make a judgment on the validity of the postulates proposed without the reference to the experts. For this reason most equations are provided with numbers so that readers can repeat calculations themselves. The elementary concepts postulated are relatively simple, so that the self-learners or high school graduates, who took advanced physics, should be able to follow the text with occasional reference to WEB pages.

The research progress provided the text with maximum evidence supporting the main postulates. Several simple derivations from different points of view were treated as proof of the validity of the postulates emerging. This made several disclosures spurious and the text long, but all alternatives had to be considered. Further conclusions are welcome from the readers.

Main idea postulated in the book

In macroscopic terms, at least at atomic size, the probability of a statistical event occurrence is the source of reality in physics and in gravitation. The event occurrence is controlled by the orthogonal geometry (six directions) of space, Ampere's Law, the rule of equipartition of energy and of conservation of energy and of momentum. The quantum uncertainty creates definite quanta in orientation and in magnitude. The event pulse must be of correct direction, it must induce the velocity of light and the input must provide sufficient energy to generate a quantum at the receiving end. The model splits the velocity of light into three orthogonal components and ten equivalent quantum orientations per oscillatory plane. The components of such split velocity of light corrected for relativity and for quantum derive a new set of integer physical constants. **The integer constants and the classical SI physical constants derive essentially the same energies and momenta. The model** provides also integer values to some other physical parameters like the masses of earth and of sun using the Special Relativity rules. The integer model may be considered as a relativistic presentation of physics with some other corrections.

Three books

Author's interest in the basic physics was raised in the middle 1960-ties by the discovery of stepped decrements in the logarithm of the relaxation current in an epoxy resin specimen excited by thousands of electrical impulses [1]. This together with local high fields conduction experiments displaying also quantum effects sustained author's interest for three decades generating several publication [2, 3, 4, 45, 46, 56, 57]. Reassessment of physics turned his interest in 2002 to gravity, when true value of gravitation lead to integer relativistic physical constants with local velocity of light. It took seven years to decipher alternatives of physics worth publishing the first research book [45]. The more compact second book took another year [46]. It has shown the same type of forces between atoms and large masses. The third book disclosed reasons for dimensional inconsistencies, attributed vectorized thermal forces for mutual attraction and repulsion based all events on probability arising from space geometry. It was not released for sale. The fourth book is the corrected third book with two new physical constants, disclosure of kinetic rules inducing new dimensions and the source of energy in the Universe. Although classical SI physics derive essentially correct energy values almost the same as the integer cubic space model, they are both based on incorrect fundamentals. These were found in the last 13st year of the research.

Method

Classically research follows experimental or well proven relations. The author did not want to be swayed at the very beginning by views expressed by famous scientists. Presently the starting positions were the quanta in relaxation and in the high field conduction experiments, the set of the international physical constants and inadequacies in physics. Any numerical fit found between parameters was followed by the search of the logic of the relation. Consequently the research deals mainly with the numerical data based on the strong belief that the particles in space do not understand any complex mathematics. For example the author was convinced and he has shown that a radiation particle carries clear information about its energy. Applying relativity and quantum correction clarified many problems. The spherical integer geometry with virtual particles was used in the first four years of the project, and then replaced with more effective integer cubic geometry. The natural units of physics were searched for several years until it was realized that all events are orientation, geometric and structural probability occurrences. Thus the final conclusions expanded the view that the world is probabilistic following the probabilistic waves explained by Max Bohr [55], supported by Albert Einstein and having uncertainties proposed by Werner Heisenberg.

Appeal for assessments

A dozen of years of research were spend by the author solely on quantum, relation, high field conduction and gravitation generating an important advancement of science, possibly with minor errors or misinterpretations. Major advancements or changes in science are not welcome in the orthodoxy of the old Universities and imperfect corrections are treated as heresy and forgotten for decades. Truth can be a disappointing even destructive for many, but it is still needed.

Because of frail health of the author and of his age, he is searching researchers for rapid multi-discipline assessment of his postulates from different fields of science. Such views would be included in the next reprint of the book. Contributions would be fully acknowledged. Discussers are authorized to use short extracts of the text for publication on WEB pages and researchers will be encouraged to follow author steps.

Acknowledgments

Prof. J. H. Calderwood is thanked for earlier support and for his direction made 50 years ago to include in research NaCl crystals, because the uncovered later features of quantum and of Bohr atom in NaCl formed the bases of gravitation findings. The Library of the University of Alberta is thanked for the access extending over 28 years.

Contents

List of Figures

List of Tables

1. Introduction

In the last few decades digital world has arrived; it has been disclosed, uncovered, invented, created, formed in music, in television, in computers, in telephones, in communications and possibly in other fields. Everybody can see it every day even in their ten fingers and their ten toes. Nevertheless the applicability of the restricted ten orientations and of size of quanta for the whole Universe is not generally apprehended. Physics uses many terms of physical constants to describe the reality; these terms express only slightly different aspects characterizing reality, but in fact they all refer to the same simplicity of the concept expressed by the digital world. **The model proposed deals with particles in space discovered to move with the average velocity 1/3 c_s; this is feasible because velocity expresses also pulse rate. The interacting radiations have smaller mass and momentum than the classical SI particles, because a fraction of source radiation does not interact with the receiver. These interacting particles move with the local velocity of light within atoms or molecules requiring well known relativity corrections. Simple probability rules of space applied to interacting radiations derive integer values of the physical constants.**

The rule of the 2nd Law of thermo-dynamics requires equalization of energy minimizing total energy, of conservation of momentum in all degrees of freedom and of maximizing entropy. This rule has led the author to derive several dozens of expressions enumerating accurately the earth gravitation listed in Table 5 of Section 19. These expressions are valid, because of overlap of the meaning of different physical expressions. In spite of only two opposing forces being needed to balance gravity between the sun and the earth the number of the gravitation expressions derived was huge.

Bose-Einstein radiation distribution is not considered in detail in the model. The only values considered in the calculations are the r.m.s. or mean energies and momenta of pulses of electromagnetic spectrum and of their frequencies at 6000 K, 300 K and 1 K. The model proposed by the author is based on two mean basic experimentally derived natural frequencies of vibrations of the electron of 5×10^{15} Hz and of the proton of 5×10^{12} Hz referred to 300 K, and their local values, without considering their normal distributions. Several relations uncovered are spurious, but all had to be investigated, because other researchers may draw different conclusions from them. Research lasting a dozen years indicates presence within atoms of local light velocity requiring specified relativity and quantum corrections establishing integer values of physical constants. In Table 1 the atomic values were converted to integer SI values listed now together with equally valid classical SI constants. They displayed several new aspects of the digital world of space physics and of gravity based on probability derived from the static geometry of space and on laws of physics. The book analyses Bohr atom on earth and its aggregates on sun. It assesses correctness of applying digitalization to general physics by applying the integer values of constants to numerous laws of physics and by deriving the digital physical constants directly from the Planck units.

How did this research start? In his 70 years of electrical engineering experience the author was always looking for a good numerical agreement with the basic science. Consistent numerical agreement with physical data using different formulations of the problem is the best verification of hypotheses. The author's studies of high field aging

[1], relaxation [2, 3], and high field conduction [4] and several others displayed details of forces in the basic physics and derived several physical constants. Nevertheless physics in author's eyes was incomplete, because some experimental data could not be explained; the common statement that the square of velocity converts mass to energy has to be revised, because mass cannot move at the velocity of light. Gravity pulse has mass.

The relaxation experiments of the author displayed elementary dipoles formed by a charged divalent impurity and a charged vacancy in a NaCl crystal. They modeled the basic Bohr atom structure. Relaxation experiments generated time resolved spectroscopy. While studying aging of dielectrics [1, 2, 3, 4] the logarithm of the relaxation current displayed a series of straight lines of slope decreasing in steps, displaying time constants decrementing by a factor of about two. Such value was also suggested by the Nobelist P. Anderson [11]. The dissociation energy of a dipole impurity (structure formed from a positive and negative charge) can be directly related to the Bohr atom.

The related central field of Bohr atom in its basic concept is common to all physics displaying the same features independently of its size in different media. Modeling for some decades Bohr atom structure using divalent impurities in NaCl crystals uncovered several important features, which could be applied to sun space radiation sourced from hydrogen. It took decades to realize the importance of an earlier finding. It was observed that a particle (a vacancy) with one degree of freedom had a slightly a larger energy, $0.6931 \ldots kT = kT \times ln2 = 2/3 \times 1.04 \times kT$, than the classical expression $\frac{1}{2} kT$. Measured energy expressed in electron volts of the classical SI units is 0.0179 eV versa the expected value 0.0129 eV. The measured value can be expressed by the squared relativity corrector 1.1547^2 for velocity $\frac{1}{2} \times c_s$ and the quantum corrector, ~ 1.04, deriving the value: $1/2kT \times 1.1547^2 \times 1.04 = 0.6926 \ldots kT$. Thus presence of a local velocity $\frac{1}{2} \times c_s$ in electrical field has a very reasonable experimental background. Velocity of light is the residue velocity of several vectorial sums of $1/4 \times c_s$ local velocities in orthogonal orientations. These data were covered earlier author's books [45, p.180, 3 and 2] and in Equation 6 of Appendix. Please note $1/4 \times c_s \times 1.1547^2 = 1/3 \times c_s$. The relation $1/3 \, kT \times 1.154700538^2 \times 1.1111 \ldots \times (1+1/80) = 1/2 \, kT$ is an example of accuracy obtained, when applying relativity corrections to an experiment. $1/80$ is really the uncertainty of Planck radiation term 0.01 J $\times 1.1111^2$, while 0.04 in 1.04 are four 0.01 J terms because direction has not been specified.

Initially local velocities were considered to be located within the atom, but it is better to consider them to be the components of light velocity in three orthogonal directions, because of independence of their occurrences in these orientations. Local velocities are affecting the values of physical constants. Different directions and velocities were later applied to the generation of gravity vectors from thermal phonons; presently the concepts developed for gravity are applied to the atomic structures. The first book by the author [45] introduces the integer concepts; it was really a research report searching for corrections to physics with many questions remaining unanswered. The second book [46] proposed a more specific model of gravitation based on one particle only, the proton converted to electrons. The book traces events within the atom from the values of physical constants corrected by relativity, while High Energy Physics does it from breaking the atoms. The present book uncovered that the Universe's parameters are in nature all probabilistic and they are all likely to have common origin, without initially displaying magnetic and electrical features. Various quantum units at atomic level are subject to probability occurrences of unity having been differently formed. The data presented demonstrate the mechanism by which several local velocities of radiation particles create the velocity of light after applying corrections involving relativity and quantum. While the value of the international gravity constant G applies well to the sun-to-earth and to similar systems, the value for large bodies at equal temperatures is different. For forces between particles of Bohr atom and for large bodies at equal temperatures the gravity constant has the value $G_s = 10^{-10}\,\text{N-m}^2/\text{kg}^2$.

The book deals primarily with interactions obeying Bose-Einstein Statistics with local velocities of light requiring physical constants to be subjected to some unusual effects caused by the Special Relativity Theory rules. Nature eludes us because all SI measurements in Physics relate to particles moving with velocity of light or its very small fraction, while physical constants proposed in the model refer to local velocities of light. The proposed presence of local light velocities introduces numerous corrections to several expressions. After corrections, this approach uncovered a set of physical constants, which are nearly all integer values. They represent geometry of space and accurate probability of creating specific events. They are derived from Plank's units in the same way as the classical SI units are obtained.

Many units in scientific use, including atomic units are not SI units, but designated BIPM and accepted for use with SI units. The book does not differentiate them and treats them as SI units. The model postulated is derived from the digital geometry of space and the gravity force has been found to be a derivative of other forces. The data demonstrate that **Bohr atom has three independent oscillators**, not two used often in the classical SI system (International System), the electron and the proton. The use of two oscillators is the cause of the prolong difficulties in understanding the gravity problems.

The present state of physics is extremely well described by the recent book by S. Hawking [63]. The author worked at establishing his own view of physics putting aside details of persuasive interpretations by Einstein, Schrodinger, Heisenberg, Hawking or Bohr. The research deals with numbers and the probabilistic aspects of the geometry of space. The model has been derived by simply guessing very acceptable values for a wide range of phenomena in physics including gravitation. These are events existing without the intervention of an observer. Uncertainties of single experiments do not apply to the whole world. The author believes that there is clear reality, cause, effect, probability and uncertainty. Research presented here indicates that the world is very exact for very long time returning to the mechanistic concepts believed by Kepler before the arrival of quantum mechanics. Most important deductions were arrived after researching the topic for over 10 years, with all significant analyses carried in that period remaining in the text.

The book provides more evidence validating the model for atomic interactions, testing it by well known laws of physics and by simple explanations of the principles involved. It is suitable for readers deeply interested in science and in the physical reality, because the number of simple equations used may be too tedious for others.

In the classical SI system the world is analyzed accurately in the way as it appears. In the integer model distance and time for particles moving at velocities observed by different observers need corrections according to the rules of Special Relativity. At the basic level there is only one particle, the electron moving with local velocity in three orthogonal directions. Proton is its creation. Really it remains the egg and hen question. Electron is subjected to space quantization and to relativity and to quantum corrections arising from the local velocity of light. This follows equal mathematical probability options for the linear velocity of motion along the axis established by the source and the observer. According to the Special Theory of Relativity particles having rest mass cannot reach velocity of light. Since energy and so mass is transferred from the sun the particle must move with local velocities smaller than the velocity of light c. The electron natural frequency of oscillation plays little role in the model of gravitation.

Integer values of physical constants require existence of a static electron charge e; a cubic space charge value e_s, is equal about 2/3 of the classical SI e value after the application of relativity and quantum correctors listed subsequently:

$$e_s \approx 2/3 \times e/\text{correctors} = 2/3 \times 1.6022 \times 10^{-19}/1.1111^{1/2} \times 1.04^{1/3} = \mathbf{1.0001 \times 10^{-19}} \text{ C} \qquad (1.1)$$

Due to equivalency of units the error expresses uncertainty of Boltzmann constant k_s.

To validate an event the science requires energy and momentum to have the same value for all observers including the stationary ones. The range of events in nature atomic units is limited. They comprise electron moments converted to proton and vice versa with statistics and probabilities expressing orientations, repetitions rates, density rates and velocities.

The simple kinetic interactions between particles, are expressed in 1D by balls on a pool table. Data presented displays them as kinetic in character, and as such they have no dimensions of length or of time. This is claimed to be the feature of all atomic physics. Hence the classical SI system has its limitations in the same way as the integer cubic model in SI presented in the book. In elastic kinetics the variables of time and of length are expressed as unity parameters. The integer values of physical constants with their cubic geometry quoted in Table 1 was necessary to uncover and clarify several experimental results.

Expressing atomic kinetic interaction in terms of SI units requires adding unity units in equations to fit dimensions or replacing the sign of equality by equivalence symbol, $\equiv$, hiding this way dimension inequality. Absence of dimensions of time and of distance establish the electron charge to be e_s = (atomic volume converter 1×10^{-30})/ε_{os} = 1×10^{-19} C. It also creates 1/10 equivalence between atomic and SI numerical terms. Hence:

Unity event $= 1$ eV $\equiv 1$ W $\equiv 1$ N $\equiv 1$ J $\equiv 1$ W/m $\equiv 1$ N/m $\equiv 1$ J/m $\equiv 1$ W/m^2 $\equiv 1$ N/m^2 $\equiv 1$ J/m^2 (1.2)

Above unity event generating unity signal per orthogonal orientation is at 300 K the vectorized Planck radiation term **0.01** J, which over average velocity of three particles of radiation derives the gravitation constant G_s expressing also energy $0.01/(1/3 \, c_s)$ = $\mathbf{10^{-10}}$ J-s per kelvin.

The true relations are expressed in atomic terms. Rotation converts electron volt units of atomic cubic integer model into joules by multiplying the product of the size converter 10^{20} and electron charge 10^{-19} derives factor of ten. Hence a 10 J SI event is **1 J** atomic event, as in c. g. s. system with unit current 10 Amperes.

Physical constants are normalized and so usually the energies derived from them are per kelvin. Energies derived from the sun radiation are measured on earth and so they refer to 300 K.

The energy density value on earth at 300 K, published on WEB, peaks at 1.55×10^{-5} W/cm² representing for six orientations of space corrected for relativity and for quantum the integer model value of

Equiv. black body radiation $= 6 \times 1.55 \times 10^{-5}$ W/cm²$\times 1.1111^{1/2} \times 1.04^{1/2} \times 10^4 \equiv 0.99972 \approx$ 1 W pulse, attributing 1 W/m² at 300 K and expressing unity probability of an event   (1.3)

The other value quoted on WEB is 4.8 W/m² expressing 1.2 W/m² for each of 4 directions reduced to 1 W/m² because in the model 20% of source radiations are not recorded by the receiver.

In comparison sun radiation at 6000 K (or 10000 K as indicated further in text) using integer space unit SI constants and using model units is expressed by:

Sun radiation power density $= \sigma_s T_s^4 \times 1/1.11111 \times 1.04^{0.12} = 3.000087 \times 10^9$ W/m²   (1.4)

The infrared quantum energy pulses measured on earth generated by sun power amount to 3 W, J, eV or 2 W, J or 2 eV, if electron contribution is neglected.

The Universe displays integer parameters, derived after valid corrections are applied. The values discussed may be only components of physical measurable values in direction; they may or not be relatively and quantum corrected; they may be mean values, r. m. s. values, per pulse, per second, per phase, per K, per 300 K, per 6000 K, per quantization pathway in one direction, per plane or in six orientations. Energies may be per orientation or per plane, in classical SI units, in integer model SI units or in natural atomic units. The interactions have to be looked at both in the atomic natural units and in SI units to build a valid gravitation model. Calculated energy pulses can be physical entities, component of energies forming vectors, gravitation vectors of energy or multidirectional oscillations. This leads to a very large number of numerically accurate estimates of the earth gravitation force in Table 5 and for planets in other Tables, while only two values are needed. Different meaning in SI units may be expressed probably in one atomic nature statement. **The true relations are expressed in relations between the momenta of particles at the nature atomic level; they have to be converted to SI energy units.**

Energy in SI units of joules multiplied by e_s is expressed in electron volts and expresses energy per charge event; conversely energy in election volts divided by e_s is expressed in joules. The basic unit of length in SI unit of one meter is 1.0×10^{10} larger than the atomic size of Bohr atom, the unit of length at the atomic level. The linear SI size converts to atomic size by 1.0×10^{-10} multiplier, the size converter for area is 1.0×10^{-20} and for volume and energy it is 1.0×10^{-30}. Probabilities of energy events occurrence are subjected to volume probability. Events measurements at 300 K encompass the term T = 300 K in several expressions. Some SI parameters express atomic sizes, others require a conversion. At times energies have been subscribed with "SI" or "at" to identify them clearly. So unity values, 1 J atomic and 1 eV atomic are often referred to as 1 Jat and 1 eVat, while J in SI as JSI. **All analyses in the book derive essentially the same numerical energy results for the integer model and for the classical SI system, establishing veritable evidence for the integer relativity model validity.**

The fundamental pulses of 1×10^{-27} J are responsible for balancing gravitation between earth and sun. Pulses are attributed with rate [K⁴] units of energy [J], mass [kg], power [W], power density [W/m²], momentum [kg-m/sec] or force [N].

Several phenomena in gravity and in physics textbooks are explained by the well accepted space quantization by Stern and Gerlach introduced in 1922. It is shown on Figure 1, for $l = 2$ showing that due to quantum requirement radiation traveling along one plane can take place only in one of five orientations, if the source-receiver propagation axis is established. It minimizes the event to the value of Planck constant h_s restricting its appearance to five directions. Equal moments are radiated in the five directions of the plane and 1/5 of the momentum appears in the forward direction. The source radiates initially momentum in two opposite directions along the established source-receiver axes. It is now suggested that usually a total reflection takes place from the back mass the reflected pulse converting the generated chains of waves into a single pulse. Figure 2 with 5 Planck constants h_s in each direction expresses energy or momentum input of 10 h_s for 1 h_s in the output direction establishing the probability of the event at 1/10 for one interaction in any three planes. To produce a vector requires sequential orientations and the formation of the total probability of an event due to space quantization expressed by the product of probability of the events in two planes, 1/10 x1/10, as in Equation 10.1 or in three planes, 1/10x1/10x1/10, as in Equation 2.5.

The signal carried in the beam along the source receiver axis comprises 5 (or 10 if reflections are included) linear oscillations, the cosine components of the original radially directed 3D atomic radiations. The average or r. m. s. sun energy or linear momentum of these vectorized oscillations expresses 1 J ≡ 1 eV ≡ unity probability reference per plane value at 300 K on earth ≡ graviton pulse 1×10^{-10} J at 300 K at atomic level. Earth generates **2 J** for 3 planes in the absence of pure electron oscillations at 300K. The unity value of **1 J** is derived from integer Stephan-Boltzmann constant 5×10^{-8} divided by 5 quantum orientations of Stern and Gerlach and multiplied by local velocity of light 1×10^8 m/s

Gravity radiation in the integer model from earth is attributed to naturally distributed longevity of proton caused by rotation of earth around the sun. All atomic energies and moments contribute to the formation of the one dimensional signal proceeding along the source-receiver axis, with exception of those at right angle to the above axis. Input energies contributing to the signal in one plane are decreased by factor 1.2 expressing input fraction 0.8333 . . ., inputs in two planes by factor 1.2^2 and fraction 0.6944 . . ., and inputs in three planes by factor 1.2^3 and fraction 0.57877037 . . . Such reduction of radiation energy is more simply considered in the orthogonal axes than in the atomic spherical configuration. The formation of the radiation signal along the source-receiver axis can be now expressed in terms of one dimensional expression considering only components of photons that generate forces. The term reflection is used, because the pulses from the back surface of the atom change their phase and being delayed convert the classical SI wave generated in both directions into a unidirectional pulse.

The cubic approach has been used a century ago by in Reyreigh-Jeans formula in expressing infrared radiation. Now with the cube representing the Bohr atom its wavelength λ is twice the atom diameter 2×10^{-10}m. Bohr atom is covered centrally by $\frac{1}{2}\lambda$.

With three orthogonal directions corrected for relativity, the output of 1 eV requires the input of the integer model Hartree energy of 30 eV, the total energy of Bohr atom exactly, exceeding slightly the SI value of 27.2 . . . eV. They are related by quantum and relativity corrections.

If one parameter requires the multiple of 10^{-3} or of 10^{3}, the product of two parameters such as voltage and current displays values 10^{-6} or 10^{6}. Since each represents probability the product of three parameters displays multipliers 10^{-9} or 10^{9}. The orthogonal directions of space produce radiations vectors expressed at the atomic level by whole numbers. They start from the unity event of 1 eV, expressed equally well by other unity values; the unity event of the recorded signal times inverse probability for each orientation of the plane 10x10x10 and times 6 orientations of directions the total product form the sun temperature of 6000 K.

Let us consider the Maxwell pulse with both voltage (and electric field) and current (or magnetic field) moving in phase, as shown in several textbooks. Only mean or r. m. s. values of the pulses are considered without their distributions. Components in phase generate real power expressed by vectors, components out of phase produce oscillations generating photons and heat carried in the radiation pulse. Each of the vector pulses oscillates solely in the direction of progression and pulses are subjected to probabilities arising out of quantum restrictions shown in Fig.2. Elastic kinetic interactions between radiation particles establish dimensions displayed by the balls on a pool table. The probability that one of ten optional directions is directed to the receiver is 1/10; the probability of two sequential pulses discharging in the same direction is $1/10^{2}$ and the probability for all ten pulses is $1/10^{10}$. Although the changes proposed are radical in form, the integer model and the classical SI energies remain equal. The postulated model supplements the SI system with different observers, who discover the probabilistic and relativistic origin of the Universe.

Particles in High Energy Physics have a very short life. They proceed fast and need high energy to break the atom, before energy is divided between the different degrees of freedom. In contrast the world around us had billions of years to reach semi-stable equilibrium. Atomic force radiation output of 1 J is derived from radiation output of 1 J reduced by value of 1×10^{-10} of the international gravity constant G_{s}, in two dimensional displays $(\mathbf{1x10^{-10}})^{2/3}$ divided by the corrector of quantum and of relativity $K_{rel}^{1/2}$:

Atomic force $\boldsymbol{F_{ats}} \equiv \frac{1}{2}\mathbf{x}\ (1\mathbf{x}10^{-10})^{2/3}/(1.15470^{1/2}\mathbf{x}1.04^{1/20}) = \mathbf{1.0005\ x10^{-7}\ N} \approx \mathbf{1.000x10^{-7}\ N}$ (1.5)

For an electron circulating around the positive proton, at velocity much smaller than velocity of light, the force of attraction by the proton is $\mathbf{1x10^{-7}}$ N. This is measured by the observer moving towards the atom with velocity $\frac{1}{2}$ $\mathbf{x}$ c as explained by the author in his Reference [45]. **It is proposed that events occur with the respect of the local centre of energy, that may be the earth, the sun, the planet, their couple, the mass of the Universe, the proton or the radiation event**; these masses becoming the reference or the observers of the event.

Space quantization with rules of probability of the event occurrence in specified direction is applied to the geometry of the configuration with relativity and quantum corrections. Axes of the source-observer system are fixed by various options with the initial unity signal. Identity sign, ≡, is used, in equation deriving the new integer values or checking their validity. The value is often absolute and not subject to approximations. Dimensional display expresses sequential probability and a variety of power indexes applied to the parameters: ½, 1/3, ¼, 2/3, ¾ and 1/12. Some vectors are emphasized by **bold** printing.

The common factor in all three books is a new set of physical constants given in Table 1, which differs from author's earlier presentations [45, 46] in attributing integer values also to the mass of earth, to the mass of the sun, to the distance between them, to the linear earth velocity, adding two new modified electron Boltzmann and Planck parameters and missed nine integer physical constants. The model postulates that the elementary vectors of radiation are longitudinal pulses expressing vector forces transferred in one dimension (1D). The source may be rotational at the source, but only its longitudinal variations are transmitted to the receiver including the component of the orbital electron spin. These are not the classical stationary SI particles. The particles in space move at average velocity $1/3$ c_s of their original classical SI magnitudes of mass and momentum incremented by the relativity correction 1.1111 . . . or its derivative and by the radiation leaking at the source at right angle to the source-receiver axis. Most masses, energies and momenta are subjected by arbitrary chosen quantum corrections required to generate integer physical constants and parameters displaying measurement accuracy limits expressed by Boltzmann constant or its multiplier or by Heisenberg Uncertainty (H. u.) and the number of degrees of freedom called DOF. These are the physical constants and parameters of our visible reality based on elementary events expressing unity probability derived from the geometry of the orthogonal geometry of space. The third column in Table 1 lists some approximate correctors for classical SI units.

Cubic integer relativity corrected constants are essentially rounded up values of physical constants in SI units, which are attributed in several cases with the absorption factor of about 2/3. Cubic integer constants are subscribed with *s*, such as h_s, h_{s2D}, G_s, v_{Hms} and for sun-to-earth gravity 2/3 G_s. The System International constants, the classical terms h, G and v_{Hm}, listed also in Table 1, are derived from analyzing data from recognized sources. **The original quantum model derives their own atomic units** and dimensions, not SI units and the model atomic units require conversion to SI units.

Space quantization is demonstrated on videos for two orientations on WEB pages. The proposed probabilistic application of the ten space quantization to three planes of the proposed integer model of space is shown in Fig. 3. Sequential radiation discharges along the chosen source-receiver axes have equal chance of occurring in any sequence in any direction. To obtain a signal all thirty discharges have to occur in the same specific orientation subject to sequential probability rules and to quantum correction generating from the initial 30 J or 30 eV signals the minimum energy pulse progressing in one specific direction with variance of ~3/2 k_s:

$$(1/30)^{1/30} \times 1.111 \ldots \times 1.04^{1/5} = (1 - 0.00017) \text{ eV} \qquad (1.6)$$

In the absence of laboratory facilities the set of internationally recognized physical constants seemed an ideal starting data for retirement research, with spherical geometry being followed for the first 4 years of this research. To derive required relations with spherical geometry required using virtual energies for corrections [45, 46].

The reality is expressed by the measurements obtained from hydrogen atoms, which may differ from radiation generated by the sun. Other groups of radiation waves created by human ingenuity in different technologies are not discussed here. **The valid classic SI system with its limitation does not preclude presence of another slightly different valid system incorporating relativity, quantum and other corrections.**

SI physics has only a few exact expressions and exact space parameters and not so long ago radiation was represented by infinite waves. **The analyses of earlier author's experiments [3] concluded that eingen functions of the Maxwell equations for the non-steady state express pulses instead of monochromic waves.** This interpretation was first suggested to the author by GUO, T. and GUO W., who carried similar research [59]. Hence the proposed radiation current pulse I_o is a transient decreasing exponentially and having equal alternating and direct components in the display [4]. This allows expressing unidirectional gravity force, which is the force of atomic attraction, both by using oscillating or by direct values. Such pulse displayed as current I_{os} contains the same information about quantum event as a decrementing half of the wave function or wave packet. Modern physics replaces pulses by wave packets.

It is suggested that **the other half of the wave function is the reflection carrying no further information and bringing some confusion by introducing the infinite wave representation of a particle.** Proposed physical constants of the integer model express parameters of space power, of energy, of momentum, of frequency and of the relaxation time constant. They are all the derivatives of the wave function, which generates velocity of light. The amplitude of current I_{os} pulse is localized by the minima and the maxima of wave energy densities; the wave length defines particle size, so that there is no conflict between the wave and the particle representation. **The model also suggests that a constant flow of pulses displaying energy and force must have a source of energy.**

Energy theories produce valid, extensive and detailed results. However, integer energy equations may be generated by a number of differential equations and so these theories are an incomplete description of nature. To display quantum energy flow rate must be the same along the 1D route from the source to the receiver. To achieve equal rates particles must slow down or accelerate at the boundaries. The mathematics of the latter process displayed experimentally has been described by the author in [3, 5].

Quantum displacement relaxation current observed in NaCl crystal is the evidence that all the results in a solid dielectric are statistical in nature. This is confirmed by the results of Tamura et ales [7], who used well separated single electrons for their display. The author uncovered in classical gravitation a corruption of the values of the measured physical constants, caused among other factors, by the absorption of only 2/3 of the initial signal. This is caused by the absence of electron frequencies from the cool earth and an incorrect assessment of energies in the absorbed sun radiation pulse. The resonance limits interactions between the radiation and receiver to those of the same frequencies or sub-frequencies. This property is used extensively in analyses of atomic structures and of elements in analytical chemistry.

The trend to reach equipartition of energy among all degrees of freedom, representing **a semi-stable equilibrium of minimized energy, is considered in the current model to be the most significant activator of the particle and of the pulse conduct.** This is the 2-nd Law of Thermodynamics in action. Although in simple space the number of degrees of freedom per particle per K can be expressed by 17 the nature attempts to create more degrees of freedom to distribute its energy in a more efficient way. In atomic terms equipartition of energy means equipartition of momenta in all degrees of freedom, which at the end of the project was found to be a huge number.

The basic gravity forces are proportional to T^4 and are three dimensional. The earth force of gravity repulsion is caused by rotation and acceleration. It is one-dimensional and it comprises components of variables along only one interacting axis. The gravity forces of sun-to-earth, of earth-to-moon and of sun-to-planets were calculated in models using both SI constants and the proposed uncorrupted physical constants, the two producing accurate features of the gravitation physics.

For over ten years the research results presented in the book were and remained dimensionally incorrect until kinetics were introduced in the last few months. In the string theory, Smolin [8] and Schwarz [36] explained that dimensions were still missing and undefined in physics.

The proposed model of gravitation, otherwise gravity, involves a large number of simple concepts, equations and corrections, their number obscuring the basic principles of the model. Elliptical orbits of planets proposed by Ptolemy in the 2nd century and by Copernicus in the 16 century, are represented now by one circular distance L_{earth} between the planet and the sun. The elliptical paths of planets described in the three laws of Kepler are neglected in the model. They do not affect the proposed representation of gravitation by the physical constants. Similarly the elliptical paths of electrons in Bohr atom do not affect the average energy. The basic law of gravitation with the universal gravitation constant G was discovered by Newton in 1684. The value of G was first measured in 1798 by Cavendish in an experiment using a torsion balance, deriving the force of attraction between the two masses.

The **force of attraction, the gravitation, must be balanced by a force opposing it**. In the Cavendish experiment that force is provided by a spring. In the same way the gravity force between the celestial bodies must be balanced by force acting in the opposite direction. Until the gravitation of planets was considered in great detail everything pointed to the gravity force being opposed universally by the centrifugal force of the planet, rotating around the sun. Centrifugal force expresses linear kinetic energy of earth over distance and so it is the outward gradient of energy displayed round the circle with earth being the reface. However, several texts in physics treat centrifugal force as virtual, or acting in the central direction, calling it centripetal and equating it to the gravity. This view treats the sun as the reference.

Other expressions in the model indicated clearly that the ratio, of **the body mass to electron mass, represents the number N of electrons comprising the body.** Different reflections were studied uncovering accurate parities between various physical constants. These lead to several dozen of reasonable accurate identities predicting correct value of the force of sun to earth gravitation expressed in terms of currents created by the velocity of the mass of the body. The source of gravitation remained unclear until the end of the project. In that sense the three books presented form sequential research reports describing the improved comprehension of physics examining topics numerically step by step. The author believes in the paramount importance of accurate numerical description having one defined meaning and confirmations describing all possible aspects of the problem. Single equation has infinite number of multiplier solutions and can be manipulated. The author rejects the nihilistic view of indeterminacy expressed by some physicists. Indeterminacy is often expressed by the probability of energy density in the electron cloud around the proton. **Events are expressed in the book by unity probability, while non-events by an absolute zero.** We see around us an infinite number of events occurring each with probability of unity [39]. We cannot express them by fractions. That is statistics. Nonevents are not observable.

The postulated source to receiver geometry is deterministic. The event is measured with accuracy derived from Boltzmann constant or Heisenberg uncertainty (H. u.), the event is real and present, but its accuracy is limited. It is later postulated that measurement uncertainty may in some cases become huge, distorting our reality.

Although the terms attraction and repulsion are used in the texts, they are misnomers. **With two celestial bodies both are attracting each other until the time the reference body is chosen; at this point the forces of the other body change directions seemingly displaying a repelling force, but the two forces are directed toward each other. The body undergoing energy decrement is the proper description of the process of radiation causing attraction without introducing controversies.**

The **physics of Bohr atom express energies solely in the terms of the electron orbits, while the model deals with energies of electrons, protons and (proton+electron) pairs.** It took long time to find justification of the two apparently different presentations. The classical and the quantum physics are without any doubt essentially valid and accurate. New theory requires bringing the model to agreements with most presentations of the classical and of quantum theory. **It is suggested the classical atomic SI system uses often electron mass for the equivalent mass of electron and proton.** The integer model derived several valid relations indicating that at 1 K and at 300 K Planck constant h_s **applies to infrared frequencies of proton** and of (proton+electron). At the end of the book the proposed integer physical constants were applied to the established numerous rules of physics and were found to fit them well. Integer physical constants were also all derived from Planck units in the same way as some SI unit are.

Most scientists may consider it ridiculous to propose another set of physical constants in view of overhelming success of quantum physics. It is postulated now that classical SI values neglect relativity and quantum corrections obscuring several fundamental processes in physics. Values of the physical constants proposed are all product of the geometry of the space and of different probabilities. They are creating a variety of signal structures having all the same initial source. Electron momenta in the far past created protons and hydrogen; **now the vectotized infrared radiation of the solar system is attributted to gravitation.**

The total gravitation force displays Mandelbot (fractal) effect by showing as elementary infrared radiation of protons. Neglecting role of electrons limits the accuracy of proposed gravity model in the sun to earth planetary system to a few parts per million. The quantum value for the sun temperature 6000 K is the number of degrees of freedom of space (DOF) listed in Equation 2.64, essentially seventeen to power three, **expressing the inverse probability of the sun occurance of quantum unity event in one plane:**

Unity pulse from the sun mass $\equiv (2 \times 10^{30}\,\mathrm{kg})^{1/6000} \times 1/1.04^{1/4} = 1.00181$ JSI or 1 eVat   (1.7).

with deviation 18 k_s, which is $\sim 2k_s$ per each of ten quantum positions per plane (Figure 3).

The integer model SI quantum value per K is the model Boltzmann constant 1×10^{-23} J expressing 0.0001 J/C on division by e_s. The probability of unity event with 17 degrees of freedom is 1/17 per K. Each degree kelvin adds another degree of freedom so for 6000 K the probability is $1/17 \times 1/6000$, because each degree is a unity event or rate. The three planes introduce power of three to probability. Corrected for relativity and quantum the probability derives energy pulse appearing on earth from unity pulse on sun and expresses also the product of the model energy of Boltzmann constant and local velocity 1/3 c_s:

Probability per second of unity sun event appearing on earth $\equiv$
$k_s \times (1/3\ c_s) = (1/17 \times 1/6000)^3 \times (1.11\ldots)^{1/2} \times 1/1.04^{1/6} = 1.00019 \times 10^{-15} \approx 1 \times 10^{-15}$ J-m/s   (1.8)

Since probability of unity expreses 1 J, the above value expresses also WSI or eVat. Referred to 300 K the value becomes the minimumu energy pulse for 3 planes 3×10^{-13} J.

Integer values in Table 1 are based on the cubic space charge of electron **1×10^{-19}** C, derived from different interprettation of Millican experiment; electron moving at local velocity **1/3 c_s = 10^8** m/s carries the radiation energy **10^{-14}** J. The unit of time of one second is taken to be 1/(86400) part of the solar day. The unit of length of one meter is the distance passed by radiation in **$1/(3 \times 10^8)$ seconds.**

The unit of current of one ampere in the integer model is the same as in the clasical SI system with the reference held in Paris; it is defined by the force of attraction **2×10^{-7} N** acting between two thin conductors per meter length one meter appart carrying one ampere. The force of attraction for the integer model is smaller by the relativity correction 1.111 ... coresponding to the average local component of the velocity of particles of **1/3xc_s** times quantum corrector $1.04^{9/20} = 1.017806$. Hence the proposed value for μ_{os} = **1×10^{-6}** [N-A^{-2}] times $1.111\ldots^2 \times 1.04^{9/20}$ = **1.256551×10^{-6}** [N-A^{-2}] compares well with the classical SI value **1.256637×10^{-6}** [N-A^{-2}]. These choices produce the whole set of model physical constants with the exception of the gravitation constant.

Let us consider forces between two parallel conductors carrying equal currents, one conductor being at 300 K and the other at 6000 K. The cool conductor generates energy derived essentially only from the oscillations of protons and (proton+electron) pairs, while the hot conductor generates according to the equipartition rule equal energies of oscillations generated by protons, (proton+electon) pairs and electrons. This causes the decrease of the force of attraction to a factor of ~2/3 and for this reason the integer model gravity constant for bodies at equal temperatures **$G_S \approx 3/2 \, G = 1 \times 10^{-10}$ N-m^2/kg^2.**

The model is based on an extended interprettation of the Millican experiment, which originally attributes electron with a charge 1.602×10^{-19} C. In that experiment earth gravity force, acting on minute oil particle of mass m_e charged with a small number n of electrons is balanced again the radiation field E attracting charge q of these electrons:

$$m_e \times g = n \times q_{e^.} \times E \qquad (1.9)$$

Table 1. *Derived integer values of physical constants in cubic space gravity converted to SI units versa classical physical constants SI units.*

Name or its inversion	Relativistic Integer value [8-9]	SI value	Notes ~ Integer value/SI value
No = 6/Boltzmann constant k_s,	6×10^{23}	6.023×10^{23}	Q
Hartree's Energy U_{Hs}	30 eV	27.2138 eV	1.1111
Atomic Velocity of light c_s	$1/111 \times 1/2 \times \sqrt{(1/\mu_{os}\varepsilon_{os})}$ $=3 \times 10^8$ m/s	2.99792×10^8 m/s	$1.04^{1/56}$
Proton radiated mass equivalent m_{ps1D}	1×10^{-27} kg		
$(m_{ps3D})^{1/2}/1.11111^{1/2}$	3×10^{-14}		
$(m_{ps3D})^{1/3}$ (3D)	1×10^{-9}		
Proton density mass m_{pds}	2×10^{-27} kg	1.6726×10^{-27} kg	$1.1111 \ldots$ $^2/1.04^4$
Electron mass m_{es}	1×10^{-30} kg	9.10938×10^{-31} kg	$1.1111 \ldots ^2 \times$ $1.04^{1/2}$
$(m_{es})^{1/2}$	1×10^{-15}		
$(m_{es})^{1/3}$ (3D)	1×10^{-10}		
Electron mass reduced $m_{es}m_{ps1D}/(m_{es} + m_{ps}1D)$	0.9990×10^{-30} kg		
Bohr atom geometric mass $\sqrt{(m_{es} \times m_{ps1D})}$	$31.6227766 \times 10^{-30}$ kg.	39.022 kg	$1.1111 \ldots ^2$
Electron charge e_s	1×10^{-19} C	1.6022×10^{-19} C	$1.1547^{1/2}/1.2^3$
Electron rotating velocity v_{es}	1×10^7 m s^{-1}		
Infrared proton frequency v_{Lms} at 300 K	5×10^{12} Hz	$4.23392485 \times 10^{12}$ Hz	
High electron frequency v_{Hms} at 300 K	5×10^{15} Hz	$2.46606102 \times 10^{15}$ Hz	
Time s 1 s_s = $1 \times 1.11111111^{1/2}/1.04^{1/20}$ = 1.052027461 s 1.0 s			
Mass of sun m_{sun}	2.0000×10^{30} kg	1.9891×10^{30} kg	Q(quantum correction)
Mass of sun, linear one direction	1×10^{10} kg		
Mass of earth m_{earth}	6.0000×10^{24} kg	5.9742×10^{24} kg	Q
Mass of moon m$_{moon}$	7.3477×10^{22} kg		

Distance of earth to sun L_{earth}	1.5000×10^{11} m	1.4961×10^{11} m	Q
Distance of earth to moon L_m	3.84400×10^8 m		
Circular velocity of earth v_{earth}	30000 m s^{-1}	29786 m s^{-1}	Q
Thermal power of sun (6 sides, 3 planes)	3×10^{30} W	2.334×10^{30} W	
Circular velocity of moon v_{moon}	1022 m s^{-1}		
Circular sun area $A_{cisunr} = \pi \times r_{sun}^2$	1.5000×10^{18} m^2	1.52179×10^{18} m^2	Q
Circular earth area $A_{cirea} = \pi \times r_{earth}^2$	1.278005×10^{14} m^2		
Earth eq. circumference	40.000×10^6 m	40.075×10^6 m	Q
Reference temperature of earth T_N	300 K	293.15 K	
Linear velocity of earth v_{earth3}	30000 m/s	29876 m/s	Q
Permittivity of vacuum $\varepsilon_{os} = 1/5 \times I_o$	1×10^{-11} Fm^{-1}	8.85418×10^{-12} Fm^{-1}	$1.1111\ldots$
Permeability of vacuum μ_{os}	10^{-6} NA^{-2}	$4\pi 10^{-7}$ NA^{-2}	$1/1.1111\ldots^2$
Planck's constant $h_s \equiv m_{es} v_{es} / (1/3c_s) \equiv k_s (m_{es})^{1/3}$	1×10^{-33} J-s	6.62607×10^{-34} J-s	$3/2 \times 1.04^{3/20}$
h_{ps} proton radiation energy	1×10^{-33} J		
Planck's constant h_{selec} for electrons	1×10^{-36} J-s	should be 6.62607×10^{-37} J-s	
Boltzmann's constant $k_s \equiv m_{es} v_{es} \equiv h_s \times 10^{10}$	1×10^{-23} J	1.3806×10^{-23} J	$1/1.1111\ldots^3$

Electron Boltzmann's constant $k_{selec} \equiv h_{selec} \times 10^{10} = 1 \times 10^{-26}$ J is energy generating muo10$_s$
$$1 \times 10^{-26} \times 1/10 \times 1/10 = 1 \times 10^{-28} \text{ J}$$

Equivalent electron $\equiv 3(m_{es}\, m_{ps})^{1/2} \times 1.111\ldots^{1/2} \equiv 1.000000000 \times 10^{-30}$ J

Avogadro's unit of force $F_{atoms} \equiv (G_s \varepsilon_{os})^{1/3} \equiv 10^3 G_s$	1×10^{-7} N	0.82363×10^{-7} N	$1.1111\ldots^2$
Radiation (space) energy per m^3 (1D)	1×10^{-28} J/ m^3		
Muo10$_s$	1×10^{-28} kg	1.881×10^{-28}	$1.2^2/1.04^{1/10}$
Bohr's atom dimension $2a_{os}$	1×10^{-10} m	1.0584×10^{-10} m	$1/1.1111\ldots^{1/2}$
Gravity constant G_s for sun to earth with $2/3\ G_s$	1×10^{-10} N-m^2/kg^2	0.66742×10^{-10}	$G_s \equiv (\mu_{os}/\varepsilon_{os})^2$

$G_s \equiv h_s/k_s = 1.111\ldots m_{ps} 1D \times c_s^2\ G_s \equiv 10\varepsilon_{os}/1.111\ldots^2 = 10\varepsilon_{os}/1.234567901 = 81\mu_{os}^2 = $ (Klitzings constant)^2

Bohr's magneton μ_{Bs}	1×10^{-23} JT^{-1}	0.9274×10^{-23} JT^{-1}	$1/1.2^3 \times 1.1547^{1/2}$
Stephan-Boltzmann constant σ_s	5×10^{-8} W-m^{-2} T^{-4}	5.6705×10^{-8} W-m^{-2} T^{-4}	$1/1.11\ldots$

$\sigma_s = 5$ W/m^2 per unit of local velovity 10^8 amounts to 10 W/m^2 with reflection

Mean earth temperature $T_{earthmeans}$	$h_s c_s \varepsilon_{os}/e_s^2 = 300$ K	NTP 293.15 K	Q
Sun surface mean temperature $T_{sunmeans}$	$20\ h_s c_s \varepsilon_{os}/e_s^2 = 6000$ K	5780 K	$1.0328/1.04^{1/7}$
Earth acceleration	10 m/s^2	~9.81 m/s^2	
Rydbergs constant	1×10^7 m^{-1}	1.0973731×10^7 m^{-1}	$1/1.1111\ldots$
Atomic force H/Rydberg constant	1.0×10^7 V/m	1.097×10^7 m^{-1}	$1/1.1111\ldots$
Klitzings constant $h_s/e_s^2 \equiv 9\mu_{os}$	1.0×10^5 J-s-C^{-2}	$0.25812.8\ldots \times 10^5$	$1.2^5/1.04^{19/20/}$
Josephson'sconst is $2_{es}/h_s$	2.0×10^{14} C/(J-s)	2.4180×10^{14}	$1/1.2 \times 1/1.04^{1/5}$
Coulomb's constant	$1/12\varepsilon_{os} = 8.333\ldots \times 10^9$	$1/4\pi\varepsilon_o \approx 8.987565$	$1/1.15470$

Space frequencies $v_{ovector}$	3×10^{-17} Hz	and	$v_{os} = h_s c_s \varepsilon_{os}/e_s =$ 3×10^{-17} Hz
Fine structure constant α	$\alpha_s = 1/137.067$	$1/137.035989$	
Fine structure constant with cor. relative. and quantum $\alpha_s' = \alpha \times 1.333 \ldots \times 1.04^{7/10} = 0.01000063$ eV			
Graviton energy	$g_{se}V = 1/60 =$ $0.01666 \ldots$ eVK^{-1}	$g_{seJ} = 1.666 \ldots \times 10^{-35}$ JK^{-1}	10^{30}
at 300K with electron/orien.	5 J/$\sqrt{3} \times 1.04 = 3.002$ J	5×10^{-33} J/$\sqrt{3}$ $\times 1.04 = 3.001 \times 10^{-33}$ J	
Geometric mass of sun and earth	$\sqrt{2} \times 10^{27}$ kg	3.44721×10^{27} kg	$1/(1.2^5 \times 1.04^{1/2})$
Number of electrons forming mass of earth	6×10^{54}		
Linear, surface and volume probability	$Pr_p, Pr_l^2, Pr_l^3 = P_v$	$10^{-10}, 10^{-20}, 10^{-30}$	
Power of sun beam to earth	4×10^{29} W	3.981×10^{29} W	Q
Spin S_s	per orientation $1/6\ h_s$, total $h_s = 10^{-33}$ J-s	$h/2\pi = 1.05457 \times 10^{-33}$	$1.1111 \ldots^{1/2}$

Relativity correctors for: $1/3\ c_s$ 1.11111; $(1/2\ c_s)^{1/2}$ 1.07547; $(1/2\ c_s)^2$ 1.1547^2; $1/4\ c_s$ 1.0328; $1/2\ c_s$ 1.15470; $(1/3\ c_s)^{1/2}$ 1.054093; $1/2\ c_s$ 1.15470; $(1/2\ c_s)^3$ 1.5396; $(1/2\ c_s)^4$ 1.7778;

Physical constants of integer cubic units of space expressed as approximately corrected SI values of physical constants and other data

$h_s \equiv 3/2 \times h\ 0.9939 \times 10^{-33}$ J-s; $m_{pds} \equiv 1.2\ m_p \approx 2.00 \times 10^{-27}$ kg; $1.666/[1.1547^2 \times 1.1111(1+1/80)]$ $= 1.000$; $k_s \equiv k/(1.15470^2 \times 1.0328 \times 1.04^{1/15}) = 1.0000 \times 10^{-23}$ J; $G_s \approx 3/2G = 1 \times 10^{-10}$ N $\equiv h_s / k_s$; $\sigma_s \equiv \sigma(1-1/10)/1.04^{3/5} \approx 5 \times 10^{-8}$ W/ m^{-2}; $\mu_{os} \equiv 1.04^{1/2}\mu_o/K_{rel} \approx 1.1111 \times 10^{-6}$ N A^{-2}; $\varepsilon_{os} \equiv \varepsilon_o$ $K_{rel}/1.04^{1/2} \approx 1.00 \times 10^{-11}$ J-m^{-3}; cube side $a_{os} \equiv 2\ r_o \times 1/(1-1/10)^{1/2} \approx 1.00 \times 10^{-11}$m;

Minimum H. u. $= h/(2\pi \times 1.11111^{1/2}) \approx h_s/10$. Charge, energy and force decrement p.u. factor due non-interactive radiation at 90°: for one plane $1/1.2 = 0.833 \ldots$, for two planes $1/1.2^2 = 0.6934$, for three planes $1/1.2^3 = 0.5787$ and for two pulses $1/1.2^4 \approx 0.4822$.

In 1923 Millikan derived electric charge to be $q_e = 1.59 \times 10^{-19}$ C, while the present value is 1.602×10^{-19} C.

   The author claims presently that 1923 Millikan electron charge interacting is decremented by the factor 1.2^3, increased by relativity corrector of $1.154790^{1/2}$ for the electron moving with velocity $1/2\ c_s$ and by quantum corrector $1.04^{19/200}$ deriving in the experiment model value 1×10^{-19} C with the quantum uncertainty of the measurement at $\frac{1}{2}\ k_s$:

$$(1.602\ 1773 \times 10^{-19})/1.2^3 \times 1.0745699 \times 1.04^{19/200} = 1.000045 \times 10^{-19} \text{ C} \qquad (1.10)$$

The uncertainty of the event is 0.000045×10^{-19} C $\equiv \frac{1}{2}\ k_s$; converted to H. u. it is $\frac{1}{2}$ $h_s = \frac{1}{2} \times 10^{-33}$ W/m per wave on division by proton frequency 1×10^{12} Hz in one pathway at 300 K. Although in the past the author used different correctors there is no doubt that the charge calculated by Millican using his modern figure 1.602×10^{-19} C is the derivative of 1.00×10^{-19} C. The Universe of integer relativistic constants seen by **an observer moving**

with velocity of mass of proton ½ c_s does not require relativity correction, 1.1547½, or any arbitrary quantum correction, because they move at the same velocity.

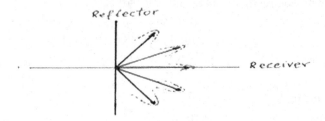

Fig.1. Five forward probability pulses. Distributions sketched applies to all vectors in all Figures. The integer values apply only to average or r. m. s. quantities. For six directions there are 30 quantum orientations. Spherical structure has 34 quantum orientations.

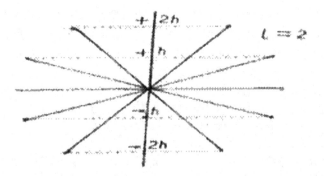

Figure 2. Space quantization in ten orientations per plane generating probability $1/10^{10}$, with 1/10 expressing probability of one step process for one of three planes. The two vertical projections of the electron orbital angular momentum do not count because they are at right angle to the progression axis and they decrease energy and forces received by factor 1/1.2=0.833 . . . per plane. Distribution of values around the mean value and direction due to uncertainty are not considered in the model.

The observer identifies the no intereactive (missing) moments perpendicular to the 1D source-receiver axis obtained in the calculation, from the measured value of 1.602 1773 $\times 10^{-19}$ C; the observer derives also the integer charge 1.00 $\times 10^{-19}$ C identified with energy of 1 J in Equation 1.2 and 2 J of Planck Radiation Law at 300 K. This value is derived in Equation 10.45. Similarly the classical static Boltzmann constant $k = 1.3806 \times 10^{-23}$ J is decreased by the factor 1.2^3 and increased by the relativity and quantum corrections to derive k_s value:

$$k_s = 1.3806 \times 10^{-23} \text{ J}/1.2^3 \times 1.11 \ldots^2 \times 1.04^{1/3} = 0.999 \times 10^{-23} \equiv 1.0000 \times 10^{-23} \text{ J} \qquad (1.11)$$

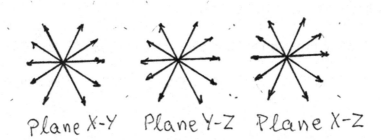

Plane X-Y   Plane Y-Z   Plane X-Z

Figure 3. Thirty space quantizations are atomic level in three planes generating radiation probability $1/10^6$ per event in the required direction

On the other hand classical proton SI mass m_{pd} is increased by relativity and quantun correction

$$m_{ps1D} = 1.672 \times 10^{-27} \times 1.11\ldots^2/1.04^{4/5} = 2.0004 \times 10^{-27} \text{ kg} \tag{1.12}$$

The integer model electron charge e_s in one dimensional display (1D) derives relativity corrected triple gravition constant:

$$(e_s)^{1/2} = (1 \times 10^{-19})^{1/2} = 3.1627766 \times 10^{-10} = (3 \times 10^{-10} \times 1.11\ldots^{1/2}) = 3 \times G_S \times 1.054092553 \tag{1.13}$$

And SI electron charge
$$= 1.6021773 \times 10^{-19} \text{ C} \approx (3 \times G_S \times 1.111\ldots^{1/2})^2 \times 1.2 \times 1.1547^2 \times 1.04^{1/29} \text{ C} = 1.602166 \times 10^{-19} \text{ C}. \tag{1.14}$$

is derived more accurately from charge deposition on electrodes in the electro-chemistry. It is suggested that electron momenta in space radiation and in the liquid solutions have the same local split velocities requiring in the above estimate the same corrections generating the static model charge value 1×10^{-19} C.

In atomic integer terms **the static charge is the proton linear momentum with the mass moving at the average component velocity of the group:**

$$e_s \equiv m_{ps1D} \times 1/3 \times c_s = 1 \times 10^{-27} \times 1 \times 10^8 = 1 \times 10^{-19} \text{ kg-m/s} \tag{1.15}$$

Electron charge $e = 1.602 \times 10^{-19}$ C measured by C.Millican in 1923 corrected for relativity and for quantum has a value $e_s = 1.000 \times 10^{-19}$ C leading to the whole set of relativistic integer space physical constants. The value 1.000×10^{-19} is also the proton charge momentum. The list of integer values of physical constants differs from the classical SI values by incorporating probability factors of the orthogonal geometry of space, the relativity corrections required by the local velocities of light and arbitrary quantum corrections.

To conserve momentum the electron radiates in two directions sending in each:

$$e_s = 1/2 \times 1.602177 \times 10^{-19} \times 1.2 \times 1.11 \ldots^{1/2} / 1.04^{1/3}$$
$$= 1.00014 \times 10^{-19} \, C = 1.0 \times 10^{-19} \, C + 1/6 \, k_s (\text{H. u.}) \tag{1.16}$$

Similar identity applies to proton mass in integer cubic units $m_{pds} = 2 \times 10^{-27}$ kg

$$1.672623 \times 10^{-27} \times 1.11 \ldots^2 \times 1/1.04^{4/5} = 2.00118 \times 10^{-27} \, \text{kg} \approx 2 \times 10^{-27} + 1 \times 10^{-30} = 2001 \text{ electrons} \tag{1.17}$$

The model uncovers the electron charge e_s to be really equivalent electron discharging during radiation in both directions proton frequency oscillatoty momentum:

$$(2m_{es} \times 5 \times 10^{12}) \equiv 1/E_{sspaceSI} \tag{1.18}$$

The $E_{sspaceSI}$ value is expresses in Equation 2.2.

The exactness of velocity of light corrected for relativity demonstrates also the exactness of the proposed integer values of ε_{os} and μ_{os}:

$$c_s = 1/\sqrt{(\varepsilon_{os} \times \mu_{os})} \times 1/1.111 \ldots^{1/2} = \sqrt{E_{sspaceSI}} \times 1/1.11 \ldots^{1/2} = 1/\sqrt{(10^{-11} \times 10^{-6})} \times 1/1.111 \ldots^{1/2}$$
$$= 3 \, 1622 \, 7766 \times 1/1.111 \ldots^{1/2} = 3 \, 0000 \, 0000 \ldots \text{ m/s} \tag{1.19}$$

The electron charge e_s discharges also its own electron frequency as oscillatory momentum $2m_{es} \times 5 \times 10^{15} \equiv 1 \times 10^{-14}$ kg-m/s. Millican experiment derives the model electron charge of 1×10^{-19} C, if the presence of the local (i.e. imaginary or split) velocity $\frac{1}{2} \, c_s$ of the proton is accepted as a fact. **With electron taking two values of moments, most values of physical constants have also two values. It requires hundreds of equations to display a novel, a more comprehensive, nearly fully comprehended but more transparent and improved view of relativistic physics and of relativistic gravity.**

Textbooks of physics (eg. A. Beiser [34]) derive Bohr atom orbit at 3.3×10^{-10} m $= 2\pi \, a_o$ obtaining Bohr atom radius of $a_o = 0.5252 \times 10^{-10}$ m. In the cubic space model π is replaced by 3; the application of correctors of 1.1111 . . . for relativity and of $1.04^{1/4}$ for quantum derives:

$$a_{os} = a_o \times 3/(1.111 \ldots \pi) \times 1.04^{1/4} = 0.4999 \times 10^{-10} \equiv 0.5 \times 10^{-10} \, \text{m} \tag{1.20}$$

Table 2 gives the approximate relation between classical SI values of physical constants and the integer values proposed by the cubic space model. The major differences appear to be the result of the 1.111 . . . relativity corrections, a quantum reflection and the 3/2 factor in places.

The basic unit of angular momentum of SI is replaced in the integer space model by relativity corrected input of ten quantum space quantization events with accuracy of 5 in 10^4 parts. It expresses also SI power of 1 W/m³ converted by size to 1×10^{-30} W/m³ atomic units and to vectorized form by three steps of probability of orthogonal orientations $1/10^3$:

$$h_s \approx h/2\pi \times 10/1.111 \ldots^{1/2} \text{J-s} \equiv 1 \, \text{W/m}^3 \times 1 \times 10^{-30} \times 1/10^3 = 1 \times 10^{-33} \, \text{J-s} \tag{1.21}$$

$\equiv k_s$ x linear probability $=1 \times 10^{-23} \times 1 \times 10^{-10} = 1 \times 10^{-33}$ J-s.

Thus in the cubic system $2\pi \equiv 6$ and h_s expresses quantum pulse in one of six orientations.

There are no static charges in the Universe. There are only momenta of electrons and of protons contained within specific space boundaries and oscillating or proceeding with limited number of mean vibrations or mean velocities. **The true interactions between particles are kinetic with no units of distance or time. They are expressed on the atomic scale of nature. The chance of an event at the atomic size level are much smaller than at SI level.** To convert to the model SI system of integer cubic space units, the multiplying size converters are: for linear dimensions 10^{-10}, for area 10^{-20} and for energy, mass and volume 10^{-30}. Furthermore the atomic nature system uses momentum versa energy of the classical system with the unit being the charge e_s versa 1 kilogram requiring multiplying by system converters e_s and c_s. The classical SI values are ten times larger than integer model atomic values expressed in SI units. The relativity corrector is displayed by the early choice of 1 m = $1/(40 \times 10^6)$ of the model circumference of earth expressed by in the following identity:

Earth circumference:
40 000 000 m $\equiv$ velocity of light x squared relativity correctors for $\frac{1}{2} c_s \times 1/10$

$$\equiv 3 \times 10^8 \times 1.1547005^2 \times 1/10 \text{ m/s} = 39\ 999\ 997.3 \tag{1.22}$$

Nature is elusive, it hides its details. The next identity still lacks full proper explanation. It is accepted for its accuracy. Relativity corrected equitorial earth circumference equals the squared sun temperature:

$$40\ 000\ 000 \text{ m}/(1.111\ldots) = 36\ 000\ 000 \equiv 6000^2 \text{ K} \tag{1.23}$$

Earth circumference, 40×10^6 m, with π replaced by 3 over 1.15470^2 equals numerically the electron rotational velocity 1×10^7 m/s. Now it is also the exactness of the next relation, which validates it:

$$(1 \times 10^7)^{1/2}/(1.111\ldots)^{1/2} \equiv 3000.000000 \text{ K} \tag{1.24}$$

The corrected product of sun radius $(1.5 \times 10^{18}/\pi)^{1/2} = 6.90988300 \times 10^8$ m, 10 quantum sun temperature 10000 K that include energy of electron oscillations and permiability of space μ_{os} is numerically equal to the earth radius:

$$\text{sun radius} \times 10 \times 10000 \text{ K} \times 1 \times 10^{-6} \times 1/(1.111\ldots^{1/2} \times 1.04^{1.5} \equiv \text{earth radius} \tag{1.25}$$
$$6.3778 \times 10^6 \text{ m} \equiv 6.378 \times 10^6 \text{ m}$$

The above identity implies that units of μ_{os} are $N^{-1}A^2$.

The energy 50 eV is derived from the Bohr space diameter $a_{os} = 1 \times 10^{-10}$ m and stomic force; the total energy of Bohr atom structure of 50 eV divided by relativity correction 1.15470^4 and reflection $(1 + 1/10 \times 2/3)$ obtains 30 eV:

Atomic force 1×10^{-7} N$\times a_{os}/1 \times 10^{-19} = 50$ eV and $50/1.15470^4$ x $(1+1/10 \times 2/3) = 30.0000$ eV
$$(1.26)$$

Circular area increases with the square of the linear dimension. Taking the square root of the rounded up circular area of the sun $\sqrt{1.5} \times 10^{18}$ m^2 over (the product of the average local component of the velocity of the particle $1/3$ x c_s and the square of the relativity correction for $1/3 \times c_s$) times quantum correction and linear probability $1/10^{10}$ derives 10 m-s expressing several parameters:

$A_{cisunr}/(1/3 \; c_s)$ x corrections =
$\sqrt{1.5} \times 10^{18}/[(1/3 \times 3 \times 10^8 \times 1.111 \ldots ^2) \times 1 \times 10^{10}] \times 1.04^{1/5} = 9.99835$[m-s]$\equiv 10$ eVat $\equiv 10$ J $\equiv 10$ W
$$\equiv 10 \text{ kg-m} \equiv 10 \text{ N} \qquad (1.27)$$

The value of sun force of attraction is derived using two steps of quantization, impulse (momentum, power, energy) generated per kilogram, temperature to power four, velocity of light and quantum and relativity corrections. The value obtained is within the limits of accuracy equal to the classical SI reference force of Equation 6.1:

$F_{Gs} \equiv 10$[N-s-kg^{-1}]$\times 6000^4$[K^4]$\times 3 \times 10^8$[m-s^{-1}]$\times 1/10^3 \times 1.111 \ldots /(1.1547 \times 1.04^{2/5}) \equiv 3.5432 \times 10^{22}$ N
$$(1.28)$$

Using Special Relativity derives its own exact value of gravitation force:

$F_{Gs} \equiv 1$ J$\times (3 \times 10^8)^3 \times 1.11 \ldots \times 1/10^3 = 3$ J$\times (10^8)^3 \times 1/10^2 = 1 \times 10^{17} \times 1000^2 \times 1/10 = 3 \times 10^{22}$ N $\quad (1.29)$
where $1 \times 10^{17} \equiv (\varepsilon_{os} \cdot \mu_{os})^{-1}$ and $1000^2 \equiv (3 \times 300 \text{ K} \times 1.1111)^2$.

The most important parameter in the model is the velocity of light c_s. Particles of radiation move with local velocities $1/3 \; c_s$ and relativity correction has to be most often applied for that value.

The values discovered now are: local velocities of light $\frac{1}{4} \; c_s$ for proton, $\frac{1}{2} \; c_s$ for electron, $1/3 \; c_s$ for the average velocity for three proton oscillators, electron and (proton+electron) pair and $1/10 \; c_s$ for the rotational velocity of electron in the Bohr atom. Velocity of light is derived now from the probability of the discharge in any one direction in the geometry of the six orthogonal orientations; the event is described by unity and the resultant formation by addition of three vectors components of light velocity is shown in Fig. 3. Furthermore in any of the three planes space quantization requires having at least the Planck constant value h_s. Addition of components of light velocity moving in orthogonal directions produce velocity of light, while the sequential probability requirement that the resultant signal is in the correct direction decreases very significantly the resultant vector value.

The SI system is a practical system based largely on man made artifact units, which can be very accurately reproduced, which are convenient to use, but very large in relation to the most elementary events in science. The first standard of the length of 1 meter established by the French Academy Sciences in the year 1791 was $1/10^7$ the distance of the equator to one of the poles, a quarter of earth circumference of Equation 1.18. It remains a perfectly valid historical scientific reference because it produced the total energy of Bohr atom proposed now, although the electrons were then unknown:

Total energy of Bohr atom $\equiv 3 \times 10^8/1 \times 10^7 \equiv 30$ eVat $\equiv 30$ J $\equiv 30$ W   (1.30)

The units representing factual science law are the current of one ampere, expressed by the force of 2×10^{-7} N per meter between thin parallel conductors, which is likely to be replaced soon by an equally valid number of charged particles. The second basic parameter of the model, the energy value of 1st term of Planck radiation pulse 0.01 J is assumed to express the permittivity of space $\varepsilon_{os} = 1 \times 10^{-11}$ J /m^3 moving at the average velocity of particles of Bohr atom times one step quantization:

Radiation pulse $= 0.01$ J $\equiv \varepsilon_{os} \times 1/3 \; c_s \times 10$ (1.31)

Energy 0.01 J is linearly dependent of temperature and referred to 300 K it becomes 3 eVat, converting to 1 eV or 1 J per direction, if energy of electron radiation is included.

In 1889 one SI meter was replaced by marks on platinum-iridium rod kept in Paris and finally it was defined by the wavelength of light emitted krypton gas. Krypton is a noble gas carrying features of hydrogen.

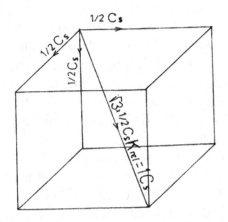

Figure 4. Relativity corrections applied to the product of three vectors moving at half of the velocity of light ½ c_s in three planes at 90° to each other produce the velocity of light c_s; component local velocity is indirectly measured in electron spin. All velocities ½ c_s start from the center of cube, which expresses also center of each square. The resultant is the relativity corrected vector sum of light velocity. Hydrogen atoms tend to make clamps, with 3x3x3 =27 facing each of 6 directions forming a cube with 54 units per clamp as found in Chapter 11. Proton-electron pairs of unity Planck mass move at ¼ c_s and with three planes each have 9 contributors in 3 orthogonal orientations. They convert mass to energy, are quantum corrected with 2/3 factor neglecting independent electron energy.

The product consumes exactly the energy of the new proposed constant of Chapter 21:

$$E_{sspace} = 1/(\varepsilon_{os} \mu_{os}) \equiv 1 \times 10^{17} \text{ J} = 27 \times (1 \text{ Planck mass} \times 1/4 \times 3 \times 10^8)^2 \times 2/[3(1+1/80)]$$
$$= \varepsilon_{os}/muo10_s \text{ mass in integer cubic units} = 1 \times 10^{-11}/1 \times 10^{-28} \text{ kg} = 1 \times 10^{17} \text{ J.}$$

In the second step of velocity conversion ½ c_s to c_s, the output becomes 1/10 of the initial input. The output is the geometric mean mass/m of all the planet-sun couples expressing the same gravitation energies of Table 4. With input in one direction of 1 W times squared local velocity derives exactly:

$$1\text{ W} \times (1 \times 10^8)^2 = 1 \times 10^{16}\text{ J/m.}$$

Presently the size of Bohr atom is derived from the above energy of the atom of 30 eVat over relativity correction 1.111 . . . being the atomic force 1×10^{-7} N and the Bohr atom radius a_{os}:

$$F_s \times a_{os} /1\times10^{-19} = 30/1.111 \ldots = 27.1262\text{ eV} \tag{1.32}$$

Above value differs from SI Hartree value by the quantum value $2 \times \sqrt{2}\ k_s T_{300}/e_s = 0.0848$ eV. Other cubic space constants follow from these simple geometries, probabilities, and space quantization; corrections were usually guessed both for relativity and for quantum. **A congregation of particles having energy distributed is subject to equipartition energy rule.** While in atomic interactions energy equipartition is a very fast process, for galaxies it takes millions of years.

The number of expressions used for energy is overwhelming and together with the number of equivalent meanings they lead to 96 expressions of the apparent gravity or repulsive forces for the earth and for the sun or thermal radiations. Units of linear elastic kinetics have only two important units expressing momentum, magnitude and rate; SI system can replace those two components by different values in SI units.

The energy of an excited Bohr atom is divided between electron, proton and (proton+electron) pair each component oscillating independently at different frequencies having mean value and distributions.

At sun temperature, T sun = 6000 K, the value $k_s T_{6000}/e_s \times$ corrections = $1\times10^{-23} \times$ $6000/1\times10^{-19} \times 1.2^2 \times 1.1547 \times 1.04^{1/16} \approx 1.000011$ eV = 1 J is the unity energy event value of the sun pulse, becoming 3 eVat for the three oscillators and $3 \times 3 \times 1.111 \ldots = 10$ J for three planes. Average radiation quantum energies of 1 J and 1 eV at 300 K are unity probability individual events at atomic level present in all radiations. Values less than one at 300 K are probabilities based on unity energy input. The corresponding uncertainty for the earth signal at 1 K is 0.05 eVat expressing five 0.01 eVat elementary energies within the atom, reduced to 0.001 eVat by space 1/10 linear probability. Energy of 0.01 eVat is the first quantum energy step per K expressed by the Planck radiation law.

The law of equipartition requires equal energies to be distributed in all degrees of freedom. Hence the photon energy gradient of the sun radiation derived below is equal to gravitation force. Gravitation force acts in one plane requiring 1/3 multiplier and 6_s does not incorporate energy of independent electron oscillations. On average sun radiates with 1000 K per each of six orientations:

$$1/3 \times 6_s T_{1000}{}^4 A_{cirsun} \text{ xcor.} = 1/3 \times 5\times10^{-8} \times 1000^4 \times 1.1547^{3/2} \times 1.1111 \times 1.04^{7/10} = 3.5429\times10^{22}\text{ N} \tag{1.33}$$

From the sun the hydrogen pulses from three orientations facing the receiver are created by three orientations so that nine 1 J energy pulses corrected by relativity 1.111 . . . deliver in total 10 J, which is 1 eV/atomic event at the atomic level. The two stage process increases velocity of the equivalent electron representing (prorton+electron) pair moving at 1.4 c_s four times = 2 x $\sqrt{3}$ x 1.1547. For the molecule H_2 the output is doubled to 20 J.

It is suggested that, that while on earth the Bohr atom is spherical (besides other shapes, which are not considered) the signals from the sun indicate at a cubic shape due to very high gravitation pressure around the sun. Equations assuming both shapes generate valid data. This book, with calculations including integer space physical constants, displayed clearly contraction and extension of several parameters recorded by different observers as required by the rules of Special Relativity.

The research can be criticized by dealing with a signal, which does not exist on earth since very high temperature is needed to sustain atomic hydrogen. The classical SI units use the same procedure.

Above introductory notes should persuade the reader to look at the strange model seriously, because it adds to the total knowledge of science. Gravity was initially investigated hoping it would provide another well recognized constant for uncertainties in relaxation. By pure mistake the gravity was expressed by an unusual way. Then new formulations in gravity of the first two books [45, 46] required reexamination of bases of whole physics. The early research covered many topics and data from the first three years are not even mentioned. The physics postulated comprises several independant findings, each needing justification, so the ideas cannot be developed sequentially. The book needs to be read more than once. It covers research progress of eleven years. The book was written to be understood by anybody interested in physics, or by followers of hazard, since nature uses statistics of the game of dice. This is engineer's view of the elementary physics and of the gravity, in which integer numbers, all derivatives of unity probability, are the most important characteristics of the reality. There is no calculus involved, only powers of ten and probability concepts. An open mind is needed, without beliefs in dogma. Close attention has to be paid to the values of constants used in equations. Calculations were done on commercial calculators displaying 10 to 12 digital places, which use approximations generating unspecified minor errors. Every reader can make his own mind on the validity of the concepts postulated without the need to be dependent on an expert opinion. WEB pages including WikiEncyclopidia are the simple and update references.

To be acceptable the postulated forces in the model must derive accurate experimental values, have correct directions, clear points of applications and be applicable to all possible range of situations. These conditions are satisfied in most identities proposed, but in several cases the use of the chosen corrector is not fully understood. Some important results may be displaced because several findings were deduced at the final book review. The model expresses simple physics in terms of the Bohr atom. Expressions derived for sun-to-earth gravitation were applied with variable success to all planets. The model uses the center of energy interaction to be the reference frame; for observers of individual radiation-mass interactions occurring near body's surfaces the references are the earth and the sun.

There were earlier suggestions that the Universe is Digital by Konrad Zuse (Digital Physics, 1967), by P.A. Wahid (The Divine Expert System, 1998), by Seth Lord (Physical Review Letters, 2003), by S. Wolfram (Science, 2002). They dealt mainly with the communications and compared the operation of the Universe to a computer program.

Preliminary resume

1. **Ampere's Law plays a principal role** in generating gravitation forces. Elastic kinetic atomic interactions between particles lack units of time and of distance. The model of physics and of gravitation proposed derives the probability values for generating, the minimum possible quantum unity values of various parameters. These parameters are time, length, mass and charge and are derived from the orthogonal structure of the Universe. They need a sequential probability of occurrence of events required to generate radiation. The probabilities values for events are: one sequence of radiations in orthogonal directions 10^{-3}, three such sequences 10^{-9}, linear conversion from SI to natural atomic 10^{-10}, volume conversion 10^{-30}, electron charge conversion from SI to natural atomic $10^{-10} \times 10^{-9}$, minimum energy expressing SI unity event in natural atomic units 10^{-14}, Planck infrared constant $10^{-10} \times 10^{-9} \times 10^{-14}$. The charge value from Millikan experiment establishes the physical event having unity value and probability; non-events have zero probability. Multiples of these units create the Universe. The gravity constant G_s is the integer cubic SI force density 1 N-m^2/kg^2 converted to nature atomic units by linear probability.

2. Above probabilities and Planck units derive both the integer values of cubic space constants converted to SI units and the classical SI units; the elementary units are for both systems one meter, one kilogram and one second together with one of the electromagnetic parameters, which according to the model proposed is event probability derivative from the orthogonal properties of the space.

3. Bohr atom has **three oscillators: electron, proton and (electron+proton)** pair.

4. **There is no independent gravity force.** The gravity force attraction of sun-earth couple is the original sun wave converted into several half wave lengths by the internal atomic reflection converting the wave into a pulse. Unspecified laser action may be responsive for force formation. Electron frequencies play no role in sun-to-earth in gravitation. The missing mass in the Universe comprises mainly its cold bodies.

5. **The mass of sun generates 1 W/(kg-K) per kilogram-kelvin** vectorized power at the surface of earth.

6. Gravitation balance is obtained by the **planet attracting the sun with the same force as the sun attracts the planet.** Attraction converts to repulsion, if the planet or center of is mass is treated as reference. The planet rotating around sun generates an outward directed centrifugal force. Rotation induces continuous inward acceleration, which is an alternative description of the reason for the appearance of the repulsive force balancing the sun gravitation. If in doubt rotate stone on a string. The events in rotating fluids equating energy densities do not apply.

7. Elements of atomic and radiation particles move with **velocities** or **local component velocities 1/10 c_s, 1/3 c_s, 1/4 c_s, 1/30 c_s, and 1/2 c_s** and require relativity and quantum corrections to express their true component values. Protons move with velocity 1/4 c_s.

Two sets of two protons in orthogonal orientations add velocities vectorially generating two $\sqrt{2}$ x 1/4 c_s = 0.35355339 c_s pulses. These two repeat this process generating $\sqrt{2}$ x 0.35355339c_s = ½ c_s. Three such pulses generate velocity c_s shown in Figure 5. Thus there are twelve original pulses pulse, which need to satisfy probability requirement to be radiating in specific orientation. The first step can be obtained also with three pulses and nine in total. The product of these probabilities is the chance of radiation being created with velocity c_s. Three components electron, proton and (proton+electron) move with local velocities 1/3 c_s which, by expressing the pulse rate, form the only one velocity of light c_s. They move in 3 forward orthogonal axes and generate one dimensional c_s moving along the long axis of the cube created by independent space orientations. Orthogonal component velocities **1/3** c_s mutually null components at right angle to c_s, all along the radiation path from the source to the receiver. Additions of light velocity components **1/3** c_s generating c_s are possible because the measured velocity expresses a pulse rate. Component local velocities uncover several new aspects of physics; proton is a nulled sum of moments of 2001 electrons; mass of 1 kg contains 10^{30} electrons; international gravity constant 1×10^{-10} is the conversion of the SI force of 1 N unit to natural atomic units; uncorrected age of the Universe is 10^{10} years. Main elements of kinetics of physics and of gravitation can be explained by the spherical or by cubic geometry using frequencies of electron 5×10^{15} Hz and of proton 5×10^{12} Hz and their local values 10^{15} Hz and 10^{12} Hz without considering their distributions.

8. The new aspects of **the digital world of relativistic space physics and of relativistic gravity is based** mainly **on the static geometry of space, on the probability of the formation of 1D signal, on the local components of light velocities and on the laws of physics**, versa **the classical SI system based on the measurements of static particles moving at the velocity of light.** SI system and classical physical constants express the real experimental values of energy and momentum comprising imbedded relativity and quantum corrections.

9. Space quantization was established by O. Stern and W. Garlic in 1921. The integer cubic model proposed by the author applies it to six orthogonal orientations each having five positions. With reflection five becomes ten position for one plane and 30 quantum radiation orientations for radiation from a spherical surface and 32, 34 or 36 quantum optional radial radiation positions from the sphere. Usually present reflections convert radiation spread in three planes to a directed pulse having thirty times (30 J) the energy of a quantum value of 1 J.

    Relativity correction of 1.1547005 indicates that the presence of local velocity of ½ c_s is not invention by the author. Although the author identified ½ c_s presence, its invert $1/(\sqrt{3} \times 1.1547005) = 0.500000016$ has been used by Goldsmith and Hollenbeck in their 1925 measurements of electron spin:

Electron spin = $\sqrt{3}/2$ x $h/2\pi \approx 1/1.1547005$ x $h/2\pi \approx 0.1378\ h$.   (1.34)
Every physics textbook quotes this equation and hence unknowley recognizes the presence of the component velocity ½ c_s.

10. Proton is the mass of 10^{30} electrons, with celestial body of mass kilogram contained the number N_e electron N_e=(mass kg)/$1\mathrm{x}10^{-30}$ and with velocity v_e generating current $-N_e v_e$ generating classical electromagnetical force expressing correct repulsion forces of four planets.

11. The Universe uses sub harmonic frequencies and its geometric factors to establish the distances between planets. The Universe used to be called Newton's clockwork Universe and it appears now to behave like one, but it is subjected to a statistical, and not to an absolute probability.

12. **After the application of relativity and of quantum corrections the calculations derive effectively the same energy values of parameters using either the classical SI units or integer space relativistic units converted to SI.** Several integer model values express more veritable values of H. u. than SI values.

13. Elementary radiations involve the conversion of electron mass to energy by the square of the velocity of light, with the expression needing rearranging.

14. The picture of crossed hands on the front page is not ornamental; it displays 5 quantum orientations in each directions of space, which have originated bone growth in these preferential directions in one plane with 4 limbs indicating four directions generating 20 orientations. Sequences of three or two pulses are needed to generate by wave functions the velocity of light. Mass referred to 300 K with probability 1/40 derives 1 J pulse with relativity and quantum correction propagating growth:

$$(300 \text{ K})^{1/40} /1.1547 \equiv [1 - 1/(20 \text{ x } 40)] \text{ J}. \tag{1.35}$$

Consideration of such unclear ideas for eleven years concluded that proton instability delivers free sources of energy responsible for some aspects of gravitation, for the age of the Universe and for processes of growth at the tips of grass.

15. The SI force between two one farad charges, Bohr radius apart, is according to classical Coulomb Law $F = 1\mathrm{x}1/(4\pi\varepsilon_o \mathrm{x} a_o) = 1/(4\pi\mathrm{x}8.85418\mathrm{x}10^{-12}\mathrm{x}5.291772\mathrm{x}10^{-11}) = 1.69840\mathrm{x}10^{20}$ N with energy $F\mathrm{x}1.6022\mathrm{x}10^{-19}\equiv27.2112$ eV=Hartree energy. In integer SI cubic model units, π is replaced by 3; then relativity 1.111 . . . correction is applied generating an integer value $F'=1\mathrm{x}1/(12\mathrm{x}\varepsilon_{os}\mathrm{x}a_{os}\mathrm{x}1.11 \ldots) = 1\mathrm{x}1/(12\mathrm{x}1\mathrm{x}10^{-11} \mathrm{x}0.5\mathrm{x}10^{-11} \mathrm{x}1.111 \ldots) = 1.5\mathrm{x}10^{20}$ N; F' multiplied by quantum and two relativity correctors $1.0328\mathrm{x}1.11 \ldots^{1/2}\mathrm{x}1.04$ derives $F=1.69832 \mathrm{x}10^{20}$ N generating an excellent agreement with the classical F since correction 1.0328 is not that accurate. This illustrates the validity of the model postulated and its relation to the classical SI system at the fundamental level.

    The classical Coulomb Law for F does not foresee the absence of electron frequency at the lower temperatures; this decreases F by the factor 2/3 deriving the true atomic force, $F'=1$ N per oscillator, the value required in Equation 1.2 by using 2D converter $1\mathrm{x}10^{-20}$. The classical Coulomb Law defines F accurately at high temperatures with three oscillators of Bohr atom radiating energy: the electron, the proton and the (proton+electron) pair. To bring F to unity at 300 K requires 2/3 factor, relativity and quantum corrections F x 2/3 x $1/(1.11 \ldots$ x $1.04^{1.2})$ x10^{-20}= 1 - 0.00008 N ≈ 1 J - $1k_s$. **Boltzmann constant in relations using SI units expresses measurable uncertainty,** because k_s =h_s x1x10^{10} =1x10^{-23}=1x10^{-33}x1x10^{10}.

16. The products of two macroscopic parameters such as E and B or I and V are needed to obtain power real. The same powers should be derived by atomic parameters ε_o, μ_o or/and by other products of physical constants such as atomic masses and frequencies establishing energies, forces or powers. Although generated photons were interfering in the research, the eleven years long unspecific analyses obtained outstanding results.

17. The model of physics, of gravitation and of Universe has been described with different names: integer, digital, relativistic, cubic, space, probabilistic each term expressing different features of the model proposed.

18. Above resume excludes important findings of the last six months. Some assertions and unimportant calculations are repeated several times, because it will not be easy to persuade conservative colleagues that such large amount of simple valid data remained hidden in SI system. The bases of the whole physics has been updated. The concept of time is lucking meaning except for recording past events in the evolving Universe.

19. Equal probabilities of event occurrences in the directions of two proton pulses are needed to increase their velocity (rate) of arrival; The process takes place in three planes with two non interacting pulses. It derives the number of progressing and reflected pulses at 1/(3x1/6x1/6) - 2 = 10.
The relation expresses 5 space quantizations of Stern and Garlich in probability terms.
The model using integer relativistic physical constants deals primarily with events outside the simple atomic structures of hydrogen and helium. It does not consider the standard atomic model of the High Energy physics.

2. *Nature events expressed by digital cubic geometry; model physical constants values derived from Planck units.*

By attributting the rules of physics to the individual observable electron and proton events the model sees the present world as formed from a primevial mist. The mist is taken to be an act of first Holy Creation or residues of later localized explosions. The cubic model will be used to consider basic signals of the Universe from electron, proton and hydrogen. **The space is attributted with three dimensions of space in six orthogonal directions, each dimension (plane) having 10 quantum orientations and one dimension of time for all. Electrons and proton generate events associated with oscillations generating energy pulses moving through atoms or molecules at local velocities expressed by a large fraction of the velocity of light.** Velociry of light has one value and it has constuctive components generating specific features of physics caused by the independance of events in six directions. The stretching oscillation is a two or three dimensional event of probability unity having equal probability of moving in both directions of five space quantizations in three planes. It's reality is defined by the probability of generating a pulse along the source-receiver axis in a direction of the receiver determined by the geometry of its formation.

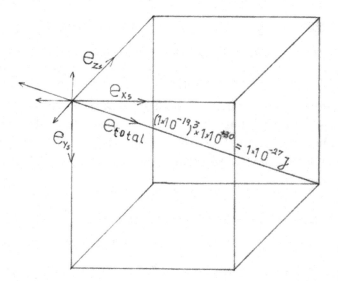

Figure 5. Summed X, Y and Z components of electron vectorized charges. If probability of e_s appearance per indicated direction is 10^{-19} probability across the cube is 10^{-57}. Charge e_s may belong to electron or proton. Factor 10^{30} converts atomic size to SI size. Component of light velocities generate energy pulse $1.111 \ldots e_s \times 1/3 \, c_s \times 3 \equiv 1 \times 10^{-10} \equiv G_s$. $Gs/1 \times 10^{17} \equiv 10 \times 10^{-28}$ J $= 10$ muo10_s masses in cubic integer SI units.

Electrons, as explained subsequently, aggregate forming protons; protons form hydrogen and other atoms. Increasing entropy decreases the energy of particle's events to unity quantum 300 K expressing minimum possible value per oscilator.

Radiations are likely to be initiated by disturbance or regular collapse of proton. Particles move locally with 1/2, 1/3 or 1/4 velocity of light controlled by equal probability of event occurance in 30 directed orientations of space and by a structured event generating the velocity of light. Relativity and quantum corrections are inherent properties of the reality; relativity and quantum corrections are absent in the classical practical SI system. **The spherical or cubic mass of 1 kilogram radiates in one plane mass, energy, momentum or power of 10 N-s/kg-K² of SI units. In one orientation at atomic level it expresses one 1D signal of 1 N-s/kg-K². In three planes this value becomes 30 eV equal to the dissociation energy of Bohr's atom, but there are 36 quantum orientations with 32 or 30 interacting. The sun radiation on earth corrected for quantum and twice for relativity in 32 quantum orientations generates per pulse:**

$$(6000 \text{ K})^{1/32} \times 1.04^{2/5} / 1.15470^2 \equiv 1 \text{ J (eV)} - 0.00019 \text{ eV}(\equiv 2k_s) \qquad (2.1)$$

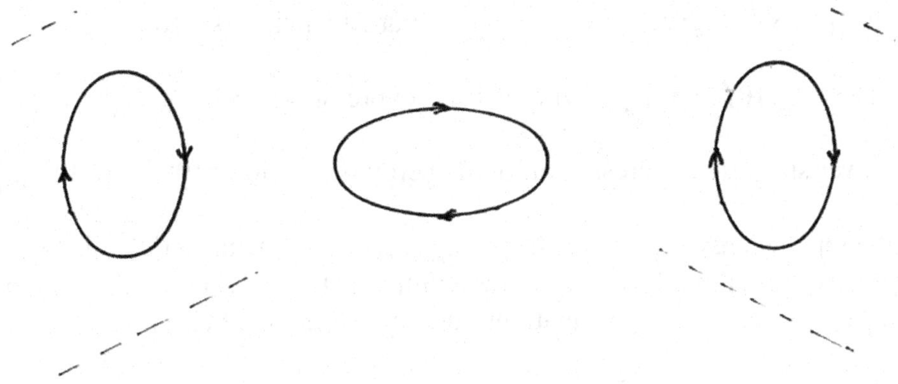

Figure 6. 2001 electrons rotating at three orthogonal planes forming the proton.

The model postulales that electrons form proton mass by chance combinations of radiating momentum as shown in Fig. 6. They interact with each other nulling their momenta to form the mass of the proton as explained in Chapter 13. The momentum of Bohr atom created by $3 \times 666 + 2 = 2000$ electrons generates the proton mass of 2×10^{-27} kg. Proton generates pulses moving through space at the statistical energy density 1×10^{-28} J/m³ expressing a muo10$_s$ on atomic scale.

The basic requirement of physics is that the valid laws of physics must apear to be the same for all the observers. According to the proposed integer model radiation particles from the edge of the Universe are hydrogen converted to radiation energy moving at the velocity of light. The model postulates that electrons generally move with local component velocity $1/2 \; c_s$, protons move with local velocity $1/4 \; c_s$, while radiation particle carry both components with average local velocity $1/3 \; c_s$. The velocity c_s and $1/3 \; c_s$ are related to the two constants of space permittivity ε_{os} and permiability μ_{os}. The relation deriving space pulse $E_{sspaceSI}$ is:

$$E_{sspaceSI} = E_{s1kgSI} = 1/(\varepsilon_{os}\mu_{os}) \equiv 1 \text{ kg} \times c_s^2 \times 1.111\ldots = (3 \times 10^8)^2 \times 1.11\ldots = 1.000\ldots \times 10^{17} \text{ J} \qquad (2.2)$$

And earth gravity force in one of 36 orientations

Vector $E_{searthSI}$ = 3.5438 x 10^{22} x$1.04^{5/12}$/36= 1.000608 x 10^{21} N or J. $\qquad$ (2.2a)

Where c_s is actually $\sqrt{3}$x1/2 c_s x 1.1547005. Product $\varepsilon_{os}\mu_o$ and ratio ε_{os}/μ_o derive interesting energies:

$$(\varepsilon_{os}/\mu_{os})^2 \equiv 1.0\text{x}10^{-10}\,\text{J}^2 \equiv G_s, \qquad \sqrt{(\varepsilon_{os}/\mu_{os})}\text{x}1/1.111\ldots^{1/2} \equiv 0.003\,\text{J} \qquad (2.2b)$$
$$\mu_{os}/\varepsilon_{os} \equiv 1.0\text{x}10^5\,\text{J} \qquad\qquad \varepsilon_{os}/\mu_{os} \equiv 1.0\text{x}10^{-5}\,\text{J}^{-1}$$
$$\varepsilon_{os}\mu_{os} = 10^{-11}\text{ x }10^{-6} \equiv 10^{-17}\,\text{J}^{-1}.$$

Square root of space permiability $\sqrt{\mu_{os}}$ generates the elementary 1D energy pulses of space of 0.001 J; five of each (ten together) generate Plank's 1st radiation term 0.01 J:

Elementary radiation SI pulses at atomic level = 0.001 J $\equiv (\mu_{os})^{1/2} \equiv (\varepsilon_{os}/10)^{1/4}$ $\qquad$ (2.2c)

These are the pulses of Figure 3 creating in one plane Planck's 1st radiation term:

0.01 J$\equiv(\mu_{os})^{1/3} \equiv (\varepsilon_{os}/10)^{1/6}$ and ε_{os} moving at 1/3 c_s expressing ε_{os} x1/3$c_{s=}\,\mu_{os}^{1/2}$ $\qquad$ (2.d)

Raised to power six, 0.01^6, expresses infrared energy per cycle 1x10^{-12} J=1/10^{12}=$\varepsilon_{os}/10 \equiv \mu_{os}^2$.

Thus ε_o is the input energy for 10 quanta and μ_{os} is square of quantum; that is the probability of light velocity formation. Please note the component 3 x 1.111 $\ldots^{1/2}$ = 3.16227766 = $\sqrt{(1\text{x}10^7)}$ appears several times in equations and $\sqrt{\varepsilon_{os}} \equiv 3\,\mu_{os}$ x **1.111** $\ldots^{1/2}$.

The product of $E_{sspaceSI}\,h_s$ is 10^{17}x$10^{-33} \equiv 1\text{x}10^{-16}$ J$\rightarrow$ x1/1x$10^{-19} \equiv$ **1000 K in the sun and earth pulse expressed in several expressions.**

In the book the relativity corrector is placed beside velocity of light. The arrow $\rightarrow$ indicates that the value has been moved to the right for the next transaction.

Although local velocities of electron, proton and (electron+proton) pair are one third of the velocity of light, they add to the real velocity of light. The parameter 10^{17} has the following relations to mimum energy 1x10^{-14} J$\equiv$1x10^{20}x1/(1.0x10^{17})^2 J, to Planck constant h_s= 1 J/(1.0x10^{17})^{2}x10 =1x10^{-33} J-s, and permittivitty $\varepsilon_{os} \equiv (h_s^{1/3})$ J.

Using the integer physical constant E_{s1kgSI} and the relativity correction for 1/3 c_s derives a new parameter of energy pulse per plane:

$$E_{sspaceSI} \equiv (\varepsilon_{os}\text{ x }\mu_o)^{-1} \equiv 1\text{ kg }x\,c_s^2\text{ x Relativity cor.}= (3\text{x}10^8)^2\text{ x }1.111\ldots = 1\,x\,10^{17}\text{ J} \qquad (2.3)$$

Planck constant $h_s \equiv 10\,/E_{sspaceSI}^2$; h_s expresses J-s at atomic level, multiplying by 10 x 10^{30} derives Planck's 1st radiation term vectorized energy at the SI level

$$\mu_{os}^{1/3} = (\varepsilon_{os}/10)^{1/6} \equiv 0.01\text{ J, at atomic integer }\mu_{os}^{1/2} = e_s\text{ x }(\,1/3\,c_s)^2 \equiv \textbf{0.001 J} \qquad (2.4)$$

Please note that 2D (surface) converter $10^{-20}/e_s$=$10^{-20}/10^{-19} \equiv$ 0.1 J is unvectorized energy.

The space radiation energy 1×10^{-28} J/cycle per orthogonal orientation is the muo10$_s$ mass. It is the minimum energy $h_s/e_s = 1 \times 10^{-14}$ J squared expressing two pulses displaced by 90°: $10^{-28} \times 10^{-28}$; $(h_s/e_s)^2 = (1 \times 10^{-33}/1 \times 10^{-19})^2$ and it expresses muo10$_s$ mass and the electron mass of five quantization pathways converted to energy per cycle of electron frequency:

(2.5)

$$(h_s/e_s)^2 = 5m_{es} \times c_s^2 \times 1.111 \ldots/v_{Hms} = 5 \times 10^{-30} \times 9 \times 10^{16} \times 1.111 \ldots/5 \times 10^{15} = 1 \times 10^{-28} \text{ J/cycle}$$

Actually the integer cubic space value derived now in integer model SI units is 1×10^{-28} W/cycle, but the equivalence of units allows us to consider this to be energy. **Energy 1×10^{-28} J is the SI muo10$_s$ mass expressed in static state by Equation 17.11.** In the classsical integer SI units the value of space energy is 10^{-26} J/m^3 refers likely to 300 K per orinettaion. These two values represent also minimum energies $(10^{-26})^{1/2} = 10^{-13}$ J and $(10^{-28})^{1/2} = 10^{-14}$ JSI discussed subsequently. In the postulated model 1×10^{-28} WSI expresses power of the radiation pulse at the atomic level and not the inherent power of space, like Casimir effect. Value of 1×10^{-28} kg is the muo10$_s$ mass in the model integer units and expresses force.

The energy 1×10^{-28} JSI in one dimensional display of $(1 \times 10^{-28}$ J$)^{1/3}$ J, corrected for relativity for ½ c_s and for quantum, expresses along one of 5 pathways of space quantization the value 1×10^{-10} J of the International Gravity constant; that value expresses also linear probability or the linear converter of SI units to the atomic units:

$$1/5 \times (1 \times 10^{-28})^{1/3} \times 1.1547^{1/2} \times 1.04^{1/16} = (1 - 0.00001) \times 10^{-10} \text{ J} \qquad (2.6)$$

Atomic force $F_{ats} = 1 \times 10^{-7}$ N applied in three orthonogal orientations as shown in Figure 6 generates the International Gravitation constant G_s for bodies at equal temperatures of 6000 K, expressing triple tenfold decrease of sequential probability of the event occurance:

International Gravitation constant $G_s \equiv F_{ats} \times 1/10 \times 1/10 \times 1/10 = 1 \times 10^{-10}$ N $\qquad$ (2.7)

Power index is drivative of probabilty conversion explained in the two earlier books [45, 46]. It may or may not change the units. The above basic value of the model expresses also the one dimensional display of the electron mass:

$$(1 \times 10^{-30} \text{ kg})^{1/3} \equiv 1 \times 10^{-10} \text{ kg} \qquad (2.8)$$

Due to one step quantization $1/10 \times h_s = \mu_{os}/1.111 \ldots \times 1 \times 10^{-28}$ J-s. $\qquad$ (2.9)

Boltzmann constant expressed by other constants

$$k_s = 1/10 \times h_s/\varepsilon_{os} = 1 \times 10^{-34}/1 \times 10^{-11} = 1 \times 10^{-23} \text{ J}. \qquad (2.10)$$

Using relativity corrector of $1.111 \ldots$ for $1/3$ c_s, derives density of the space signal:

$$1.111 \ldots \varepsilon_{os} c_s^2 \equiv 1.111 \ldots/\mu_{os} \equiv 10^3 \times 10^3 \text{ JSI} \equiv 1 \text{ MW/m}^2 \qquad (2.11)$$

The product of electron charge e_s and velocity of light derives ε_{os} implying that ε_{os} is current. Since in Equation 1.15 e_s has been found to be momentum ε_{os} remains energy:

$$\varepsilon_{os} = e_s \times 1/3 \ c_s = 1 \times 10^{-19} \times 1 \times 10^8 \equiv 1 \times 10^{-11} \ \text{J/m}^3 \tag{2.12}$$

According to integer model the local velocity 1×10^8 m/s expresses product of two pulse rates generated by temperatures of $\sqrt{3} \times 6000 \times 1.0392 = 10000.03$ K so that: 10^{-11} J/m^3

$$\varepsilon_{os} = e_s \times 1/3 \ c_s = 1 \times 10^{-19} \times [10000 \ \text{K} \times 10000 \ \text{K}] = 1 \times 1 \times 10^{-11} \ \text{J/m}^3 \tag{2.12a}$$

Product of pulse rates is the inverse of the probability product, which decreases the value of the resultant. In $\varepsilon_{os} = G_s \times 1/10$, G_s can be expressed by the product $(1/10^5 \times 1/10^5)$, where 10^5 J is the earth pulse and the term in brackets is the product of probabilities.

The constant ε_{os} can be expressed by the product of pulse rates or of probabilities. In atomic terms $\varepsilon_{os} = 1 \times 10^{-11}$ expresses momentum of radiation pulse moving through space; divided by E_{s1kgSI} it derives energy in integer SI units:

Power (equal muo10$_s$ mass) at 1 K $= \varepsilon_{os}/E_{s1kgSI} = 1.0 \times 10^{-11} \times 1 \times 10^{-17} = 1 \times 10^{-28}$ J/s-m^3 (2.12b)

At 300 K per meter of velocity of light 1×10^{-28} J/s-m^3 $\times 300/3 \times 10^8$ m/s $\equiv 1 \times 10^{-34}$ J, which is minimum energy at the atomic level $1 \times 10^{-14} \times 10^{-30} = h_s/10$. (2.12c)

Energy in eV for 3 oscillators $\equiv 1 \times 10^{-28}/10^{-19} \times 10^8 \times 1/10 \times 3 = 0.03$ eV/s (2.13)

Energy moving at 300 K $= 0.03$ eV $\times 300 \times 1.111 \ldots = 10$ eV (2.14)

At SI level that energy varies with the square of temperature and comprises the invert of linear probability $= 1 \times 10^{-28}$ J/m^3 $\times 300^2 \times 1 \times 10^{10} \times 1.111 \ldots = 1 \times 10^{-13}$ JSI. (2.15)

With sun reacting only with 1/2 of its mass and in one plane 1/4 of its temperature per orientation generates

Energy moving at 6000 K $= 0.03$ eV $\times 1500 \times 1.111 \ldots = 50$ eVat (2.15a)

At SI level energy $\varepsilon_{os} = 1 \times 10^{-28}$ J/m^3 $\times 3000^2 \times 1 \times 10^{10} \times 1.111 \ldots = 1 \times 10^{-11}$ JSI $= 1 \times 10^8$ eVat $\equiv 1/3 \ c_s$. (2.16)

suggesting that **1/3 c_s (eV) expresses a unity energy pulse of the sun in one orientation**, which in specific direction becomes $1 \times 10^8 \times 1/10 \times 1/10 = 1 \times 10^6$ eV $= 1$ MeV pulse.

Earth pulse is taken as pair of quantum corrected pulses in wave function formation increasing 4 times by relativity and quantum correction and displaying H. u. of $6kT$:

$$4 \times (\sqrt{3} \times 300 \ \text{K}) \ X \ (\sqrt{3} \times 300 \ \text{K})/1.04^{59/60} \equiv \mathbf{0.99983 \times 10^6 \ eV} \equiv 999.83 \ \text{K} \approx 1000 \ \text{K} \tag{2.17}$$

The products of most wave function formations in the book are marked by X indicating the process of the velocity c_s formations.

With cor. (correction) $3\sigma_s T_{300}{}^4/\text{cor.} = 3 \times (5 \times 10^{-8}) \times 300^4/[1.2(1+1/80)] = 1000.00 \ldots \text{W/m}^2$

$$(2.18)$$

Power from earth at 300 W/m^2 $\equiv$ 300 K: $\sigma_s T_{300}{}^4 \times 2/3 \times 1.111 = 300.00 \ldots \text{W/m}^2$ $\qquad$ (2.19)

Thus in above Equations $\sigma_s T_{300}{}^4 = \sqrt{\sigma_s T_{300}{}^2} \times \sqrt{\sigma_s T_{300}{}^2}$. $\qquad\qquad$ (2.19a)

Dividing by 300 power from earth of 1 W/m^2 $\equiv \sigma_s T_{300}{}^3 \times 2/3 \times 1.111 \ldots = 1.00 \ldots \text{W/m}^2$ (2.20)

Three 300 K added in orthogonal orientation relativity corrected $=\sqrt{3} \times 300 \times 1.1111 = 1000$ K. Energy of 1.0 MeV is actually the relativity corrected electron mass converted to energy $\equiv$ 1000 W/m^2 so there is more than one way to obtain that power density or temperature:

With $m_{es}xc_s{}^2 \times 1.111 \ldots = 10^{-30} \times 9 \times 10^{16} \times 1.11 \ldots = 1 \times 10^{-13}$ J $\rightarrow x1/1 \times 10^{-19} = \mathbf{1\ MeV_{SI}} \equiv 1/\mu_{os}$ defining also μ_{os}.

$$(2.21)$$

When proton mass is converted to energy, the relativity corrected (cor.) pulse energy *is.* (2.22)
$\qquad m_{ps} c_s{}^2 \times 1.111 \ldots = 1 \times 10^{-10}$ J $\equiv G_s$ converting to eV $1 \times 10^{-10}/1 \times 10^{-19} \equiv 1 \times 10^9$ eV $= \mathbf{1\ GeV_{SI}}$.

During the conversion of electron, of proton, of atom or molecule into radiation, conservation of momentum is satisfied because momenta appearing in the two opposite orientations add to zero. A source of energy is still needed. Conversion of proton mass delivers energy of graviton. These oscillating pulses convert to vectors by the probability of being space quantized $(1/10 \times 1/10)^3$ converting two 1 MeV input thermal photon pulses into 2 J vector forces.

It is suggested that similar process converts also electron clouds into protons and into the hydrogen, two atoms forming molecules. They create the space radiation 2.725 eV discussed below.

Term $m_{ps}c_s{}^2$ is proton momentum $m_{ps}c_s$ converted by c_s to SI units. In the integer model $c_s{}^2$ has the wave function format of two pulses generating another pulse at the velocity of light generating energy per wave of the space frequency $c_s{}^2 \times 1.1111 \ldots /1 \times 10^{17}$ = 1 J. However two pulses in wave function format, as indicated in other parts of the book, generate $\sqrt{2}$ larger pulse, which has to be relativity and quantum corrected and multiplied by $\sqrt{2}$ because interacting radiation is in one orientation of one plane. The total more accurate vector pulse is: $2c_s{}^2 \times 1.111 \ldots^2 \times 1.154700^2 \times (1+1/80) = 2.999997 \times 10^{17} = 3 \times 10^{17} \rightarrow \times 1 \times 10^{-19} = 0.03$ eV or J with uncertainty 300 k_s, expressing k_s per degree kelvin. The variance is k_s per each k_s in 30 orientations. Converted with relativity correction to 300 K derives 10 eV and 30 eV for three planes.

All gravity radiations are vectors, because they move in specified direction and oscillate only in one orientation. However term energy does not specify the vector format and its vectorized format is reaffirmed at times.

The product of radiation energy and E_{s1kgSI} derives vectors of energy 10^{-14} J x 1 x 10^{17} J-s $\equiv$ 1000 K $\equiv$ 1000 W/m^2. Two such pulses in wave function format generate 1 MeV vector pulse of Equations 2.21-22. Energy of 1 MJ divided by 1×10^8 derives Planck radiation term 0.01 J.

Radiation energy of proton mass is also $(\varepsilon_o \mu_o) \times 1 \times 10^{-10} \equiv \mathbf{1 \times 10^{-27}\ J}$ $\qquad$ (2.23)

It expressing energy of ten muo10$_s$.

Energy per second is 1×10^{-17} J x 1×10^{17} J-s $\equiv$ 1 W.

Wave function of energy $\varepsilon_{os}\mu_{os}$ X $\varepsilon_{os}\mu_{os}$ = 10^{-17} X 10^{-17} = 1×10^{-34} J = $h_s/10$ is expressing in SI classical units ½ of the crossed $h = h/4\pi$ (2.24)

Vectorized energy 1×10^{-14} J is produced by the 1×10^{-17} pulse. It is 3 proton momenta at 600 K converted to 600 K:

Vector energy $1\times10^{-17}=1\times10^{-14}$ x1/10 x1/10 x1/10 $\equiv 3k_sT_{600}^2/(1.1547^{1/2}$ x $1.04^{0.13})$ (2.25)

And $1\times10^{-14} \equiv \varepsilon_{os}$ x1/10 x1/10 x1/10.

Radiation peaks at 2.82 kT which times $1.1111^{1/2}\times1.04^{1/4}=3.001829$ k_sT is giving the model integer value. Energy 1×10^{-15} JSI is one of 10 local space quantization energy components comprising the minimum input energy pulse of 1×10^{-14} JSI. At 6000 K $3k_sT_{6000}$ = 1×10^{-15} JSI is one of 10 local space quantization energy components comprising the minimum energy pulse of 1×10^{-14} JSI $\equiv \sqrt{(}$muo10$_s$ integer mass), i.e. $(1\times10^{-28}$ kg$)^{1/2}$.

The two infrared red pulses produced by radiation mass of the proton 1×10^{-27} kg and by the pair (proton+electron) generate effectively the same energy each moving at velocity $1/3$ c_s:

$$1\times10^{-27}\text{x } 1/3 \text{ x } 3 \times 10^8 \equiv 1 \text{ eVat},$$

producing in 3 planes $\qquad 1\times10^{-27}\text{x}3\text{x}10^8 = 3 \text{ eVat}$ (2.26)

These vectorized energy E_n values are derived also by energy density equation in which energies densities per meter cube of electrical fields E^2 and magnetic field B^2 expressed per atom by constants ε_{os} and μ_{os} with $10^{-10}\times10^3$ converters and other correctors:

$$E_n/volume=1/2 \text{ x } [1/(c_s^2\varepsilon_{os})^2+\mu_{os}^2] \text{ x}10^{-10}\text{x}10^3/(10^{-19}\text{x}1.11\ldots^2) \equiv 0.99985 \text{ eVSI/m}^3 \quad (2.27)$$

The value 1 eVat has been now derived from several experiments of relaxation and high field condustion in NaCL crystal carried out by the author [2-4]. Naturally occurring calcium divalent impurities in the crystal (actually beside the crystal) combines with the crystal vacancies forming free dipoles modelling well the Bohr atom. The energy of the closest association is 0.88 eV, which times relativity and quantum correctors 1.11 . . . x$1.04^{0.6}$ =1.001 eVat. The average NaCl vacancy mobility derived by Whitman and Calderwood [12] from data of **40 researchers** is 0.78 eVat ±0.066 eV, that value being effectively the authors' jump energy of U_j = 0.88 - (8x0.0109) = 0.79 eVat. The energy 0.78 eVat is the lattice energy of NaCL over quantum corrector $1.04^{0.7}$. Corrected by relativity energy 0.79 eV derives 0.79 x $1.2/1.1111^{1/2}$ = 0.9993 eV $\approx$ 1 eVat in one plane. Because earth does not generate pure electron radiation only 2/3 of 1 W = 0.667 eVat of sun pulse is balanced by thermal radiation corrected for relativity 1.1547 to 0.77 eVat, fitting also [12] data. The energy of Bohr atom of 10 J expresses above 0.78 eV as 10 eV/($1.2\times1.11\ldots^{1/2}\times1.0328^{1/2}$) = 7.77 eV, factor 1.2 being unvisible component of energy discussed subsequentley.

Both ε_{os} and μ_{os} refer to energy transmitted in one plane at 1 K. **The Planck Radiation law derives the value 0.01 J. On converting to 300 K it becomes 3 J for three oscillators**. Oscillations of proton and of the pair, (proton+electron), produce in author's interpretation the well known, doublets in frequencies, because the two masses differ by 1/1000.

The Stephan-Boltzmann constant σ_s has been undervalued in its importance in physics. Two pulses $\sqrt{10}$ x ($\sqrt{2}$ x6000)^2 divided by the squared local velocity and corrected for quantum correction derives energy of 3 J from the sun including electron oscillation:

$$[\sqrt{10}x(\sqrt{2} \; x6000)^2]/(1 \; x10^8)X[\sqrt{10}x(\sqrt{2} \; x6000)^2]/(1 \; x10^8) \; x1/\sqrt{3} \; x1.04^{3/50} \equiv 3.000035 \text{ J or eV} \tag{2.28}$$

where $\sqrt{10}/(1.11111^{1/2}) = 3.00000000$ J is sum of energies of electron, proton and the pair, with variance $k_s/3$. Temperature here is considered to be the pulse rate.

The $\sqrt{3}$ larger resultant vector probability 1/1000 appears along the longest axis of the cube, as in Fig.5. Multiplied by relativity correction K_{rel}=1.1547 derives probability value of 1/500, making it 2 eVat for 1000 eVat input in one direction (1/6 x6000).

The first experimental data consistent with the cubic integer model were provided by Volta over hundred years ago in the measurements of the gradient of potential in the atmosphere and some final calculations are at the end of the Appendix. The force pulses of 1 N/m are intensified to 88 V/m per orientation of the cube by the presence of water below and above the measuring range generating ($\sqrt{3}$ x 88) V/m. Two pulses in one form of the wave function format generate the field gradient $\sqrt{2}$ x ($\sqrt{3}$ x 88)^2 = 32855 V/m $\approx$ 30 kV/cm just sufficient to break air in the presence of a free electron. Electron streamers convert to the arc of the lightning, while changing orientations in 50-100 meter's steps. This is how nature confirms the integer model assertions.

The ratio of volumes of the cubic Bohr atom to the spherical Bohr atom is

$$(1x10^{-10})^3/[4/3 \text{ x } \pi \text{ x } (0.5291772 \text{ x}10^{-10})^3] \text{ x}1.04^{1/6}/1.2 \text{ x } 1/1.1547^2 = 1.00035 \tag{2.29}$$

With quantum and relativity corrections the volume of the cubic Bohr atom deviates from that of its spherical presentation by the uncertainty expressed by 0.00035x10^{-30} x 30 eV = 1x10^{-32} J/m^3 $\equiv$ 10 h_{as}; this expresses H. u. of energy h_s in ten quantization directions. In expressing the distance traveled in terms of frequency and of wavelength several corrections have to be made. The frequency has five pathways of space quantization shown in Fig. 1 and only one has the right direction. The pulse moves with velocity 1/3 c_s, its probability of forming a vector is 1/1000 and the chance of the event in linear atomic units is 10^{-10} smaller than in SI units.

Equipartition of energy is taken to mean that different forms of energies are equal, expressed by oscillation and by direct progression. The energy 1 x 10^{-28} J/m^3 is the Bohr atom energy expressed by twice atomic force F_{atom} times radius 1/2 x10^{-10} m deriving 2x1x10^{-7}x ½ x1x10^{-10} = 1x10^{-28} J, which divided by volume probability derives energy of Planck Radiation Law 0.01 J. Corrected for relativity, **energy 1x10^{-28} J expresses muo10$_s$ mass** in Equation 17.11. Muo10$_s$ is the particle responsible for forces between particles within the atom expressing 10 quantum components of proton mass. Energy 0.01 J of one orientation discharging from the whole atom in 32 orientations, three planes and relativity corrected derives unity value with 5 k_s variance:

$$\text{Energy} = 32 \text{ x } 3 \text{ x } 1.111 \ldots^{1/2} = (1 - 0.000514) \text{ J or eV per atomic event} \tag{2.29a}$$

The muo10$_s$'s energy density 1×10^{-28} J/m^3 in one of five quantization pathways expressed in 1D by power index 1/3, corrected by 1.07456 for relativity, corrected for local velocity ($\frac{1}{2}\, c_s$)^2 and converted to SI units by linear probability 10^{10} with quantum corrector derives the value 1 eV with accuracy 1 in 10^5:

$$E_n/volume = 1/5 \times (1 \times 10^{-28})^{1/3} \times 1.07456 \times 10^{10} \times 1.04^{1/16} \equiv 0.99998 \text{ J/m}^3 \approx 1 \text{ J/m}^3 \qquad (2.30)$$

Please note that $1/5 \times (1 \times 10^{-28})^{1/3} \times 1.07456 \times 1.04^{1/16} = 1.000002877\times10^{-10}$ is the accurate linear probability. The variance $\sim3\times10^{-16}$ J/m^3 is high, but per K it expresses $10e_s$ that is e_s per orientation.

The above energy represents a vector produced by electron subjected to the atomic force of $F_{as} = 1\times10^{-7}$ N/m moving over half of the Bohr atom diameter, $a_{os} = \frac{1}{2}\times10^{-10}$ m. This expresses energy in integer SI units, the value in atomic terms requires multiplication by 1×10^{-10} linear probability to express the atomic event displaying the source of 1×10^{-28} J/m^3 energy from the Bohr atom, the muo10$_s$ mass. The multiplication of two minimum energy terms 1×10^{-14} J (or probabilities) displaced by $90°$ generates two vectors displaced from the origin by velocity of light producing the pulse progressing at c_s, the expression taking wave function form:

$$1/2\, F_{as} \times a_{os} \times 1 \times 10^{-10} = 1 \times 10^{-28} \text{ J per pulse} \equiv (1 \times 10^{-14}) \text{X} (1 \times 10^{-14}) \text{ J} \qquad (2.31)$$

the product deriving the constant $1 \times 10^{-14} \times 1 \times 10^{8} = 1 \times 10^{-6}$ W $= \mu_{os}$ per second per plane.

The pulse in SI integer units in four directions of the geometry of space generates energy $2\, F_{as} \times a_{os} = 1 \times 10^{-17}$ J, which at $E_{sspaceSI} = 1 \times 10^{17}$ J-s at 1 K has gradient 1 J^2; the square should be neglected, because it is product of two pulses forming another.

Since radiation energy expresses (mass x c_s^2), the mass of space/m^3 is the radiation pulse in one direction at the atomic level per kelvin:

$$1 \times 10^{-28}/(9 \times 10^{16} \times 1.111\ldots) = 1 \times 10^{-45} \text{ kg/m}^3 \qquad (2.32)$$

For ten directions atomic energy density becomes 1×10^{-44} kg/m^3. Now $h\varepsilon_{os} \equiv$ Min. energy $1 \times 10^{-14} \times 10^{-30} \equiv 1 \times 10^{-44}$; energy in one orientation $\frac{1}{2}\,(600K)^4 \times 6\times10^{24}$ kg $\times3\times10^8/1.04^{3.9}$ $= 1.00092 \times10^{44}$ multiplied by 1×10^{-44} derives unity pulse. The assumption that local velocities of radiation particles is $1/3 \times c_s$ derives correct radiation energy gradient in

$$1 \times 10^{-44} \text{ kg/m}^3 \times (1/3 \times 3 \times 10^8)^2 = 1 \times10^{-28} \text{ kg-s}^2/\text{m} \qquad (2.33)$$

Multiplied by invert of volume probability, 1×10^{-44} kg/m^3 $\times1\times10^{30}$, the gradient derives again the minimum SI particle energy 1×10^{-14} J. This means that values 1×10^{-44}, 1×10^{-27} and 1×10^{-28} express units of the integer cubic model SI system.

Actually it is suggested that the format of Equation 2.33 is misleading. In the wave function format energy 1×10^{-28} J moving forward is created by the invert of probability 1/10 from the line of 10 orientations in two planes the product of two pulses from two planes:

$$(k_s \times 1/3\, c_s \times 10) \text{X} (k_s \times 1/3\, c_s \times 10) = (1\times10^{-14}) \text{ X } (1\times10^{-14}) = 1\times10^{-28} \text{ J} \qquad (2.34)$$

Radiation pulse 1×10^{-27} J expresses one half of (proton+electron) pair mass and when accelerated to velocity of light and corrected for probability delivers energy of the gravity constant, 1×10^{-10} J:

$$G_s = 1 \times 10^{-27} \times (3 \times 10^8)^2 \times 1.111\ldots = (3 \times 10^{-14} \times 3 \times 10^8)X(3 \times 10^{-14} \times 3 \times 10^8) \times 1.111\ldots^{3/2} = 1 \times 10^{-10}\, J \quad (2.35)$$
$$\equiv 10\varepsilon_{os.}$$

The above value refers to atomic dimensions; SI units require multiplication by the volume probability 10^{30} generating 1×10^{-15} kg/m³. So there are three classical SI values to be considered:

Proton power generated per cycle $k_s/(1/3\ c_s) = 1/5 \times 5 \times 10^{12} \times 1 \times 10^{-27} \equiv 1 \times 10^{-15}$ W or J   (2.36)

Electron radiation generates power $1/5 \times 5 \times 10^{15} \times 1 \times 10^{-30} \equiv 1 \times 10^{-15}$ W or J   (2.37)

and electron mass $1 \times 10^{-30} \equiv (1 \times 10^{-15}\, J)^2$

The radiations generating forces observed in NaCl experiments [3] involved the whole Bohr atom expressed by the equivalent electron oscillating with proton practically at frequency 10^{12} Hz involving convertion of the whole mass to radiation.

Values 1×10^{-15} W or J apply to 10 local pathway, so that the minimum energy 1×10^{-14} J transmitted value remains valid. Above values in 1D presentation are the cube of Equation 2.37 $(1 \times 10^{-15}\, W)^3 \equiv 1 \times 10^{-45}$ kg/m³ and 1×10^{-15} W time volume converter 10^{-30} deriving energy at the atomic level. This value is derived from the rules of probability in orthogonal geometry generating a progressing pulse having significantly smaller energy than its components in three orientations.

The basic premise of the model is the equipartition of energy rule. The Bohr atom carries three independent oscillators produced by electron vibrating at the mean frequencies of 5×10^{15} Hz locally represented by 1×10^{15} Hz, proton at 5×10^{12} Hz and (proton+electron) pair at about 4.995×10^{12} Hz, usually not differentiated in the model from 5×10^{12} Hz. Frequency parameters are expressed by the same values applicable both to SI and atomic interactions. The momenta of three 10^{-15} kg/m³ masses add, with converters from atomic to SI being 1×10^{15} and 5×10^{12}, and generate the following Planck Law energies per second, one of violet frequency and two of infrared frequencies:

$$3 \times 1 \times 10^{-15} J \times (1 \times 10^{15})^2/(1 \times 10^{15}) \equiv 3\ eVat\ \text{and}\ 2 \times 1 \times 10^{-15} J \times (5 \times 10^{12})^2/(5 \times 10^{12}) \equiv 0.01\ eVat \quad (2.38)$$

In three planes the product of 3 eV with correctors of relativity and quantunm $3^2 \times 1.1547^3/1.04^{11/25} = 13.619$ eV. Please note again **that energy values E_n derived by the integer space model and by the classical SI are effectively equal.**

The electron frequency per space quantization pathway is

$$v_{Hms} = E_n/h_s e_s = 10.2/(1 \times 10^{-33} \times 1 \times 10^{-19}) = 1.00033 \times 10^{15}\ Hz \quad (2.39)$$

While the corrected square of local self frequency of electron moving with proton at ¼ c_s derives 5×10^{15} Hz:

$$(¼\, c_s)^2/(1.11\ldots \times 1.04^{15/48}) \equiv 5.0008 \times 10^{15}\ Hz \quad (2.40)$$

For three oscillating planes and time relativity correction for $1/3\ c_s$ the energy is

$$3 \times 3 \text{ eV} \times 1.11 \ldots = 10 \text{ eVat} \tag{2.41}$$

When using wavelength it is necessary to distinguish incoherent radiation from radiation subjected to quantization forming vector forces and also from radiation of normalized values per kelvin and from values measured at 300 K. In cubic space unit π of several SI expressions is replaced by 3.

Integer Boltzmann constant is the minimum energy expressed in integer SI units converted by linear converter to atomic level:

$$\text{Integer Boltzmann constant } k_s = 1 \times 10^{-13} \text{ J} \times 1 \times 10^{-10} = 1 \times 10^{-23} \text{ J} \tag{2.42}$$

Boltzmann constant can be derived in terms of Avogado's number=6.022137 $\times 10^{23}$ of carbon atoms in 12 grams of carbon. Inverse vectorizarion inverses grams into SI kg and radiation in one direction converts 12 to 6. Hence k_s expresses the probability of mass of 1 kg carbon generation power or energy 1W or 1 J

$$P = 6/6.022137 \times 10^{23} = (1 - 0.003878) \times 10^{-23} \tag{2.42a}$$

The electron radiation input energy 1×10^{-13} J per cycle of electron frequency 5×10^{15} Hz units in five quantization pathways is exactly the energy per unit volume of space expressing again muo10$_s$ value: $\quad 5 \times 1 \times 10^{-13}/5 \times 10^{15} = 1 \times 10^{-28} \text{ J/m}^3 \tag{2.43}$

Electron charge is space energy moving at the velocity of light in three planes with 1.111 . . . relativity corrections:

$$e_s = 1 \times 10^{-28} \times 3 \times 10^8 \times 3 \times 1.111 \ldots = 1 \times 10^{-19} \text{ C} \tag{2.44}$$

Electron charge is negative proton charge and there is tendency to forget the presence of two 'electron charge' frequencies.

Space energy momentum is unity:

$$(1 \times 10^{-28}) \times c_s/e_s \times 3 \times 1.111 \ldots = 1 \text{ kg-m/(s-C)} \tag{2.45}$$

Mass of 1 kg moving at local velocity of ½ c_s converted to natural atomic units by multiplying by volume probability derives 10 Boltzmann constants:

$$1/3 \times c_s \times 1 \text{ kg} \times \text{vol.probability} \equiv 1/3 \times 3 \times 10^8 \times 1 \times 1 \times 10^{-30} = 1 \times 10^{-22} \text{ J} \equiv 10 \, k_s \tag{2.46}$$

Electron mass moving at local velocity of ½ c_s expresses energy in atomic units.

Pulses claimed to be observed from the edge of the Universe are likely to be according to the integer model proposed converted H_2 molecules into radiation comprising $6k_s$ oscillators including 2 electrons, two protons and two (proton+electron) pairs expressing energy in 3 planes referred to 300 K:

$$6k_s/e_s \times ¼ c_s = 6 \times 2 \times 10^{-27}/1 \times 10^{-19} \times ¼ \times 3 \times 10^8 \equiv 3 \text{ eVSI} \tag{2.47}$$

On earth the pulse represents well the measured:

$$3 \text{ eVSI-}5 \, k_s T_{300}/e_s \times K_{rel}^2 = 3 - 5 \times 2 \times 10^{-27} \times 300/1 \times 10^{-19} \times 1.333 \equiv 2.725 \text{ K} \tag{2.48}$$

where 5 are for six k_sT/e_s oscillators less two $1/2xk_sT/e_s$ uncertainties for two degrees of freedom. If these pulses were from an excitation state their energy would decrease on radiation expressing a decrement of energy gradient *dw/dx* meaning an attractive force. So initially the only force in the Universe would be the the gravitation force of attraction. A total conversion of such atoms into radiation proposed above would also generate a force of attraction.

Viewed from a vantage point above North Pole the earth appears to rotate in counter clockwise direction around the sun. From the same vantage point earth rotates in counter clockwise direction around its axes. The earth rotation speed creates force of rotation momentum equal to the force gravity illustating the equipartition rule of energy. This is shown by the following relations:

Force $F_{Gs} = m_{earth} v_{earth}^2/L_{earth}$ x1.1547²x1.0327¹ᐟ⁴= 3.5429x10²² N
$\qquad = m_{earth}$ x(mass velocity at 2/3 earth radius)/(2/3 earth radius)
x1.1547²x1.0328¹ᐟ⁴= 6 x10²⁴ kgx196.62²/8.5x10⁶ x1.1547²x 1.0328¹ᐟ⁴=3.5429 x 10²²N $\qquad$ (2.49)
Thus F_{Gs}/m_{earth} = 3.5429x10²² N /6 x10²⁴ kg =0.0059 N/kg ≈ 0.006 N/kg $\qquad$ (2.50)

Or introducing Equation 16.52 F'_{Gs} =4.0000 x10²⁴/1.11 . . .=0.006x6x10²⁴ =3.6000 x10²⁴N.

The value 0.006 N/kg derived several times applies also to the ratio (squared linear earth velocity, 30000²)/distance to sun:

1 kg x 30000²/1.5x10¹¹ = (196.62²/8.5x10⁶) x1.1547²/1.04¹ᐟ⁴ ≡ 0.00600 N/kg $\qquad$ (2.51)

During earth rotation **the impulse generated** by gravitation force multiplied by duration *t* (seconds/year) of earth annual rotation] is within 0.6% equal to the **reactive impulse** formed by the mass of sun (1x10³⁰ kg) and the force of 1 N/kg over the relativity correction:

$$F_{Gs} \text{ x } t = 3.5438 \text{ x } 10^{22} \text{ x } 31.557 \text{ x } 10^6 \equiv 1x10^{30} \text{ x } 1 \text{ N/kg} /1.111 \ldots \qquad (2.51a)$$

Duration of 1 year balances units in the equations. When **force is replaced by product of mass and acceleration the impulse becomes momentum:**

$$\textbf{F = m x a = m x v/t therefore F x t = m x v} \qquad (2.52)$$

Aften ten years of searching the dimensions missing in SI sytem have been uncovered. The atomic interactions within the atom and in space have to obey the laws of elestic kinetics. Kinetics display the absence of dimensions of time and of length generating for space interactions the following equivalences of dimensions and properties:

$$\textbf{Force = Mass= Power = Momentum = Impulse = Energy = Charge} \qquad (2.52a)$$
and

Time = Velocity = Frequency = Rate = Stress = (Temperature)^2 (2.52b)

the latter confirmed by Equation **18.27**

Length and time are descriptives of space and not of interactions in point location of space. It is interesting to note that Einsten introduced these units in his spacetime. These two dimensions are absent in equations of atomic interactions. Consequently proposed relations of the integer model do not need to be dimmensionally correct in SI units.

Since force is mass, atomic force to power 4 is the muo10$_s$ mass, because atomic force is formed from two 1×10^{-14} J minimum energy 1 kg pulses in X and Y axes generating wave format at c_s in Z axis and forming the second branch of the radiation loop:

$$(F_{atomic})^4 \equiv 1 \text{ kg} \times (1 \times 10^{-7} \text{ N})^4 \equiv 1 \times 10^{-28} \text{ kg, integer cubic muo10}_s \text{ mass} \qquad (2.52c)$$

Similar equivalence of impulses in one dimension over time was first observed in the author's relaxation experiments. It required balancing impulse and energy applied to extend the Bohr atom (formed by the impurity), to the impulse and the energy of relaxing thermal flux. The well known energy of dissociation 8.1 eV [Dekker 53] corrected by relativity $8.1 \times 1.11 \ldots ^2 = 10$ eVat is also the dissociation energy of the integer model. In the integer model $1/2 \times 1000 k_s T_{300} \ln 2/(1.04 e_s) \approx 9.9973$ eVat with variace $27 k_s$ expressing k_s per orientation. The earth-sun circular balance is analysed in Equation 7.4.

The gravitation constant G_s is a vectorized parameter derived also from converting an elementary SI energy density of space in three step quantization, requiring the application of the third root of the radiation value. One of the five pathways of 1D presentation of the space energy derives, after correction, the linear probability 1×10^{-10}, otherwise G_s:

$$G_s \equiv 1/5 \times (1 \times 10^{-28})^{1/3} \times 1.1547^{1/2} \times 1.04^{1/16} = 0.99999 \times 10^{-10} \equiv 1 \times 10^{-10} \qquad (2.53)$$

Above value becomes the classical G with 2/3 multiplier.

The minimum SI radiation pulse is derived by applying h_s/e_s to two dimensions requiring square root value:
$$\text{Minimun output SI pulse } (h_s/e_s)^{1/2} = (1 \times 10^{-28})^{1/2} \equiv 1 \times 10^{-14} \text{ J} \qquad (2.54)$$

The minimum energy radiation value 1×10^{-14} J is the Planck constant per electron charge

$$h_s/e_s = 1 \times 10^{-33} \text{ J} / 1.0 \times 10^{-19} \text{ C} = 1.0 \times 10^{-14} \qquad (2.55)$$

Hence, **Klitzing$_s$ constant** in four orientations, directions dealing with the Hall effect, has in integer cubic units converted to SI the value of:

$$h_s/e_s^2 = 1.0 \times 10^{-14} / 1.0 \times 10^{-19} = 100\,000 \ \Omega \qquad (2.56)$$

in the model integer units. In that form it expresses four orientations so it is about four time the value ~25 813 Ω, which was the subject the Nobel Prize in 1985.

Electron mass converted to radiation subject to relativity correction and to one step quantization derives the minimum output energy value in the model integer SI units. The relation in wave function format expresses energy as the square of two vectors:

$$h_s/e_s = m_{es} \times c_s^2 \times 1.111\ldots \times 1/10 \qquad (2.57)$$
$$= (m_{es}^{1/2} \times c_s)^2 \times 1.111\ldots \times 1/10 = (1.0 \times 10^{-15} \times 3 \times 10^8)^2 \times 1.111\ldots \times 1/10 = 1 \times 10^{-14}\ J$$

Energy 1×10^{-14} J derives fairly accurate velocity of light in:

$$c_s \equiv \tfrac{1}{2} \times 10^{-14}\ J \times 3.5438 \times 10^{22}/1.1547 \times 1.04^{3/5} = 2.997677 \times 10^8 \qquad (2.58)$$

Linear probability per second in eV is three times the electron charge or energy 3 eV:

$$1 \times 10^{-10}/(1.111\ldots c_s\, e_s) \equiv 3\ eVSI/s \qquad (2.59)$$

The following identity expresses relation between the squared temperature and velocity of light. Energy dependance on the square of temperature is expressed also in Equations 14.5, but the power index may vary with the activation energy of the process.

$$c_s \equiv (6000-300)^2 \times 10 \times 10 \times 1.04^{3/5}/1.111\ldots \equiv 2.99373 \times 10^8\ m/s \qquad (2.60)$$

Individual $1/2c_s$ oscillations are subject to distribution not discussed in the book. On some rare occasions they overcome the potential barriers surrounding the atom generating radiation. The three moments of hydrogen protons move at sun surface along X-Y-Z axes at velocity $1/4 \times c_s$. These moments generate velocity $\tfrac{1}{2} \times c_s$ obeying rules of conservation of momentum and they create equal pulses of $-1/2 \times c_s$ in the opposite direction. The waves convert to pulse in the following manner. The latter pulse, opposite to the pulse crossing the barrier, is wholly reflected within the atom, its polarity is reversed and so after some delay it cancels the forward vector pulse of $1/2\ c_s$. It lags the forward pulse by a span of $3/2 \times a_{os}$ representing the time lapse of $3/2 \times a_{os}/(1/2\ c_s) = 1 \times 10^{-18}$ s. That time interval multiplied by the atomic force $F_{atomics}$ and inverse of linear probability derives:

$$1 \times 10^{-7} \times 1 \times 10^{-18} \times 1 \times 10^{10} = 1 \times 10^{-15}\ J = 1/10 \times h_s/e_s \qquad (2.61)$$

Energy of sixty pulses from **the H_2 molecule** has to be used to produce the minimum elementary pulse 10^{-14} J moving at velocity c_s in 6 orientations with energy

$$60 \times 1/10 \times h_s/e_s \times 1.1547^3 \times 1.04^2 \approx 9.99138 \times 10^{-14}\ J \equiv 1 \times 10^{-13}\ J \qquad (2.62)$$

Mass of sun in one dimension is

$$(2 \times 10^{30})^{1/3}/(1.111\ldots^2 \times 1.04^{1/2}) = 1.00071793 \times 10^{10}\ kg \qquad (2.63)$$

The space is considered to have seventeen independent degrees of freedom per K as explained earlier. **Each pulse of degree of freedom requires the same energy of 1 eV**

in three orthogonal directions and contributes directly to the temperature of the sun derived below. The vector addition at 90° introduces a factor of √2. The requirements of sequential probability for the signal to remain in the same direction needs the power index of 3 to 17 DOF comprising 10 directions of quantization, 3 planes, 3 orientations of angular space and one spin. The temperature of the sun is the result of the equipartition of energy rule applied to 17 DOF corrected by relativity related to local velocities of 1/3 c_s, 1/4 c_s, 1/10 c_s and of the quantum corrector $1.04^{1/24}$:

$$T_{sun} (K) = \sqrt{2} \times 17^3 \times 1/K_{rel\ 1/3\ cs} \times 1/K_{rel1/4\ cs} \times K_{rel1/10cs}/1.04^{1/24} - 8kT/e_s\ x ln2$$
$$= \sqrt{2}\ x17^3 x1/1.111 \ldots x1/1.07456993 x1.03279559/1.04^{1/24} - 8x10^{-23}/1\ x10^{-19}\ x ln2$$
$$= 6000.032 - 0.033 \approx 6000.00\ K \tag{2.64}$$

In the above expression the energy 0.033 eV is the measurement accuracy uncertainty in S.I. units = 1.111 . . . $k_s T_{300}/e_s$=1.111 . . . $x1x10^{-23}x300/1x10^{-19}$ = 0.0333 eV. The validity of the above hypotheses is accepted because of the excellent accuracy obtained. The sun temperature expresses the minimum total energy of the seventeen degrees of freedom in its radiation corrected by relativity. In the above expression 1.0328 is the correction for the linear velocity 1/4 c_s. That value was noticed in some expressions several years ago, but it could not fit then the data. Now it applies to one directional velocity of proton.

For 3 planes 17 DOF expresses total DOF as 17^3 = 4913; with the sun power directed at earth of $1x10^{30}$ W and probability 1/4913 the power around earth surface is $3 \times (1x10^{30})^{1/4913}/1.04^{1/3} \equiv 3.003\ J/m^3$.

The presence of 17 DOF is also confirmed by the next accurate derivation of the sun reference gravitation force of Equation 6.1 from 1 of 5 quantum pathways, energy being tripled to power three by 3 planes and to power four by 4 rotational directions and by velocity of c_s and quantum corrector:

Gravitation force =1/5 x $[(17)^3]^4$ x 3 x 10^8 x $1.0328^{15/12}$ = 3.5430 x 10^{22} N (2.65)

There is one quantum event (W, J, eV, or N) per pathway and the inverted probability factors with velocity of light derive sun gravitation force, which is equal that of earth.

In the inverse direction 6000 K x 1/6 x 1/10 x 1/10 x 1/10 = 1 eV ≡ 1 W (2.66)

With atomic force of 1 x 10^{-7} N the hydrogen molecule energy seems to be

$$F_{as} x1/4\ a_{os}/e_s = 1x10^{-7}x1/4x10^{-10}/1x10^{-19} = 25\ eVSI \tag{2.67}$$

deriving Bohr atom ionization energy 25 x ½ x 1.1111 x $1.04^{1/2}$ = 13.62 eV.

With electron velocity within the atom v_e = 1 x 10^7 m/s the elementary electron power is

$$F_{as} x\ v_e = 1\ x10^{-7}x1x10^7 = 1\ W\ making\ 10\ eV\ for\ 10\ orientations \tag{2.68}$$

Identifying the unity event as probability of 1.00, 1 J or 1 W occurring on earth at 300 K from the pulse of the sun suggests the following events. **The statistics of gravitation are those of a dice, which when stopped after rotation in one plane displays probability of ¼ of the number to be displayed in the required direction.** If this has to be repeated three times the total probability is $(¼)^3 = 1/64 = 0.015625$. The pulse of 6000 K corrected by relativity of 1.1547 of the proton during pulse formation moving with velocity $½ c_s$ and quantum delivers the pulse of energy:

$$E_n = (6000 \text{ K})^{0.015625}/1.1547 \text{ x } 1.04^{1/5} = 0.999933 \equiv 1 \text{ eV}_{at} \text{ or } J_{SI} \qquad (2.69)$$

Thus the number of different energy values is doubled depending on the choice of unity event. The calculation shows that each quantum vector in Figures 1-3 is now 6000.

The sun pulses of 6000 K are space quantized by a factor 10^{-3}. In forming 6 eVat energy they represent 1 eVat vectors in each of 6 orthogonal orientations. Converting to SI units for two planes requires multiplication by electron charge e_s over velocity of light c_s:

$$\textbf{6000 K } x1/10^3 x1x10^{-19}/3x10^8 \equiv 2 \text{ } x10^{-27} \text{ J per pulse} \qquad \textbf{(2.70)}$$

Above energy is an elementary sun pulse expressing the energy of a proton delivered in radiation. The 2/3 factor derives correct ionization energy of hydrogen atom from sun discharging 1500 W/m² pulses in four orientations:

$$1/4 \text{ x } 6000 \text{ x } 2/3 \text{ x } 1/10^2 \text{ x } 1.15472 + 6kT = 13.62 \text{ eVat at 300 K} \qquad (2.71)$$

Velocity $1/3c_s$ appears to derive the elementary energy $3x10^{-14}$ J in three planes $\qquad (2.72)$

$$1/3 \text{ x } c_s \text{ x } c_s \text{ } x10^{-30} \equiv 3x10^{-14} \text{ J} \equiv 3x10^{-14} \text{ eV; corrected } (1/1.2^3 \text{ } x1.04^{1/17}xc_s)^2 x10^{-30} \equiv 3x10^{-14} \text{ J.}$$

Energy of five quantization paths each carrying $1x10^{-14}$ J over the distance $0.5x10^{-10}$ m adds to

$$5x1x10^{-14}J/0.5x10^{-10}m = 0.001 \text{ N, J, eVat} \qquad (2.73)$$

Energy of 0.001 eVat time 10 relates to the value of the input energy per quanta measured some 30-40 years ago by the author at 293 K in a divalent impurity of NaCl crystal representing Bohr atom:

$$\sqrt{3} \text{ } F_{Gs}a \text{ } x10 = ½ \text{ } k_s T_{293}/e_s \text{ x } 1.04^4 = 1 \text{ } x10^{-23}x \text{ } 293/10^{-19} \text{ x } \ln2/1.04^4 = 0.0174 \text{ eVSI} \quad (2.74)$$

Referred to 1 K $0.0173/(1.04 \text{ x } 300 \text{ x } 1.11 \ldots) = 5 \text{ x } 10^{-5} \rightarrow \text{ x } 1 \text{ x } 10^{-19} = ½ \text{ } x10^{-23} \text{ J} = ½ \text{ } k_s$. **Thus at 1 K the uncertainty is one half Boltzmann constant.**

The interaction between the sun and the earth differs from that occuring between bodies on earth, because the temperature of the two bodies is different. The hydrogen radiation particles from the sun due to the equipartition of the energy (and momentum) rules contain equal energies (momenta) of proton oscillations, electron oscillations and (proton+electron) oscillations. Radiation from earth has negligible amount of electron oscillations and its particles contain only oscillations of proton and of (proton+electron) pair. Consequently the International Gravity constant in the integer model proposed, 1 x10⁻¹⁰ N-m²/kg² which represents bodies at equal temperatures has value about 3/2 time the classsical SI value.

In SI units Planck Length L_p according to E. Tryton [Encyclopedia of Physical Science and Technogy, p. 780], is 4 x10⁻³³ cm. In proposed relativity corrected integer space units Planck Length

$$L_{ps} = [G_s h_s/c_s^3]^{1/2}/1.111 \ldots$$
$$= [1 \times 10^{-11} \times 1 \times 10^{-33}/9 \times 10^{24}]^{1/2}/1.111 \ldots = [1/9 \times 10^{68}]^{1/2}/1.111 \ldots = 3.00000 \times 10^{-33} = 3h_s \quad (2.75),$$

while the value in Table 2, express 30 quantum positions and it is 30 times smaller at $l_{ps} = 1 \times 10^{-34}$ m. It is invert of velocity of light to power three corrected for relativity and another $1/c_s$ multliplier to convert from natural units to SI units:

$$l_{ps} = 1 \times 10^{-34} \, \text{m} \approx 1/(c_s^3 \times 1.111 \ldots^2) \, c_s \quad (2.76)$$

The mass of earth is represented by the number of electrons; the number of electrons equals the ratio mass earth/electron mass; the linear velocity of earth defines the earth current. The basic interaction of earth is that of a mutual inductance, compensating magnetically totally the flux in two dimensions produced by the sun radiation. The respective formulas are:

Hydrogen atom in Balmer series from n =3 to n = 2 radiates in SI units 2.85x10⁻¹⁹ J.

Divided by electron charge (1.602x10⁻¹⁹ C) represents 1.7790 eV. Divided by $K_{rel}^4 = 1.778$ equals 1.006 eV. This calculation agrees with the statement by William Houston, Nature 117, p. 590-59 (24 April 1926)]. (2.77)

One dimensional (1D) quantum Bose-Einstein energy distribution, applicable to photons derived compactly by D. Ingram [30]:

$$w_{Lm}(v_{Lm}) = hxv_{Lm}[exp(hxv_{Lm}/kT) - 1], \quad (2.78)$$

provided the smallest quantum energy step w_{Lm}/e eV displayed in the earlier author's experiments [3] as:

$$w_{Lm}/e = hxv_{Lm}/e = kTx\ln 2/e = 0.017510 \, \text{eVSI} \quad (2.79)$$

Electrical parameters are vectors, expressing parameters that are already space quantized. Electrical field $E_{SI} = 1 \times 10^7$ V/m in a Bohr atom can be identified with the atomic force per unit charge, F_{atSI}. In space at atomic level $H_{SI} = I$ because $H_{SI} = 1/a_O = 1A/0.5 \times 10^{-10}$m times 1×10^{10} while $B = \mu_{os} \times 1 \equiv 1 \times 10^{-6}$. Because of the absence of time and of distance variables, at kinetic atomic level, the product of the local velocity of light times the vector product of two parameters, **ExB** $= 1 \times 10^7 \times 1 \times 10^{-6} = 10$ W, J or eV. The value 1 W, J or eV per orientation 1 J $\equiv$ 1 W is identified with **the radiation pulse of 1x10⁻²⁷ J expressing mass of 10 muo10s**. The wave function format generates a pulse at the velocity of light. This is also the integer model Planck mass of Table 2 $m_{ps} = 1 \times 10^{-8}$ kg converted to atomic units on multiplying by 1/3 x c_s =1 x 10^8 m/s deriving 1 W. This value represents 3 eVat with 1 eVat pulses in three planes. The temperature of 6000 K should be read as 3 *x* 2000 K, and 2000 K per orientation $\equiv 5\sigma_s T^4$ x K_{rel}. With $F_{atomss} \equiv$ 1x10^7 V/m, current per pathway I_{os}=5x10^{-11} A [45, Equation 4.13] and

$$(\mu_{os} I_{os})^2/1.0328 \; x \; 1.04^{3/5} = 3.000008 \; x \; 10^{-33} \; J \equiv 3h_s \qquad (2.80)$$

$$e_s \; x \; e_s/\varepsilon_0 = 10^{-19}x10^{-19}/10^{-11} \equiv (m_{ps} \; x1/3 \; c_s)^2/\varepsilon_{os} = (10^{-27}x10^8)^2/10^{-11} \equiv (\; Pv)^3 \equiv \mathbf{1x10^{-27}} \; J \quad (2.81)$$

These values decrease by a factor of 20 are being referred to **300 K**. The value of 1 eVat becomes then 0.05 eV.

The energy of the radiation particle has a very specific meaning. For the cubic space model at the nature atomic level it is momentum of one component per unit local velocity:

$$e_s \; /(1/3 \; c_s \;) = m_{d\backslash ps1D} \; x \; 1 = 1 \; x \; 10^{-19}/1 \; x \; 10^8 = 1 \; x10^{-27} \; J \qquad (2.82)$$

Proton having velocity ¼ c_s requires two sequences of Figure 4 to bring its velocity of light, these processes occurring at probability 10^{-9}:

$$1.11 \ldots k_s \; c_s^2 \; x1/10 \; x1/10 \; x1/10 = 1/10^9 \; J \qquad (2.83)$$

repeating this process for three pulses finds that 1 $x10^{-27}$ is probability $(1/10^9)^3 = 1 \; x10^{-27}$

where $k_s \; c_s^2 \; x \; 1.11 \ldots = \mu_{os}$ and $(k_s \; c_s^2)^{1/3}/e_s = 10$ eV.

The quantum of radiation atomic energy 0.001 J is expressed by:

$$(\mu_{os})^{1/2} = (B_s)^{1/2} = (\varepsilon_{os}/10)^{1/4} = \varepsilon_{os} \; x \; (\; 1/3 \; c_s) = 10^{-11} \; x \; 10^8 = 0.001 \; J \qquad (2.84)$$

The above element is one of ten quantum components in one plane of 1st Planck radiation term 0.01 J. Space permittivity μ_{os} =10^{-6} is generated from ten components $\mu_{osorient}$ = 10^{-5}. Three such values with probability 1/3 generate term 0.01: 3 x (1 x $10^{-5})^{1/3}$x1.111$^{1/2}$=0.01 J

Similarly there are in ten oriented forces of $F_{atomicorient}$ = 1 x 10^{-6} values which generate

$$\varepsilon_{os} = \mu_{osorient} \; F_{atomicorient} = 1 \; x \; 10^{-5} \; x \; 1 \; x10^{-6} = 1 \; x \; 10^{-11}. \qquad (2.85)$$

The reference 1 W power is the product of B_s, one orientation atomic field 10F_{atomic} and proton frequency 1 x 10^{12} Hz:

$$B_s \; x \; 10F_{atomic} \; x \; 1 \; x \; 10^{12} \; Hz \equiv 1 \; x \; 10^{-6} \; x \; 1 \; x10^{-6} \; x \; 1 \; x \; 10^{12} = 1 \; W \qquad (2.86)$$

In integer model basic terms $1/10^9$ J expresses probability10^{-9} generating at 300 K Planck radiation term 0.01 J in a two plane process:

$$\text{The resultant converted to SI} = (1/10^9) \, (1/10^9) \, x1x10^{-20} = 0.01 \; J, \; eV \qquad (2.87)$$

The above equation clearly demonstrates that one energy component of $1x10^{-27}$ J describes other parameters, which could be considered superfluous.

Science requires the same result to be obtained independently of the velocity of the observer, so that objects seen stationary or moving at the velocity of light are recorded as having the same energies although forces observed may be different.

The energy expressed by converting mass of 1 kg moving at the atomic velocity $v_{at}=$ 1×10^7 m/s, at temperature 3000 K, light velocity, relativity corrector and integer muo10_s mass 1×10^{-28} kg expresses the first term of Planck radiation value 0.01 W:

$$1 \text{ kg } x \ (v_{at})^2 x 3000 \text{ K } x c_s x 1.111 \dots x \ m_{muons} \equiv 0.010 \text{ W.} \qquad (2.88)$$

The elementary energy 0.001 J seems to come from the electron rotation around the Bohr atom:

Unity electron mass in 3 orient: $3x1.0x(1x10^7 \text{ m/s})^2 x \ 300 \ x1.111 \dots x10^{-19} \equiv 0.001$ Jat (2.89)

Above value for 3 planes time 1.11 . . . obtains the expected value 10 eVat. Energy 3 eVat normalized per K becomes the 0.01 J of the Planck Radiation Law of Equation 2.27.

Unity probability events (1.00) on earth at temperature of 300 K with recognized source-receiver axis are measurables occurring in ten orientations of the radiation source in each of three planes. The probability of events to occur in any specific orientation is 1/10 with equal chances of pulses to occur in any orientation. So the resulting probability $1/10^3$ derives $1/10^9$ for three planes. These pulses have to be orientated in the three orthogonal directions so the resultant is across the cubic space requires total probability of
$$1/10^9 \text{ x } 1/10^9 \text{ x } 1/10^9 = 1 \text{ x } 10^{-27} \equiv 1 \text{ x } 10^{-27} \text{ W or J} \qquad (2.90)$$
sun gravitation to the expression of the force of international standard of 1 A.

In his research report [45, p.22] the author attributed earth repulsion balancing the component current 1 of the repulsive force $F_R = \mu_o I^2/2\pi$ is the image current representing field configuration in SI units. The earth current of $1.7795x10^{29} \approx 1.8 \ x10^{29}$ A is produced by earth electron number of $N = m_{earth}/m_{es}$ moving with earth velocity 29796 m/s. The original reflection force F is as that subjected to 1 m long conductor creating magnetic field B moving at velocity v_v, $F=B \ x \ v_v$. At the atomic dynamic level the expression for force of repulsion is the product of flux density produced and of the velocity of light c_s:

$F_{Rs} = B \ X \ c_s = \mu_o I/2\pi \ x \ c_s$, which on converting to SI units is multiplied by c_s so that the expression for the macroscopic repelling force becomes the exact reference

$$F_{Rs}=2/3 \ x\mu_o I_{earth}/2\pi \ x \ 1/1.04^{1.1}=2/3 \ x3x10^{-11} \ x1.7795x10^{29} \text{ A}/1.04^{1.1}=3.5436 \ x10^{22} \text{ N} \quad (2.91)$$

The force obtained using integral cubic SI units is: $F_{Rs} = 1/5 \ \mu_o \ xI_{earth}=3.6 \ x10^{22}$ N.

Magnetic permeability μ_o per pathway can thus be defined to be essentially the force of 2 $x10^{-7}$ newtons per ampere$=1/5 \ \mu_o$, also because $2x10^{-7}$N $x1.8 \ x10^{29}$ A$=3.6 \ x10^{22}$ N.

The force $3x10^{-7}$ N/(m-K) per radian is that which would be seen by the observer moving with velocity 1/2 x 29796 m/s against the earth as explained earlier [45 p. 22]. Its decrement to $2x10^{-7}$ N/(m-K) in a dynamical system is attributed to the absence of the reflection component, because the earth is moving.

In the introduction formations of the velocity of light was derived from the local velocity as a two stage process. However the application of relativity correction appears to suggest that the conversion can be achieved in one step process deriving accuracy of 2 parts per 10^6 in measurements discussed in Conclusions. The two step process and one step process are not exclusive. **It is suggested that one step process to generate velocity of light given below is needed, while radiation is started from a static proton or molecule. If the radiation is from around, the sun pulses move already with local velocity of light and are retransmitted without adding extra energy:**

$$\textbf{Three local velocities generate} = \surd 3\ x\ 10^8\ x\ 1.154700^3\ x\ 1.111\ldots x\ (1+1/80)\ x1\ \textbf{kg}$$
$$= \textbf{3.000 029}x10^8\ \textbf{m/s} \tag{2.92}$$

Adding 1 kg to the equation cleared the initial doubts about it validity. Adding different unity values can correct other identities lacking agreements in dimensions. Absence of dimensions of length and time in kinetics is the correct explanation.

Other identities:
Boltzmann constant $k_s = m_{es}(1/3\ c_s)^3/E_{sspaceSI} = 1\ x\ 10^{-30}$ kg $x\ (1\ x\ 10^8)^3\ x\ 1\ x\ 10^{-17} = 1\ x\ 10^{-23}$ J (2.93)
Equations in physics are deceptive. In the following wave function format, Boltzmann constant is generated by the product of probability of two pulses generating another pulse in the third orientation. The resultant progressing at the velocity of light comprises energy and momentum of the two:

$$k_s = (m_{es}^{1/2}x1/3\ c_s)X(m_{es}^{1/2}x1/3\ c_s)x1/3\ x\ c_s\ E_{sspaceSI} = 1\ x10^{-23}\ \text{J}.$$

Since $m_{es}^{1/2} = 1$ J/10^{15} Hz (electron) $= (1/F_{atom})^2/10 = (1/10^7)^2/10 = 1x10^{-15}$ J (2.94)
the value is pulse energy per pathway in SI units.

A recent finding expresses square root of electron charge in one orientation with the graviton constant:
$$G_s \equiv 1/3\ x\ (1\ x\ 10^{-19})^{1/2}\ /1.111\ldots^{1/2} \equiv 1.000000000\ x10^{-10} \tag{2.95}$$
The figure $1\ x\ 10^{21}$ appearing in some relations is interpreted to be the rate of vector pulses in SI units arriving on earth at 300 K expressing 1 J atomic pulse time rotational converter and inverse linear probability 10:

$$1\ \text{J}\ x\ 10^{20}\ x\ 10 \equiv 1\ x\ 10^{21}\ \text{with 30 J/pulse the force becomes}\ 3\ x10^{22}\ \text{N} \tag{2.96}$$

Several of the above expressions may have a different format that is being expressed by different parameters and corrections of relativity and quantum.

Derivation of integral physical constants of the integer cubic model from the Planck's units

The classical SI units are based on very accurate and easily reproducible physical processes in space, which not necessarily display clearly all features of nature. SI system of units is derived from Planck units discussed in detail on WEB pages in ten million references; the basic formulas are based on dimensional analyses. They fit dimensions, but require finding a correct proportionality factors. Many readers would find them incomprehensible, because of the different complex derivations proposed by scientists. A particular interesting contribution on WEB is by Xavier Borg of Braze Lab Research, "A note on h and h bar", following Plank in using unbarred h in equations. This implies applying standing waves across the particle as in the model postulated now and not around the circular path. Borg uses also 1D, 2D and 3D displays in his model.

The integer cubic SI space constants are derived from the same source as the classical SI units have, that is the Planck units. Various laws of nature define equations creating new constants, from the more basic constants. Geometry of generation of radiation in 3D space and their interactions determine their relations, as given in the main body of the book, explaining the processes involved.

The formulas used to derive Planck units in the book are those displayed on WEB pages with the details specific to the model: using the reduced Planck constant h_s; 2π is replaced by 6; the velocity of light is expressed at times by the local velocity $1/3\ c_s$ representing the average speed; relativity and quantum corrections are used, if needed to obtain likely integer values expressing probability; gravity constant of the model of bodies at equal temperatures is 3/2 of the sun-to-earth system.

Since accuracy is limited by H. u. no more than 5 digits are usually quoted. Table 2 gives SI values of Planck's constants and those of the model incorporating relativity and quantum corrections required to create a set of Planck constants having the integer parameters. Planck units and the classical SI units are based on three fundamental units: c, h and G. For the SI integer model these have values c_s, h_s and G_s. These values derive all other units similarly as in the classical SI system.

The two values of the gravity constants, G_s one for bodies at equal temperatures and the other G for the sun-earth couple creates a difficulty in deriving Planck unit values having the same values as in SI system. Applying G_s value derives integer Planck units valid for atomic relations on earth having equal temperature. The gravitation constant for sun to earth pair has value 2/3 of G_s. Planck SI units do not have integer values as shown in the first two values, but this is corrected subsequently.

The ratio of Planck masses in the classical SI units corrected for relativity and quantum differ only 1/3 of Boltzmann constant this being the limit of accuracy:

$(2.1742\text{x}10^{-8}/1.8371\ \text{x}10^{-8})\ \text{x}1/(1.11\ldots\text{x}1.04^{3/10}) =0.9987$; with gravity 10^{-10} acting in one plane its uncertainty probability $10^{-20}\text{x}(1-0.9987) = 1/3\text{x}10^{-23} \equiv 1/3\ k_s$ \hfill (2.97)

The elementary units of the model, from which others units can be easily derive, are as follows:

The local velocity of light c_s is derived from the ratio of Planck unit of length over Planck unit of time:

$$l_{ps}/t_p = 1.0015 \times 10^{-34} \text{m}/1.0015 \times 10^{-42} \text{ s} = 1 \times 10^8 \text{ m/s} \qquad (2.98)$$

Planck units derive Planck constant to be proton radiation energy

$$h_{ps} = E_{nds} t_{ops} = 1 \times 10^9 \times 1 \times 10^{-42} = 1 \times 10^{-33} \text{ J-s} \qquad (2.99)$$

Both electron charge a_s and h_{as} have been found to be derivative of radiation from proton frequency or (proton+ electron) pair oscillating essentially at the same frequency 10^{12} Hz. There is need to introduce Plank constant parameter h_{select} derived from electron oscillations having 1/1000 smaller value to keep oscillators energy values equal:

$$h_{select =} \mathbf{1 \times 10^{-36} \text{ J-s.}} \qquad (2.100)$$

The five parallel paths of 250 Ω derive resultant 50 Ω, which times 2/3 obtains 30 eV of the dissociation energy of the Bohr's atom in SI units. The relativity and quantum corrected with √2 multiplier

250 Ω x √2 x 1.11 . . .$^{/2}$ x $1.04^{1/4}$= 376.35 Ω differs from classical value of the SI space impedance $Z_{ach} = \sqrt{(d_o/d_o)} = 376.68$ Ω by only 1/3 Ω representing H. u.

The integer model SI space impedance S_{acs} /1.111 . . .$^{1/2}$ = 300.00000000 Ω and

$$Z_{cs} = \sqrt{(\mu_{os}/\varepsilon_{os})} \times 1.1547 \times 1.04^{8/10} = 376.79 \qquad (2.101)$$

The three oscillator components of the Bohr atom are: proton, proton+electron, and electron. Their individual velocities express their momenta, which at the instant of radiation oscillate in the same direction, increasing total momenta by a factor of three and the total velocity by a factor of three, forming the light velocity 3×10^8 m/s, because there is effectively only one proton mass.

Duration of one second is identified with $1/10^8$ period of 3 years, each of 365.25 days; this definition is accepted because of its accuracy of 2 parts per 10^4. The period $1/10^8$ is related with rotary momentum of the earth along one of three orthogonal planes. There are 31.5576 s per year and using the quantum corrector $1.040785^{11/8}$:

$$31.5576 \times 10^6 \times 3 \times 1.040785^{11/8} \approx 1.0002 \times 10^8 \approx 1.0 \times 10^8 \qquad (2.102)$$

Frequency with unbarred h_s, $f_s = \sqrt{(c_s^5/h_s G_s)} = \sqrt{(3 \times 10^8)^2/(1 \times 10^{-33} \times 1 \times 10^{-10})\}} = 5 \times 10^{42}$ Hz.

Multiplied by time $f_s t_p = 5 \times 10^{42} \times 1 \times 10^{-42} = 5$ cycles for 5 space directions $\qquad (2.103)$

The length of one meter is defined as the distance traveled by light in period of 1/(3 $\times 10^8$) seconds. The attractive alternative definition of the length of one meter is the ancient 1791 French Academy Sciences reference of 1/40 million's part of the earth circumference. Its validity is attributed now to the local rotation velocity of electrons in Bohr atom, its velocity of 1×10^7 m/s adds and subtracts from the linear light velocity 3×10^8 m/s. Earth diameter corrected for relativity and quantum is 1×10^7 m:

$$40 \times 10^6/(\pi) \times 1/(1.1547 \times 1.11\ldots) \times 1.04^{1/5} = 1.0002 \times 10^7 \,\text{m}$$

$$2.104)$$

Above earth radius $\frac{1}{2} \times 10^7$ m fits very well formation of potential barriers created by Bohr atom radiations involving atomic unit force 1×10^{-7} N generating $\frac{1}{2}$ eVat or 1.5 eVat for three planes in all directions. It leads to realization that orientation, sequence and equal probability are our reality. In their absence we have mash-nothingness. It is believed that the geometry generates the formation of quoted earth circumference dimension by attracting or repelling particles from the original mist over billions of years.

The finding that the model Planck units are integer values is very important in validating the model and these values are printed bold. The physical meaning of the Planck units of mass and of length was derived by the author by the usual method of adding suitable non-dimensional factors to reference equations in the final text corrections. In deriving l_p in Table 2 the dimensions of the product of the squared l_p and of the local cubed velocity of light are equal to the dimensions of product $h_s G$. Then non-dimensional factors are added generating an integer product. The use of barred h_s and local velocity used in Table 2 derives a second, different l_{ps} value.

For three planes the dimensions of the product of two point masses of earth 1 m apart and of gravity constant $G_s \times m_{earths}^2$ is equal to $h_s c_s = 3 \times 10^{-25}$ J/m after adding non-dimensional factors for relativity $1.111\ldots/1.2$, the number of electrons 10^{54} per kg and linear probability $1/10^{10}$ needed because the interactions on the right side are at the atomic level:

$$3G_s \times m_{earths}^2 \times cor. = 3N_{elkg} \times h_s c_s \times Pr_l \qquad (2.105)$$
$$3 \times 10^{-10} \times (6 \times 10^{24})^2 \times 1.11\ldots/1.2 = 1 \times 10^{54} \times 3 \times 10^{-25} \times 1.11\ldots/10^{-10} \equiv 1 \times 10^{40} = 1 \times 10^{54} \times 10^{-14}\,\text{J/kg}.$$

It should be noted that these derivations are unique to the model proposed, because they incorporate relativity correction, number of electrons per kg, N_{elkg}, and linear probability, the unknown terms in classical SI system. Some scientists consider Planck units having no function, but above derivations clearly display two physical interactions or rules, which together with velocity of light lead to the derivation of other physical constants. Converting the integer cubic space SI values of Table 2 into the natural atomic units derives the unity event probability values as in Equation 1.2:

Mass $= m_{ps} \times 10^8 = 1.000$, momentum $= m_{ps} \times 1/3 \times c_s = 1.000$, energy $= m_{ps} \times c_s^2 \times 1 \times 10^{-10}$ $= 1.000$, velocity $= l_{ps}/t_{ps} \times 1/1 \times 10^8 = 1.000$, current $= I_{ps} \times \ldots = (2 \times 10^{24})^{1/3}/(1.11\ldots^2 \times 1.04^{1/2}) = 1.0007$ etc. $\qquad (2.106).$

The research presented covers the semi-stable, effectively balanced state of radiation and moving particles of physics in the Solar system. It does not cover the matters related to High Energy Physics describing subatomic particles, except for relating in Equations 17.11-17.13 the muo10$_s$ mass to the Gravitation constant, to the radiation density of the elementary radiating particle and hence to all other integer constants.

The integer Planck length is taken to be the square of (the space frequency in one plane plus its uncertainty) so that $l_{ps}=1.0015 \times 10^{-34}$m $\equiv [1/3\ v_{os} + \Delta(1/3\ v_{os})]^2 \approx (1 \times 10^{-17})^2 + (4 \times 10^{-19})$. Hence the mass uncertainty is $4e_s$, expressing rotating of moments of protons in four orientations (Equation 1.15).

The classical Planck units listed below are not the exact values listed by Wikipedia, updated on October 30/2011, which included derivation of Maxwell Equations. The integer model Planks units were derived using the same classical expressions that are used in classical SI derivations and deviations from integer values will be re-assessed.

Table 2. *Planck units of the model and of the classical SI values*

Planck unit	Dimension	Classic Expres.	SI Value	Model Expression	Model SI value
Length l_p	Lp	$\sqrt{(hG/2\pi c^3)}$*	**1.6163x19^{-35}m**		
		$l_{ps=}\sqrt{[h_s G_s/6(1/3c_s)^3]}x1/(1.1111^2 x 1.04^{1.14}$ **0.99998x10^{-34}m**		with 2/3 G_s	0.54433 x10^{-34}m
0.0015 x10^{-34}m is 1/20 x kT is H.u.					
	$l_{ps} \approx 6l_p$ x1.04$^{3/4}$				
Mass m_p	Mp	$\sqrt{(hc/2\pi G)}$	**2.1764x10^{-8} kg**		
		$m_{ps} + m_{es=}\sqrt{[h_s(1/3c_s)/6G_s)]}x1.1111^2 x1.04^{1.14}$			**1.0001x10^{-8} kg**
	$m_{ps} \approx 2\ m_p$ x1.04$^{21/10}$			with 2/3 G_s	1.8371 x10^{-8} kg
Time t_p	Tp	l_p/c	**5.3912 x 10^{-44} s**	l_{ps}/c_s	**1.000x10^{-42} s**
					$t_{ps} \approx 8\ t_p/1.04^{9/10}$
Momentum	[LMT^{-1}]	$m_p c$	**6.5248 kg-m/s**	m_{ps} x1/3c_s x 10^{-8} x 10^{-8} =	**1 kg-m/s**
For three orthogonal directions with H. u. derived from the mass					**1. 0001 kg-m/s**
6.5248 kg-m/s x 1.5 /1.04$^{11/20}$ =10 kg-m/s within 1 part in 10^4					
or m_{ps} x velocity =1 x 10^{-8} x 1 x 10^8 = 1 kg-m/s					
Energy e_p	[L^{2}MT^{-2}]	$m_p c^2$	**1.9561 x 10^9 J**	m_{ps} x1.11111 xc_s^2	**1 x10^{10} J**
Local energies per plane or per quantization path are leas important here					
				$e_{ps} \approx 5\ e_p/1.04^{11/20}$	
Power	[L^{2}MT^{-3}]	c^5/G	**3.6283x10^{52} W**	c_s^5/G_s (G_s=10^{-10})	**9x10^{50} W**
3.6283x10^{52} W 2/3 =2.43x10^{52} W; power per plane (1/3c_s^5/G_s)					**3x10^{50} W**
Force	[LMT^{-2}]	c^4/G	**1.2103x10^{44} W**	c_s^4/G_s (see Eq.18.33)8.1x10^{43} N	
Integer powers are local per plane 3x(1/3c_s^4/G_s); 3x3/1.1111=8.1 (Eq.18.33a)					**3 x10^{42} N**
Planck temperature	Kp	T_p x $m_p c^2/k_B$ = 1.41676 x 10^{32}		$m_{ps} c_s^2/1.1111 k_{Bs}$	**0.01 Kp**
Density	[L^{-3}M]	$d_{p=} m_p/l_p^3$	**5.1550 x10^{96} kg/ m^3**	m_{ps}/l_{ps}^3	**1x10^{96} kg/m^3**
				$d_{ps} \approx 1/5\ d_p/1.04^{3/4}$	
Charge	q_P	$\sqrt{(4\pi\varepsilon_0 hc/2\pi)}$	**1.8755 x10^{-18} C**	$\sqrt{(2\varepsilon_{os} h_s x1/3c_s)}$	**2.000x10^{-18} C**
	$q_{Ps} \approx q_P$ x1.04$^{1.7}$				

Current	I_P	$I_P = q_P t_p$	**3.4789 x 10^{25} A**	$q_{Ps/} t_{ps}$	**2 x10^{24} A**
Proton	E_n	$E_{n\,=}\,m_p\,c^2$	**1.9561x10^9 J**	$E_{ns\,=}\,1.1111 m_p\,c^2$	**1.0 x10^9 J**
Voltage	V_p	E_n /q_P	**1.04295 x10^{27} V**	E_{ns} /q_{Ps}	**5 x10^{26} V**
Impedance	Z_p	V_p /I_P	**29.9779 Ω**	V_{ps} /I_{Ps}	**250 Ω**
Planck constant	h	$E_n t_p$	**1.054631 x10^{-34}** J-s	$E_{ns} t_{ps}$	**1x10^{-33} J-s**

h_{ps} derives its exact model value, while h_p is ~1/6 of its h SI value

Velocity		l_{ps} / t_{ps}	**10^{-34}/ 10^{-42}**	**1x10^8 m/s**
Space vectorized frequency		$v_{osvector} = \sqrt{t_{ps}}$		**1.00x10^{-21} Hz**
Momentum	[LMT^{-1}]	$[1x10^{-8} \times 1x10^{-34}/1x10^{-42}]$		**1.0 kg-m/s**

3.   *Formation of Bohr atom radiation series and sun data*

In quantum theory energy, momentum, charge and other parameters of physics do not have a continuous range of numbers. The author observed this feature in his relaxation experiments [2, 3, 4, and 57] and he described it also in the Appendix. The smallest SI energy at 1 K is the Bohr atom radiation derived by Planck's Radiation Law as 0.01 J. It converts to 3 J ≡ 3 eV on multiplying by 300 K, because most physical constants are per one kelvin. The above energy is divided equally between the oscillators of the Bohr atom. It is suggested that for three oscillatory planes and relativity correction 1.111 . . . the energy becomes 10 J in SI unit. The above reference energy is equal in six orthogonal directions. The electromagnetic vector signal expressing radiation moving in one dimension to the receiver at the velocity of light has to comprise all 10 J energy of 3 orientations. Since radiation has initial probability 1/6 total probability of three such events is generating signal corrected by relativity and quantum $10^{(1/6)\uparrow 3}/1.1547$x $(1 + 1/60) = 1.00000 - 0.00041$ J or eV, which displays a very acceptable variance of ~ 1/2 k_s. At atomic level the force is 10^{-10} smaller and the value $G_s = 10^{-10}$ J expresses the graviton incorporating independent electron oscillations existing on the sun. The 10 J expresses ten 1 J units. It is the product of 1 J signals in four optional orthogonal orientations per plane and three components of velocity 1/3 c_s forming the 1 J pulse, less two non-interacting radiations energies, discharged at right angle to the propagation direction. Smallest pulse energy is 0.01 J/K. Experimental thermal radiation values display Bose-Einstein statistical distribution. The root mean square of the sum of the signals components orientated along the source receiver axes comprising all frequencies, wavelengths and colors expresses the effective force, moment, mass and energy values of Table 1. These are 'effective' measurables often exact values, which may not exist, but sustain balance of nature for billions of years. The gravity force acts both at the molecular level and at the celestial level.

Energy vectorized by space quantization in atomic events occurs with the same probability in all directions, but it moves in one forward direction, if reflected internally. Reflection originates probably from the whole obstructing mass at the back, so the forward pulse may comprise components from the recent past representing semi-stable situation. Total probability of the event is the product of sequential probabilities required to generate the event. With equal chance of occurrence of the quantum event in any of the three directions of three planes the chance of the event occurring in a specific direction is 1/10 x1/10x1/10 = $1/10^{-3}$ as shown in Figures 3. Since an event is the product of two parameters formatted by wave function the resultant probability becomes $1/10^3$ x$1/10^3 = 10^{-6}$ per plane. The three vectors formatted in three planes are subject to further probability of occurring in specific source-receiver axis. They generate a resultant having probability of 10^{-9}. The output value for the input of.1 J expressed in one dimensional presentation becomes:

1 J x $(10^{-9})^{1/3}$ = 1x10^{-27} J. Density of 1 MJ/m^2 of Equation 2.9 times 10^{-6} derives 1 J/m^2. It is suggested that Bohr atom comprises in two planes 10 quantum oriented masses of muo10$_s$ of mass 1x10^{-28} Kg (integer cubic SI units) generating the total proton mass of 2 x10^{-27} kg. Radiated in one direction 1 x10^{-27} x1 x10^{30}= 1000 J SI units.

The creation of a vector of above probabilities moving in 1D using either classical SI units, integer model SI units or natural atomic units expresses simply probability. On converting from SI to atomic units converters have to be used to express true atomic relations to SI units.

Orientation display for a number of physical parameters, mainly energies and moments, has three forms called 1D (one dimensional), 2D (two dimensional) and 3D (three dimensional). Display 1D refers to parameter in one of local pathways of three orientations. Display 2D refers to parameter being formed by vector additions and probability product of the two 1D parameters. It is 1D value with power index ½ applied to it. Energies or moments of components are added in each operation. Display 3D refers to parameter being formed by vector additions and probability product of three 1D parameters. It is the 1D value with power index 1/3 applied to it. Changing D display may or may not change dimensions (units) of the parameter.

Thus electron charge e_s = 1 x 10^{-19} C expresses 1D.

Electron in 2D: $\qquad e_s^{½} \equiv (1 \times 10^{-19})^{½} = 3 \times 10^{-10} \times 1.11\ldots^{1/2}$ m $\qquad\qquad$ (3.1)

is expressed by three Bohr atom dimensions $2a_{os}$ or three gravitation constants G_s corrected by relativity correction.

Electron e_s in 3D display becomes decremented μ_{os} corrected by relativity and quantum:

$$e_s^{1/3} = (1 \times 10^{-19})^{1/3} = 0.464159 \times 10^{-6} \times 10^{-10} \approx \mu_{os}/(1.2^3 \times 1.111\ldots^3 \times 1.04^{1/4}) = 0.464176 \times 10^{-6}\ NA^{-2}\ldots$$

$$(3.2)$$

The sun interacts with earth only with 1x10^{30} kg of its mass, which vectorized by the probability 1/10^3 derives 1x10^{27} kg vector. The 2D display of sun mass 1x10^{15} kg times minimum input pulse power 10^{-13} W/kg derives 100 W inclusive of the electron power. Power of 60 W is derived from 100 W by applying relativity correction and excluding electron power, by factor 2/3. Finally 50 W is obtained by 1.2 radiation input reducer, expressing the power of sun infrared energy pulse of proton and (proton+ electron) pair:

$$(1 \times 10^{30})^{1/2} \equiv 1 \times 10^{15}\ kg \rightarrow \times 10^{-13}\ W/kg = 100\ W \rightarrow \times 2/3/1.11111 = 60\ W \rightarrow \times 1/1.2 = 50\ W \quad (3.3)$$

In 2D display $(1 \times 10^{30})^{1/2} \equiv 1 \times 10^{15}$ kg expresses sun mass reduced by probability.

Energy or power may refer per oscillator, per plane, per orientation, per degree or refer to 300 K or 6000 K, to different fractional excitations between zero level and dissociation level expressed in Table 3. A mix of these frequencies creates white color.

In space, static mass is always oscillating and such oscillations have to express one of the space quantization orientations. Energy events requiring probability of occurrence proceeding through three orthogonal directions are subject of probability 1/10^9. With Planck energy 1x10^{10} J obtaining one elementary unit of energy or momentum requires linear converter of 1x10^{-10}: 1x10^{10} J x1x10^{-10} $\equiv$ 1 J, eV. Surface converter becomes 1x10^{-20} and volume converter 1x10^{-30}.

Physical constants do not display the difference between high energy photons oscillating in all orientations and the low energy vectorized radiation decreased in 1D by a probability factor 1 x 10^{-10} into a vector. All values had to be considered.

According to the model Pashen radiation series for H$_2$ hydrogen atom have long wave lengths displaying oscillations of one proton at 1x10^{12} Hz within the atom. Balmer series having twice the frequency at n = infinity is related with mobility of two protons forming a pair. Lyman series with its short wavelengths reflects oscillations of electrons at 10^{15} Hz. Bracken series expresses oscillations of (electron+proton) pairs. Pfunk and Brackett series represent other combinations of electrons and protons oscillating together. From n =1 to n = ∞:

Lyman line delivers energy: $3 \times 1.0967 \times 10^{-9} \times 3 \times 10^8 \times 1.04^{1/2} = 1.00667$ eVSI   (3.5)

Pashen line energy: $1/3 \times 9.87 \times 10^{-9} \times 3 \times 10^8 \times 1.04^{1/2} = 1.0067$ eV SI   (3.6)

Value of 1 eV comes also from other sources [William Houston Nature 117 590-59 (24 April 1926)]. Hydrogen atom in Balmer series from n=3 to n = 2 radiates in SI units 2.85×10^{-19} J. Divided by electron charge (1.602×10^{-19} C) delivers 1.7790 eV. Divided by $K_{rel}^4 = 1.778$ derives 1.0006 eV

In earth gravitation the interacting unity gravity force of sun generated by three oscillators (3 eV$_{at}$ ≡ 3 N) per kilogram of body mass and diminished by linear probability to 1×10^{-10} is further decreased by 2/3 to 2 eV$_{at}$ by the absence of electron radiation from earth. In hydrogen atom the energy transfer of 1 eV$_{at}$ per plane is reduced to 0.001 eV$_{at}$ because of the electron/proton mass ratio, but both estimates start at atomic level with unity force being generated from unity mass.

If the proposed integer space cubic constants have merit they should predict correct hydrogen radiations frequencies of different series. One form of the classical Bohr atom energy expressed in integer space cubic units with relativity and quantum corrections is:

Energy = $k_s e_s^2/(a_{os} h_s) \times (1/n_1^2 - 1/n_2^2) \times 1.04^{1/5}/(1.11 \ldots)$;
  for $n_1 = 1$, $n_2 = 2$
Energy = $1 \times 10^{-23} \times 1 \times 10^{-38}/(½ \times 10^{-10} \times 1 \times 10^{-33}) \times (1-1/4) \times 1.04^{1/5}/(1.11 \ldots)$
    $= 20 \times 10^{-19} \times 3/4 \times 1.04^{1/5}/(1.11 \ldots) \equiv 13.60$ eV$_{at}$ (3.7)

To conserve energy the mass of proton is discharged in both directions; proton moving with other oscillator at velocity 1/3 x c_s, derives $e_s = 1/3 \times c_s \times m_{prms} = 1 \times 10^8 \times 1 \times 10^{-27}$ $= 1 \times 10^{-19}$ C ≡ 1 eV, **the charge e_s expressing the momentum of the proton particle.**

Three such momenta for proton, electron and the pair (proton+electron) generate energy 3 eV$_{at}$, which for three planes time relativity corrector 1.111 ... total 10 eV$_{at}$. That energy time $K_{rel}^2 = 1.3333$ and $1.04^{½}$ derives 13.60 eV$_{at}$.

The whole gamma of radiation energies is now obtained by manipulating values in the bracket $(1/n_1^2 - 1/n_2^2)$ according to the classical Bohr method, with n_1=1 and n_2 =2, 3, 4 etc; for Lyman series, with n_1=2 and n_2 =3. 4, 5 etc; for Balmer series, with n_1=3 and n_2 = 4, 5, 6 etc; for Pashen (i.r.) series, with n_1= 4 and n_2 = 5, 6, 7 etc; for Bracken series and with n_1 = 5 and n_2 = 6, 7, 8 etc and similarly for Pfunk (i.r) series. Using space cubic integer constants replaces 1.096775×10^7 by 1×10^7 m^{-1} x 1.111 .../$1.04^{1/3}$ in Table 3 with space cubic integer constants deriving essentially the same radiation data as using SI physical constants.

The value of 1 eV$_{at}$ was identified as the vector produced by one step quantization of 10 eV. The energy of 10 eV represents total ionization of the Bohr atom in one orientation. At room temperature Bohr generates two equal infrared frequencies due to proton and due to (electron+protone) pair oscillations delivered in quantum amounts of about 10 eV$_{at}$, the same as in the impurity model in NaCl crystal. The basic signal oscillates only in the propagation direction but it is suggested that the two oscillators are displaced in phase by 90° so that with the quantum corrector the resultant energy obtained is:

$$10 \times \sqrt{2}/1.03972 = 13.6015 \text{ eV}_{at} \tag{3.8}$$

That signal is due to the oscillations of the reduced masses, approximating closely electron mass, but frequencies at room temperature are infrared with mean close to 5×10^{12} Hz. There is no doubt that radiation energies derived using classical SI units are correct. Nevertheless the author is certain that the **radiation wave must carry the components of the energy of radiation subjected to the proposed corrections,** in spite of other explanations on WEB pages.

With quantum correction twice the Lyman frequency $f_n = 4.934\times10^{15} \times 1.04^{1/3}$ $= 4.9989 \times 10^{15}$ Hz represents the electron frequency 5×10^{15} Hz. For the first item of the Lyman series the energy is expressed in following identity including term 10^{-3} derived from Figure 3:

The **radiation wave carries its own energy of radiation because Bohr atom total energy displayed in Lyman frequency is**

$$\textbf{2.467} \times 10^{15} \times 10^{-30}/1 \times 10^{-19} \times 1 \times 10^{-3} \times 1\text{eV} \times 1.2 \times 1.04^{1/3} = 29.9936 \text{ eV}_{at} \tag{3.9}$$

with the variance 0.0064 eV$\equiv 64k_s$ that is $2k_s$ for each of 30+2 degrees of freedom of 1 eV.

Table 3. List of first radiation wavelengths of the hydrogen radiation spectra series
$\quad \lambda = 1/[1.096776 \times 10^7 \times (1/n1^2 - 1/n2^2)]$ m $\quad f_n = c_s/\lambda = 3 \times 10^8/\lambda$ s^{-1}

Lyman	1.21568×10^{-7} m	$f_n = 2.467 \times 10^{15}$ Hz	
Balmer	6.56469×10^{-7} m	$f_n = 4.5699 \times 10^{14}$ Hz	These figures could be
Pashen	18.20586×10^{-7} m	$f_n = 1.6478 \times 10^{14}$ Hz	manipulated
Brackett	4.05228×10^{-6} m	$f_n = 7.4032 \times 10^{13}$ Hz $\rightarrow$/6 $\times 1.1111^2 = 1.00 \times 10^{13}$ Hz	
Pfunk	7.4589×10^{-6} m	$f_n = 4.0220 \times 10^{13}$ Hz $\quad$ (3.10)	
Humphey	12.400×10^{-6} m	$f_n = 2.4193 \times 10^{13}$ Hz $\rightarrow$/6 $= 4.032 \times 10^{12}$ Hz	

Velocity of light appears to be derivative of the local force 10^{-7} N/m, electron frequency, 2/3 absorption and average local linear velocity:

$$\text{Light velocity } c_s = F_{atomic} \times 5 \times 10^{15} \times (2/3)/1.1111 \equiv 3 \times 10^8 \text{ m/s} \tag{3.12}$$

The first quantum term of the **Planck's radiation Law** expressing infrared radiation requires $e^{hc/\lambda kT} = 2$

$$\textbf{\textit{E}}(\lambda, \textbf{\textit{T}}) = 8\pi hc/\lambda^5 \times 1/(e^{hc/\lambda kT} - 1) \tag{3.13}$$

producing vectorized pulse energy $\quad \textbf{\textit{E}}(\lambda, \textbf{\textit{T}}) = 0.01$ J or 0.01 eV per atomic event.

SI energy of 0.01 J is 0.001 eV in atomic terms because rotational size converter is 10^{-20} and electron charge 10^{-19}:

$$0.01 \text{ J } \mathbf{x} \ 10^{-20}/10^{-19} = 0.001 \text{ eV} \qquad (3.14)$$

Energy 0.01 J may be considered to be energy per quantum pathway

$$1/5 \ h_s v_{Lms}/e_s = 1/5 \ \mathbf{x} 1 \mathbf{x} 10^{-33} \mathbf{x} 5 \ \mathbf{x} 10^{12} \text{ Hz}/1 \mathbf{x} 10^{-19} = 0.01 \text{ eV} \qquad (3.14a)$$

The product of two pulses and linear probability (1D converter to atomic terms) derives the minimum energy $1\mathbf{x}10^{-14}$ J of Equation 2.45:

$$(0.01 \text{ J } \mathbf{x} \ 0.01 \text{ J}) \ \mathbf{x} \ 10^{-10} \equiv 1 \ \mathbf{x} \ 10^{-14} \text{ J, made of 10 local } 1\mathbf{x}10^{-15} \text{ J energies.}$$

The same energy 0.01 J is obtained using integer space cubic constants converted to SI in quantum corrected Planck Radiation Law:

$E_n(\lambda, T) = 8\pi h_s c_s / \lambda_s^5 \mathbf{x} 1.04^{5/6}$
$\mathbf{x} 1/(e^{hc/\lambda kT} - 1) = 8\pi \ \mathbf{x} 1 \mathbf{x} 10^{-33} \mathbf{x} 3$
$\mathbf{x} 10^8/(60 \mathbf{x} 10^{-6})^5$
$\mathbf{x} 1.04^{19/24} \mathbf{x} 1/(e^{hc/\lambda kT} - 1)$
$. = \mathbf{0.010002} \ \mathbf{x} 1/(e^{hc/\lambda kT} - 1) = F_{Gs} a_s \mathbf{x} 10 \ \mathbf{x} 1 \ /(e^{hc/\lambda kT} - 1) \text{ J/per unit wavelength per m}^3 \qquad (3.15)$
0.01 J at atomic level becomes 0.001 J or eV.

Above value $1/3 \mathbf{x} k_s \mathbf{x} T_{300}/e_s = 0.01$ eVSI seem to indicate that Planck radiation term applies to temperature of 300 K, but the quantum value of energy 0.01 J is temperature independent, because charge expresses proton momentum proportional to temperature. Energy of one pulse is proportional to the square of the frequency and the pulse is produced by two pulses spaced about 90° in wave function format with power being proportional to T^4, as predicted by the Stephan Law. The energy per plane 1 J or eV is needed to generate energy expressing a vector directed along the source-receiver axes. The energy 3 eVat is for three oscillators. At 6000 K the energy generated displays 60 eVat expressing 50 eV times 1.2.

In the other form the first term of Planck Law $E(\lambda, T) = 2h v^3/c^2 \ \mathbf{x} \ 1/(e^{hv/kT} - 1)$ using cubic integer constants expresses energy for three planes and requires now a different treatment:

$$E_n(\lambda, T) = 2h_s v_s^3/c_s^2 \ \mathbf{x} 1/(e^{h_s v_s/k_s T_s} - 1) \ \mathbf{x} \text{ correctors}$$
$$\equiv 2\mathbf{x}10^{-33} \mathbf{x} (5\mathbf{x}10^{12})^3/(3\mathbf{x}10^8)^2 \mathbf{x} 10^{10} \mathbf{x} 1.11 \ldots \mathbf{x} 1/1.04^{7/10} \mathbf{x} 1/(e^{hv/kT} - 1) = 0.030028 \mathbf{x} 1/(e^{hv/kT} - 1) \text{ J/ m}^3.$$
$$(3.16)$$

Since the above value is also 0.030028 eVat per atom or event, the variance 0.000028 eV is actually 1/3 of Boltzmann constant k_s representing H. u.

In most relations the sun appears to involve only ½ of its mass in the interaction. Decreased by the probability $1/10^3$, involving three sequential displays out of thirty possible in three planes of space and divided by the square of relativity for 1/2 x c_s, the mass of sun per unit surface circular sun area generates average mass density along the sun-earth axes of 5 x 10^8 kg/m²:

Mass density radiated using the integer space units =
$$= 1 \ \mathbf{x} \ 10^{30} \text{ kg } \mathbf{x} \ 1/10^3/1.5 \ \mathbf{x} 10^{18} \text{ m}^2 \ \mathbf{x} \ 1/1.1547^2 = 5.0000047 \ \mathbf{x} 10^8 \text{ kg/m}^2 \quad (3.17)$$

Above sun surface, the mass of 1 kilogram moving at the velocity of light, while subjected to relativity and quantum correction expresses the momentum of the surface mass density:

The corrected average momentum rate of 1 kg mass density along the sun-earth axes

$$= c_s \text{x1 kg/m}^2 \text{ x1.2 x } 1.1547^2 \text{ x1.111} \ldots^{1/2}/1.04^{7/24} = 5.002 \text{ x10}^8 \text{ kg/(m-s)} \quad (3.18)$$

Otherwise the average sun mass-gradient along the sun-earth axis =

$$1 \text{ x } 10^{30} \text{ kg x } 10^{-10}/(1.5 \text{ x } 10^{11} \text{ x } 1.1547^2) = 5.00000 \text{ x10}^8 \text{ kg/m} \quad (3.19)$$

The mass gradient converted by the square of light velocity, subjected to space quantization 1/1000, and relativity and quantum correction is the gravitation force accurate within 0.07%:

$$5.00000 \text{ x10}^8 \text{ x } 9\text{x10}^{16} \text{ x1/1000 x1/(1.11111}^2 \text{ x1.04}^{7/10}) = 3.5463 \text{ x10}^{22} \text{ N} \quad (3.20)$$

The 1 kg/m$^2 \equiv$ 1 kg $\equiv$ 1 J, which subjected to space quantization combining each 1/10 probability derive the Planck radiation energy 1x1/10x1/10 = 0.01 J. It represents the product *ExB* expressing energy at 1 K.

Author's experiments derived correct values of physical constants from radiation of NaCL impurities at temperature of 21 C, at which electron frequencies are effectively absent. They derived Planck constant h_s=1x10^{-33} J-s, which must relate to infrared oscillations of the proton and the pair (proton+electron), because no significant electrons radiations are radiated at 300 K. How can then science attribute electrons for delivering correct energy values? It is proposed now that electron mass is the Planck constant referred to 300 K and to three planes and multiplied by relativity corrector 1.11111 for 1/3 c_s:

Electron mass $m_{es} \equiv h_s$ x 300 x 3 x 1.1111= 1x10^{-30} kg $\quad (3.21)$

meaning that at **300 K the value of Planck constant h = 6.62607 x 10^{-34} J and h_s should be used with infrared frequencies, and not with electron frequencies.**

Further consideration discussed subsequently attribute Bohr radiation energies to the equivalent (reduced) electron mass expressing (proton+electron) pair.

Energy 3x10^8 W divided by 3x10^8 m/s required for 10 pathways is 10.0065 eV and 30 eV for three planes. The variance is 60 k_s or 6k_s per pathway with k_s per space degree of freedom. The total energy is 10.0065 eV plus 2 x 10 x 0.010002 derives the first step of 10.21 eV as indicated in the classical SI system.

The quantum value 1/3x$k_s T_{300}$/e_s is 0.01 eV. At 6000K the input energy 10 x 1/3x$k_s T_{6000}$/e_s = 2 eV and it increases to 3 eV by adding energy of electron radiation. The fraction 1/3kT times 3/2 converts 1/2xkT per degree of freedom, if electron energy takes part in the interaction. The author's quantum value measured was ½ x√3x0.01 eV, while the graviton energy [45] is 0.01x√3/1.04 = 0.01666 eV.

The velocity of light can be derived in terms of the microwave wavelength from radiation from space wavelength measured at 0.21 m and the related frequency 1.429 x 10^9

Hz (see Equation 10.9). Referring the frequency to 1 K from 2.725 K times 3/2 because of the absence of electron frequency in infrared and times quantum corrector obtains:

$$1.429 \times 10^9 \text{ Hz}/(2.725^2) \times 3/2 \times 1.04 \equiv 3.002 \times 10^8 \text{ m/s} \qquad (3.22)$$

If the pulse energy is 50 eV the power is 1.0007×10^{10} W; this value at the atomic level time linear probability represents 1 W.

The first term of $E_n(\lambda, T) = 8\pi h_s c_s/\lambda_s^5$ multiplied by surface probability and relativity expresses $8 \times 3 \times 1 \times 10^{-33} \times 3 \times 10^8/(0.21)^5 \equiv 2.035$ J $\equiv 2.035$ eV summing energy of proton and of (electron+proton) pair   (3.23)

The wave length, $\lambda = 60 \times 10^{-6} = 600000$ A $= 2 \times 10^{-10}/300 \times 10^3$, expresses directed radiation energy comprising a vector component derived from incoherent energy of Planck Law radiation $E_n(\lambda, T)$. The maximum power per K derived for $T_s = 300$ K for electron radiation incorporates 10^3 space quantization factor increasing input incoherent radiation by 10^3:

$$\lambda_e = c_s/v_{Hlms} \times 10^3 = 3 \times 10^8/5 \times 10^{15} \times 10^3 = 60 \times 10^{-6} \text{ m} \qquad (3.24)$$

For protons, the interaction is only in two dimensions with space quantization factor of 10^2. The proton velocity 3×10^7 m/s expresses both energy in two linear dimensions and in spin, each component contributing 1×10^7 m/s totaling 3×10^7 m/s and deriving the same value of proton wavelength λ_p:

$$\lambda_p = 3 \times 10^7/v_{Llms} \times 10^2 = 3 \times 10^8/5 \times 10^{12} \times 10^2 = 60 \times 10^{-6} \text{ m} \qquad (3.25)$$

Energy of the sun pulse is 500 J $\equiv$ 500 eVSI, subjected to linear quantization of 1/10 making 50 eV on reaches earth:

$$h_s \times c_s/(\lambda \times 1/2 \times k_s) = 1 \times 10^{-33} \times 3 \times 10^8/60 \times 10^{-6} \times 1/2 \times 10^{-23} = 500 \text{ J} \equiv 500 \text{ eV} \equiv 1/12 \times 6000 \text{ K}$$
$$(3.26)$$

Above relation specifies expresses maximum value of $E_n(\lambda, T)$ at 0.01 eV $\pm$ 5 parts per 10^4. The value $F_{Gs} \times a$ is exact at wave length $\lambda_s = c_s/v_{Lms} = 3 \times 10^8/5 \times 10^{12} = 60 \times 10^{-6}$ m expressing **Planck's radiation law** of Equation 3.13. Using in the above expression the wavelength $\lambda_s = 2 \times 10^{-10}$ m requires first converting it to room temperature and multiplying λ_s by 300 K to obtain 6750 A; using now 3 for π with relativity corrector 1.111 . . . derives for 3 planes energy 3×0.01 eVat.

Thousand times higher electron frequency with thousand time smaller mass derives radiation energy $1000 \times 0.01 = 10$ eVat, which represents electron energy, proton energy and (proton+electron) pair making the model Hartree energy of 30 eVat.

In six orientations the radiation energy 10^{-28} J at 300K oscillates at 5×10^{12} Hz so that the related energy

$$6 \times 1000 \times 1 \times 10^{-28} (5 \times 10^{12})^2 = 15 \text{ J} \qquad (3.27)$$

According to Equation 17.11 energy 10^{-28} J with relativity corrections is the mass of muo10$_s$ responsible for the force within the atom.

The maximum wavelength of the Planck Radiation Law distribution varies with temperatures and so does the oscillatory frequency v. At 1 K the low temperature component:

$$e^{h_s v_s/k_s} - 1 = 1, \ e^{h_s v_s/k_s} = 2, \ ln\,2 = 10^{10} \times 1/(\sqrt{2} \times 1.04^{1/2}) = 0.69337\,v_s.$$

At 1 K $\qquad v_{s1K} = 10^{10} \times 0.9999925 = 1.0000075 \times 10^{10} \approx 1.0 \times 10^{10}$ Hz $\qquad$ (3.28)

Consequently the International Gravitation Constant 1.0×10^{-10} N-m^2/kg^2 proposed for bodies at equal temperatures is energy per unit frequency; multiplied by the invert of linear probability it derives 1.0×10^{-10} N-m^2/kg^2 $\times 1.0 \times 10^{10} \equiv 1$ W or J.

At 300 K $\qquad v_{s300K} = 300^4 \times 1372 \times 1/2 \times 1/(1.07547 \times 1.0328) = 5.0036 \approx 5.00 \times 10^{12}$ Hz $\quad$ (3.29)

The value 1372 W was chosen from Equation 11.3 expresing energy discharged from a cube of 10^3 Bohr atoms on the sun $\hfill$ (3.30)

At 6000 K, $v_{s6000K} = 3 \times 6000^4 \times \sqrt{2} \times 1/2 \times 1/1.11 \ldots \times 1.04^{1/4}) = 4.9974 \times 10^{15} \approx 5.00 \times 10^{15}$ Hz.

Conversion of 20 H atoms into 5 He atoms generates residue mass, which is derived from any table of atomic elements:
Atomic weight of 20 H /Atomic weight of He - 5 = 20.16/4.0026 - 5 = 0.036726 units.

The number of 1×10^{-27} kg radiations released by 0.036726 kg mass is $0.036726/1 \times 10^{-27} =$
$\qquad$ 3.6776 $\times 10^{25}$ times energy per local particle $1/5h_s \times 1/3c_s = 2 \times 10^{-26}$ kg of pulse mass corrected for relativity:
$\qquad$ 3.6776 $\times 10^{25} \times 2 \times 10^{-26} \times 1.15470^2 = 1.0011$ J $\hfill$ (3.31)

In terms of earth parameters 1×10^{-27} kg $\times c_s^3 \times 300K \times 1.111 \ldots^2 \times 1/10 = 1.000\,000\,00$ J
$\hfill$ (3.32)
Above value is still per K. Converted expression 1 J $\times 300^2$ K$^2 \times 10 \times 1.111 \ldots = 1 \times 10^6$ J. balances well the sun pulse of 1 MeV calculated in Equations 10.20-21 and others and expressed now as
$\qquad$ 1×10^{-27} kg $\times 1000$ K $\times 10^{30} \equiv 1 \times 10^6$ J $\hfill$ (3.33)

With energy $E_n = \lambda \times v_{Lms}$, $\lambda = 1 \times 10^6$ J$/1 \times 10^{12}$ Hz $= 1000$ nm is on the low energy wavelength section of black body radiation generated by the sun (see WEB). Reduced by the non-interacting components of radiation of 1.2^3:

$\qquad$ **$\lambda = 1000$ nm/1.2^3 = 578.7 nm**, close to the λ_{max} radiation maximum $\qquad$ (3.34)

The maximimu intensity of radiation according to the classical physics textbooks is
$\lambda_{max} T = 2.898 \times 10^{-3}$ mK; it **derives 579 nm of Equation 3.34 on** division by 500, because 6000 K is reduced to 5000K by 20% non-interactive radiation. The wavelength of 1000 nm derives the classical value of the Stepfan-Boltzmann constant in:

$$\sigma \equiv 10 \times 1000 \text{ nm}/(1.2^3 \times 1.04^{1/2}) = 5.67 \times 10^{-8} \text{ W/m}^2 \tag{3.35}$$

The wavelength λ_1 is the product of Rydenberg constant of hydrogen and relativity correction $\qquad \lambda_1 = 91.117633 \times 10^{-9} \times 1.111 \ldots = 101.242 \text{ nm}$ (3.36)

generating energy $q_1 = \lambda_1 c_s = 101.242 \times 10^{-9} \times 3 \times 10^8 = 30.313 \text{ J} = 30 \text{ J(or eV)} + 10 \, k_s T_{300}$ (3.37)

Magnetic versa electric parameters

The concept that electromagnetic wave comprises progressing magnetic and electric waves converting to each other are well established in our minds. Its validity is demonstrated in traveling way experiments and explained in traveling wave theory. Nevertheless the magnetic theory of physics remains the source of controversy over several decades. The importance of the probability as the ultimate source of the numerical values does not disclose the structure of the physical event. In the eleventh year of research the relations between electric and magnetic parameters have been uncovered. The magnetic parameter is expressed by 1/10 of the first term of the Planck radiation term 0.01 J or eV, or otherwise $1/30 \, k_s T_{300}/e_s = 0.001$ J. These are energies in each of ten orientations of one plane. They are the magnetic permeability in one dimension and express the converted dielectric permittivity: $(\mu_{os})^{1/2} = \varepsilon_{os} \times 1/3 \, c_s = 1 \times 10^{-11} \times 1 \times 10^8 \text{ F/s} = 1/30 \, k_s T_{300}/e_s = 0.001$ J or eV. Converted per degree per plane and corrected for relativity the value becomes $1/3 \times 0.001/(300 \times 1.111 \ldots) = 10^{-6} = 0.000001 = \mu_{os} = (\varepsilon_{os}/10)^{1/2}$ per kelvin. Terms μ_{os} and ε_{os} are dynamic F/s and static radiation charges and not parameters of space, although they may express energy capacity of space for transmission filling the empty space. There are 32 energies of 0.01 J or eV in three planes, which time relativity correction 1.111 . . . derive their total energy at 1 J or eV. Each of these energies has ten quantum factors 0.001 J per plane and when balance is lost all components leave in one direction, and are observed by the receiver.

4. Magnetic lens condensing to earth gravitation forces of the sun

The hot sun radiates energies distributed around the mean three frequencies 5×10^{12} Hz, 4.995×10^{12} Hz and 5×10^{15} Hz. Due to equipartition of energy needed for the minimization of energy of oscillating particles, the energies of oscillations of electron, of proton and of electron+proton pair are equal. Such equipartition of energy is generally accepted in science. It represents maximum probability and entropy. Equipartition of energy was displayed in author's earlier experiments, but it was not reported, because the importance of the observation was not appreciated. WEB pages display the proof that for quadratic energy terms the mean energy is spread equally over all degrees of freedom.

Earth, being cool, generates effectively only energy of proton and (proton+electron) frequencies. Consequently the elementary pulses of earth carry per kelvin energies, which are 2/3 of those of the sun. The maximum rate of the interactions between the sun and the earth radiations takes place in the region, where the radiation densities are greatest. Their density decreases as the square of the distance from the source of radiation and their power density increases with temperature proportionally to T^4 of the source. Equating power densities from the sun and from the earth at balance distance L_b from earth

$$\sigma_s \times 6000^4/(1.491 \times 10^{11} - L_b)^2 = \sigma_s \times 300^4/(L_b)^2 \qquad (4.1)$$

The distance from earth center at which sun radiating power densities are equal to that of earth is 15500 km, representing over five earth radii. It is suggested that the sun and earth radiations are spread out and are interacting with as much as some 30000 km from earth centre. In the consequence of these interactions the resultant gravitation vectors are directed to the center of earth representing all the infrared radiation within the surface area as shown in Fig.7. The resultant gravitation vectors summing radiations from the sun and from the earth cover earth from all its sides, the beam starting along the earth plane of rotation around the sun.

Thus **vectorized infrared radiation from the sun is concentrated from much a larger area than earth surface.** Consequently electron frequency power density falling on earth is less than 1% of infrared frequency and its effects are neglected in the model. **The process of condensation of sun infrared radiations on earth is essentially the same, but reversed, as in gravity magnifying lenses deflecting beams of particle radiations.**

The late finding in Chapter 17 of Equation 17.30 hypothesizes with fair probability that sun radiation interacting with earth radiation is derived only from only $1 \times 10^{30} \times 1/30^2 \times 1/1.1111 = 1 \times 10^{27}$ kg of its mass. Justification for this statement is derived from the accurate derivation of reference F_{Gs}:

$$1 \times 10^{27} \text{ kg} \times 0.006 \text{ N/kg} \times 1/10 \times 1/10 \times 1/1.2 \times 1/1.1547^2 \times 1/1.11 \ldots^{1/2}/1.04^{1/10} = 3.5436 \times 10^{22} \text{ N} \qquad (4.2)$$

Re-examination of Equation 4.1 requires the introduction of factor 10^3 on the earth side:

$$\sigma_s \times 6000^4/(1.491 \times 10^{11} - L_b)^2 = \sigma_s \times 300^4/(L_b')^2 \times 1 \times 10^3 \qquad (4.3)$$

Solution of the quadratic equation with new balance distance from earth L_b' corrected for relativity and quantum:

$$L_b'=-1.8755 \times 10^9 \pm \sqrt{(1+5.5597 \times 10^{20})}/2 \times 1/1.11 \ldots \times 1.04^{1/2} = 0.9992 \times 10^{10} \approx 1 \times 10^{10} \quad (4.4)$$

Thus the distance at which force balance occurs L_b' is 1×10^{10} m from the earth and 14×10^{10} m from the sun the latter interpreted as $\sqrt{2} \times 1 \times 10^{10} \times 10$ expressing rotational addition of two vectors. At one kelvin the electron frequency $f_e = 1 \times 10^{10}$ Hz has wavelength $\lambda = 1 \times 10^{-10}$ m to generate:

$$\text{SI progression} = \lambda \times f_e = 1 \text{ cycles/m, becoming } 1 \times 10^{-10} \text{ at the atomic level} \quad (4.5)$$

Equation 17.50 establishers that radiation from the whole circular area of the sun delivers the gravitation force and so that beam is condensed on earth. The components repulsing the sun have to oppose the force of attraction in the direction and in the intensity. It is postulated that the earth circular area facing the sun generates the force of 1 N (1 N-s/m at atomic level) per square meter and since the propagating radiation moves at the speed of light it generates the quantum corrected repulsive force of:

$$\text{Earth repulsive force} = 1 \text{ N-s/m}^3 \times \text{Circular earth area} \times \text{velocity of light}/1.04^2$$
$$= 1 \times 1.278 \times 10^{14} \times 3 \times 10^8 /1.04^2 = 3.5447 \times 10^{22} \text{ N} \quad (4.6)$$

This is one of gravitation forces considered. At lower temperatures two oscillators of Bohr atom (proton and electron-proton pair) move with the average local velocity $1/3 \times c_s = 1 \times 10^8$ m/s in orthogonal directions. Eight digits of the quantum corrector $1/(2 \times \ln 2^2) \sim 1.04$ had to be used to obtain best accuracy. The Newton force, generated by the one dimensional display of the mass of earth $(m_{earths})^{1/3}$ kg, is 1 N-s/m per unit of vector sum of local velocities, with accuracy of 9 parts in 10^5:

$$(6 \times 10^{24})^{1/3} \text{ kg} \times 1.0406849^{1.2}/(\sqrt{3} \times 1 \times 10^8 \text{ m/s}) \equiv 1.000087 \text{ N-s/per unit local velocity} \quad (4.7)$$

The above value converts to 2 J/m on multiplying by $\sqrt{3} \times 1.154700 = 2.000173$ N-s/m and to 10 N-s/m gravitation pulse acting on mass of 1 kg summed from the 5 pathways of the space quantization, as given in Equation 10.1.

The sun attracts the earth with the same force as the earth attracts the sun. Only vectors of the same frequency or multiplier frequency interact.

Any radiation must have a supply of energy and **involve** recession (**movement**) **of charges away from the receiver and** away from the reference mass. This reduces the negative energy of the elementary dipole and releases energy in the direction of the receiver. **The movement of the reduced electron away from the receiver** oscillating at an infrared frequency takes at places at velocities of c_s, $c_s/3$, $c_s/2$ or $c_s/4$ is duly observed by the receiver. The observed shift is due to the movement of the reduced electron with respect of the centre of mass, effectively the proton or atomic mass, but not the shift of the mass of the planet.

Paul Marmet proposed that such signal is absorbed by hydrogen atoms on its way to the receiver, and each re-radiation generates a small delay. The observed signal may display infrared shift of velocities close to the velocity of light increasing with distance from the receiver.

Consequently **it may be incorrect to base the velocity of the Universe expansion on the infrared radiation shift observed**. A confirmation is required that infrared shift does occur between earth planets and is related with their mass velocities.

Since a part of the energy from the sun does not take part in the gravity interaction some of the proposed physical constants of space in the cubic geometry converted to SI units are about 2/3 of values of standard physical SI units.

The infrared aspects of the model are displayed in the three decades old author's experiments. The field applied to impurity dipoles in NaCl crystal induces an increment of the local field by a factor of dielectric constant ε_r comprising the ionic ε_{rio} and electronic ε_{rel} components. The ionic component is ε_{rio} = 5.62 - 2.25 = 3.24 units. The maximum expansion of the Bohr structure takes place at 9 steps of corrected energy of 1 eVat, as measured and calculated earlier [45, 46]. The energy of subsequent relaxation is sourced only from the infrared energy stored and radiated from the cool crystal. That energy in classical SI terms is:

$$1 \text{ eV} \times 3.24 \times 9 = 27.236 \text{ eVat} = \text{Hartree energy} + kT \text{ term} \qquad (4.8)$$

Relaxation experiments described in the Appendix displayed the relaxation current as series of straight sections decreasing their slopes in quantum steps. Second differential displays these results as pulses at increased separation due to the decrease of the stored energy. Pulses are formed from numerous reflections from the masses behind the radiator and comprise components of waves generated in both directions. Pulses are considered by the author an inherent feature of radiation and temperature their generator.

The geometry of 300 K repulsive forces suggests that earth pulse of 300 K acting in three orientations generate an effective resultant $\sqrt{300} \times 1.1547 = 600$ K indicating that energy is radiated from hydrogen like structure element and not from Bohr atom.

Some textbooks quote 20 eV for Bohr atom energy. According to the integer cubic model with SI units this represents temperature in 1D:

$\sqrt{300} \times 1.1547005 = 19.99999934$ J indicating that uncertainty is 3×10^{-25} J times 1×10^{-19} = 0.000003 = 3 μ_{so} 　　　　　　　　　　　　　　　　　　　　　　　(4.9)

5. Power of sun beam and sun radius

Parts of this section remains unchanged as originally written some years ago, with integer values applied only to physical constants. The presently derived integer model power of sun beam is 4.0000×10^{29} W with classical SI value assumed in Table 1 to be 4.0000×10^{29} W$/1.04^{7/50} = 3.981 \times 10^{29}$ W. Values quoted on WEB pages range $(3.6-4.1) \times 10^{26}$ W. These values require according to the integer model converting to 6000 K deriving range $(2.16-2.46) \times 10^{29}$ W with an average $\sim 2.3 \times 10^{29}$ W. That expresses value 3.981×10^{29} W$/\sqrt{3}$. The power 3.981×10^{29} W appears accurate because using typical correctors of the book it derives the pulse power 6 space orientations to be 6.00041 W/m³ with uncertainty $2/3\ k_s$ per orientation: (5.0)

Pulse power per space orientation$= (3.981 \times 10^{29})^{1/2}/1.11 \ldots {}^{1/2} \times 1.04^{1/16} \times \varepsilon_{os} = 6.00041$ W/m³.

Pulse generated$= 3.981 \times 10^{29}/(6 \times 1.111 \ldots) \times 1.04^{1/9} \times 1/6 \rightarrow$ muo10_s force $= 0.99535 \times 10^{27}$ J. Double division by 6 is needed because there are two pulses $3.981 \times 10^{29}/6$ in wave function format generating velocity of light and $2 \times 0.99535 \times 10^{27}$ J output. The rotational symmetry expresses power in one direction by the square root of the mass ratio, $1/3(m_{sun}/m_{es})^{1/2}$ over $\sqrt{2}$ and relativity correction. The preliminary estimate of power of the sun is

$$P_{sun} = 1/3 \times (2 \times 10^{30}/1 \times 10^{-30})^{1/2}/\sqrt{2} \times 1/1.15470 = 3.8649 \times 10^{29} \text{ W} \qquad (5.1)$$

The sun generates in six directions the force of 1 W per kilogram per plane measured at earth at 300 K. Actually only ½ of the sun mass interacts with earth. The above power P_{sun} represents the power mass of sun generated in one direction; after relativity and quantum corrections per each of three planes the beam power, generated per kilogram from pulses in wave function format (90° space displaced), is:

$$(3 \times 3.8649 \times 10^{29} \text{ W})/ (1 \times 10^{30} \text{ kg} \times 1.1547 \times 1.04^{1/10}) = 1.0002 \text{ W per kg} \qquad (5.2)$$

Since physical constants are per K, power $\sigma_s \times T_{sun}{}^4$ W/m² has to be multiplied by 6000 K, has to be divided by $\sqrt{3}$ and corrected for relativity: (5.3)

$\sigma_s \times T_{sun}{}^4 \times A_{cirsun} \times T/\sqrt{3} \times 1.1547 = 5 \times 10^{-8} \times 6000^4 \times 1.5 \times 10^{18} \times 6000/\sqrt{3} \times 1.1547 = 3.8888 \times 10^{29}$ W. Above power per unit circular sun area delivers per square meter of earth power of three planes: Sun power$/(A_{cirsun} \times$Surface of earth$) \times$ 6000 K
$= 3.8900 \times 10^{29}/[1.52178 \times 10^{18} \times 4\pi(6.371 \times 10^6)^2] \times 6000 \text{ K} = 3.007 \text{ W} \equiv 3.00 \text{ W/m}^2 \qquad (5.4)$

The sun beam power is the gravitational energy time velocity of light over Hartree energy:

$$F_G \times c_s /\text{Hartree energy} = 3.5438 \times 10^{22} \times 3 \times 10^8/27.2114 = 3.9070 \times 10^{29} \text{ W} \qquad (5.5)$$

Lack of dimensions of time and of space makes this equation acceptable, because Harare energy is really a non-dimensional factor $3^3 = 27$ representing also the proton mass and radiation of 1×10^{-27} W. The translation of the gravitation force of $F_G = 3.5430 \times 10^{22}$ N, times relativity corrections $1.111 \ldots$ multiplied by the linear factor 10^{10}, the space

65

quantization factor $1/10^3$ and quantum correction generates essentially the same value of sun power:

$$P_{sun} = 3.5430 \times 10^{22} x\, 1.111 \ldots x 10^{10} x 1/10^3\, x 1/1.04^{1/5} = 3.9059\, x\, 10^{29}\ \text{W} \qquad (5.6)$$

On the other hand the product of gravitation force and L_{earth} distance derives the energies of the sun and of earth for $2\,x$ three planes of moments. The moments divided by six times 1.11111 over $1.04^{28/50}$ linearly accelerate and decelerate the earth at 3.7755 mm/s while the earth is moving around the sun:

Integral of sun and earth energies $= F_G\, L_{earth} = 3.5430\ \times 10^{22}\ \text{N}\ x\ 1.5 x 10^{11} \text{m} = 5.3145\ x 10^{33}\ \text{J}.$
Value in one orientation $= 5.3145\ x 10^{33}\ \text{J}/6\ x\ 1.15470\ 1/1.04^{28/50} = 1.00055\ x 10^{33}\ \text{J}.$

The value $5.3145\ x 10^{33}$ N times $1/10^3$ and $1/K_{rel}^3$ derives essentially the same sun power in six directions:

$$1/10^3 x 5.3145 x 10^{33}/K_{rel}^3 \equiv 6\ x\ 3.906\ x\ 10^{29}/1.04^{1.1}\ \text{W} \qquad (5.7)$$

With earth eccentricity neglected earth circulates around spherical volume $V = 4/3 x\pi x L^3 = 4/3\ \pi\ (1.496 x 10^{11})^3 = 1.402 x 10^{34}\ \text{m}^3$. Since ε_{os} is 1D uncertainty of energy density the uncertainty for volume V becomes $1 x 10^{-11} x 1.402 x 10^{34} = 1.402\ x 10^{23}$ J. The uncertainty at Bohr atomic level with relativity correction and energy per particle $2 x 10^{-23}$ J becomes $1.402 x 10^{23} x 1.07547 x 2 x 10^{-23} = 3$ J or eV_{at}. Multiplication by volume probability derives surprisingly 1.5 of sun power of $3 x 10^{30}$ J. This is previous value $3.86\ x\ 10^{29}$ in 6 directions with relativity corrections:

$$6\ x\ 3.86\ x 10^{29}\ x\ 1.1547^2/1.0328 = 2.99\ x 10^{30}\ \text{J} \qquad (5.8)$$

Above value expressed per one type of oscillator

$$F_G x\ c_s\ x\ 1/10 = 3.5435\ x 10^{22} x 3\ x 10^8\ x 1/10\ x 1/1.11 \ldots .^{1/2} = 1.008\ x 10^{30}\ \text{W} \qquad (5.9)$$

For two orientations the above value becomes sun mass $2\ x 10^{30}$ W $\qquad (5.10)$

The thermal power $3.8630 x 10^{29}$ W times space quantization $1/10^3$ generates vector power $3.8630 x 10^{26}$ W, but with 10 possible positions in each plane the probability of taking the same position decreases by 30 for each component parameter increasing the input 900 times which for 3 planes makes with quantum corrector:

$$3.86 x 10^{26}\ \text{W}\ x\ (30^2)\ x 3/1.04 = 1.00 x 10^{30}\ \text{W or 1 W/kg per } \tfrac{1}{2} \text{ of sun mass} \qquad (5.11)$$

Thus the power generated by sun is 1 W per kilogram of its mass.

Sun power $1/2 x 3 x 10^{30}$ J times e_s derives uncertainty $1/2 x 10^{11} \text{J}/(2\pi x 5 x 10^{12}) = 0.016$ eV, J.
The same uncertainty was measured in author relaxation experiments, which were later interpreted to display models of Bohr atom.

Various derivations of the same sun power suggest strongly that treating the atmospheres of planets and of sun having multiple interactions as black surfaces in Equation 5.2 and sun constant in Equation 5.25 by $\sigma_s x T_{sun}{}^4$ multiplier is a valid method. Uniform distribution of sun radiation 3.8630×10^{29} derives a correct value of the sun constant P_s (the power density of sun measured at the earth surface):

$$3.8630 \times 10^{29} \text{ W } x \; 1/10^3 \; x \; 1/4\pi L_{earth}{}^2 = 1373.5 \text{ W/m}^2 \tag{5.12}$$

Sun constant is also given by the product $1000 \text{ K } x \; 1.1547^2 \; x \; 1.04^{3/4} \equiv 1372.5 \text{ W/m}^2$.

Actually the true value of sun density radiation on earth is 1417.4 W/m^2, $=1500/1.11111^{1/2} x 1.04^{1/10}$. Its **2/3 fraction** divided by 1.1547^2 relativity and quantum corrections **interacts with earth thermal radiations** deriving:

$$1417.4 \text{ W/m}^2 \; x \; 2/3 \; x \; 1/(1.1547^2 x 1.04^{1/4}) = \textbf{701.78 W/m}^2 \tag{5.13}$$

Please note that two 1000 K rates produced from two circular sun discs displaced by 90° generate $\sqrt{2} \; x \; 1000$ K rate $\equiv 1414.2$ J/m^2 decreased on earth by $1.04^{3/4}$ and measured as sun constant 1373.2 W/m^2. The 1000 K rate in 2D display over relativity correction derives 30 eV $\equiv$ 30 J:

$$1 \text{ kg } x \; 1000^{1/2} /1.111\ldots^{1/2} \equiv 30.000\,000 \text{ eV or J.}$$

Please note that 30 eV radiation signal is displayed in the Lyman frequency $2.467 x 10^{15}$ Hz of Table 3 when inserted to Equation 3.9.

The sun **input pulse energy** of 1 MeV of Equations 2.14, 2.9 and others is generated by the wave function formatted 30 eV pulses:

$$[(30 \, x30) \; X \; (30 \, x30)] \; x \; 1.111\ldots^2 = 1 \text{ MeV} \tag{5.14}$$

For balance sun and earth have to attract themselves with the same power density:

Sun: $1 \text{ J } x \; 1000^2 = 10^6$ units and earth $10 \text{ J } x \; 300^2 x \; 1.111\ldots = 10^6$ units.

And the value $701.78 \, x 1.04^{0.225} = 708$ is expressed by:

$$(30 \text{ X } 30)/1.111\ldots^2 x \; 1.04^{3/4} = 707.87 \text{ W/m}^2 \equiv 1000/\sqrt{2} \text{ K} \tag{5.15}$$

The interacting mass of sun over the geometric mass of the sun-earth couple derives essentially the same power 707.1068 W/m^2. That value to power eight over ($2x1.04$) derives the lowest special relativity value of the sun gravitation force on earth:

$$(707.1068)^8/(2 \, x1.04) \equiv 3.0048 \, x10^{22} \text{ N} \tag{5.16}$$

Radiation from the earth of $\sigma_s T_{300}{}^4 = 405$ W/m^2 increases by factor $\sqrt{3}$ because it progresses in three orthogonal directions deriving **701.48 W/m²**. The variation 0.3 W $\equiv 10$ kT or kT per direction represents uncertainty in three planes for rotations.

Also
$$1/6 \; x \; 1417.1^{1/4}/1.04^{1/2} = 1.0027 \text{ eV} \tag{5.17}$$

Sun power 3.8900×10^{29} W-s derives the gravitation as the force of attraction:

$$F_{Gs} = 3.8900 \times 10^{29} \text{ W} \ x \ 10^{-10} \ x \ 10^{3} \ x \ 1/1.1547^{1/2} \ x1/1.04^{13/25} = 3.5469 \ x10^{22} \text{ N} \qquad (5.18)$$

Using 10 pathways in 3 planes represents otherwise 30 eV Hartee energy indicating that the gravitation interaction occurs only in one plane, which is consistent with the centrifugal force of the rotating earth being in the same plane:

$$F_{Gs} = 3.8900 \ x10^{29} \text{ W}/3x10^{8} \ x10 \ x3/1.11 \ldots x(1+1/80) = 3.5538 \ x10^{22}. \qquad (5.19)$$

Gravitation force F_G and two invert probabilities $1/10 \ x1/10$ derives WEB value:

Infrared sun beam power $= 3.5438 \ x10^{22} x10^{2} x10^{2} xK_{rel}^{1/2} x1/1.04^{1/2} = 3.905 \ x10^{26}$ W $\qquad (5.20)$

The power of the sun comes from the last stage of the mass conversion of the sun involving iron [18], which displays relative magnetic permeability of about $\mu_r = 1489$. It is suggested that μ_r in conjunction with hydrogen to helium conversion, is the resonating source of energy, which supplies all degrees of freedom with their energy and so it expresses the attractive and possibly the repulsive force in variety of ways. However if there is any disturbance in the system it is the power supplied by the sun, which re-establishes the balance of the system.

The circular sun area $A_{cirsun} = 1.52229 \text{ x } 10^{18} \text{ m}^2 = \pi R_{sun}^{2}$ and $\qquad (5.21)$
the sun radius corrected for quantum and for relativity:

$$R_{sun}' = 6.9600 \ x10^{8}/(1.1547 \text{ x } 1.04^{1/10}) = 6.004 \ x10^{8} \text{ m}.$$

Sun radius expresses the value of $2 \ c_s = 2 \text{ x } 3 \text{ x } 10^{8} \text{ m/s}.$ $\qquad (5.22)$

This suggests presence of a barrier of potential at sun boundaries.

Kinetic energy of 1 kg in three orientations at the square of the local velocity of light per cycle of infrared frequency v_{Lms} derives the numerical value of the sun temperature:

$$3 \text{ x}(1x10^{8})^{2}/(5x10^{12}) \equiv 6000 \text{ K} \qquad (5.23)$$

The energy radiated per square meter using Stephan-Einstein constant σ_s and T_s^4, times $4A_{cirsun}$, times 10^3 quantum factor, times 6 directions and factor $\sqrt{2}/1.1111$ x2/3, times 1 kg per watt derives the mass of the sun as $2.0076 x10^{30}$ kg with 0.4% accuracy of the integer cubic space SI value: $\qquad (5.24)$

Sun mass $= 5 \ x10^{-8} \text{ x } 6000^{4} x4x \ 1.52118x10^{18} \ x10^{3}x \ (6x\sqrt{2}/1.1111) = 2.0074x \ 10^{30}$ kg

Thermal energy radiation density on the surface of the sphere around which earth is rotating is the sun constant:

$4\pi L_{searth}^2 =$ $\sigma_s T_s^4$ time sun sphere surface area $4\pi R sun^2$/surface area of earth rotation sphere

$$5 \times 10^{-8} \times 6000^4 \times 4\pi \times (6.9098 \times 10^8)^2 / [4\pi(1.5 \times 10^{11})^2] = 1375 \text{ W/m}^2 \quad (5.25)$$

The first radiation energy step of Planck radiation law is accurately derived from ½ of sun mass, which oscillates radiating in both directions 1 W/kg, from the surface of the earth sphere of rotation and over velocity of light and relativity and quantum corrections:

Sun power/surface of earth sphere of rotation $\times 1/[1.11 \ldots ^{3/2} \times 1.04^{1/6} c_s] =$

$$1 \times 10^{30} / 4\pi(1.5 \times 10^{11})^2 \times 1/[1.11 \ldots ^{3/2} \times 1.04^{1/6} \times 3 \times 10^8] = 0.0100009 \text{ W/m}^2 \quad (5.26)$$

Masses of earth and of sun express balance of the inverse ratios of their thermal emissive powers. Studies of initially incomprehensive, but accurate relations, such as one below have uncovered the 2nd new set of integer model physical constants:

$$1.04^{7/24} \times 6000^4 / 1.0004 \times 10^{30} \text{ kg} \approx 300^4 / [6 \times 10^{24} \text{ kg} \times (1.11 \ldots ^{1/2})] \quad (5.27)$$
$$1.31 \times 10^{-15} \approx 1.28 \times 10^{-15}.$$

Proper justification from the measurement of quantum value 0.693 kT and the interpretation of Millikan experiment took further eight years to uncover.

Multiplying above by $\sigma_s / 1.1547^2 = 5 \times 10^{-8} / 1.1547^2$ J/K derives energy radiated per kilogram:
~5×10^{-23} J $= 5k_s = 5$ Boltzmann constants in five pathways and **electron charge e_s is defined as momentum of proton in one direction taken to mean that it is the equivalent (reduced electron), while the source the square root of local velocity 10^4** often appearing in relation is clarified:

$$e_s = k_s \times (1/3 c_s)^{1/2} = 1 \times 10^{-23} \times 1 \times 10^4 = 1 \times 10^{-19} = 1/2 m_{ps} c_s = 1 \times 10^{-27} \times 1 \times 10^8 \text{ C} \quad (5.28)$$

Dividing by 1.1547^2 indicates that energies generated per kg of earth and per kg of sun are equal 1×10^{-15} J/kg, with sun splitting its moments in two orientations. That value multiplied by 1×10^{-19} is $h_s/10$.

The sun beam power 4×10^{29} W per unit light velocity c_s, corrected for relativity, $1.00000000 \times 10^{21}$ J/m, derives the term $E_{searthSI}$:

$$E_{searthSI} \equiv 4 \times 10^{29} \text{ W} / (1.1547^2 \times 3 \times 10^8 \text{ m/s}) = 3.000 \times 10^{29} / 3 \times 10^8 = 1 \times 10^{21} \text{ J/m} \equiv 1 \times 10^{21} \text{ J-s`} \quad (5.29)$$

The above relation uncovers that the sun beam power $(1/3 c_s)^{1/2} = 6000 \times \sqrt{3} / 1.04$ directed at earth as consequence of Special Relativity is an integer 3.000×10^{29} W taking value 4×10^{29} W due to relativity correction 1.1547^2 and only slightly varying from integer.

Sun pulse $= 1 \times 10^{21} \times 3 \times 10^{-21} = 3$ W $= 3$ eVat; power of 2 W are interacting in the gravitation balance. This value is per K, now referred to 300 K and decreased by relativity corrector 1.1547^2 the sun pulse derived is the same as in Equations 3.25 and 5.3: $\qquad\qquad 2 \times 300 / 1.1547^2 = 500$ W $\qquad\qquad\qquad$ (5.30)

Referred to 6000 K the value becomes 20 times larger generating after corrections:

Sun temperatures = 500 x 20/(1.2 x1.111 . . .³x1.04$^{1/3}$) = 5996.1 K ≡ 6000 K - 4 eV   (5.31)
The energy 4 eV represents the quantum 1 eVat in each of four orientations.

It is subsequently assumed that sun interacts with earth only with one of its beams. Circular sun has six orientations and each has five quantum directions making a total of 30 beams. In conclusion 64 a more realistic model of reality than SI has been proposed based on probability of the elementary unity event and the need to use elastic kinetic units in radiation interactions. Sun pulses interact with earth in one of four orientations of one plane each pulse being formed from three pulses in orthogonal orientations less two for non-interacting axes. The resultant can be formed in planes XY, YZ or XZ creating three more options and a total of 30 possible beams. Thus the sun power 3.9 x 10^{29} W of Equations 5.1, 5.3, 5.5 and 5.5 derives the reacting sun mass 1 x 10^{30} kg:

Interacting sun mass ≡ 3.9x10^{29} Wx 30/1.1547 x1/10 /1.04$^{1/3}$x1 kg/W≡ 1.00 x10^{30} kg (5.32)

Temperature in kelvin has been found to express pulse rate with parameter deriving often unpredictable effects. Some identities having undefined physical meaning are left below for consideration by other researchers.

The disagreement in units is immaterial, because the final conclusion shows that all SI terms express probability derivatives of the basic reference event. The model sun temperature of 6000 K expresses essentially the proton oscillations. With electron's own oscillations the peak sun temperature becomes 10000 K, which times 2/3 x 1/1.111 . . . derives the following quantum values of the wave: 1 section of 6000 K, 2 sections of 4000 K and 2 sections of 2000 K. The mean value 1000 K is derived on dividing by 5 time intervals over 1.1547^2 times 1.111 . . . and spreading into three planes. Three vectors pulses of squared 1000 K over relativity and quantum correction derive essentially the local velocity of light: 3 x 6000^2/(1.1547^2 x 1.04$^{1/8}$) = 100013817.7 m/s. The next calculations display energies derived for different observers. Two 1000 K pulse generate the resultant energy ($\sqrt{2}$x1000 K)X($\sqrt{2}$ x1000 K) ≡ 2 x10^6 eV or **1x10^6 eV per source atom or per orientation.**

Two pulses of 10000 K in wave function formation may generate the local velocity of light (10000 K)X(10000 K) ≡ 1x10^8 m/s. The formation of 10000 K pulse can be expressed by the product of four terms 1/10 x 1/10 x 1/10 x 1/10 or

$$1/2x6000^3/(1.1547^{1/2}x1.04^{3/25}) \equiv 10\ 003.34\ K \tag{5.34}$$

Temperature 10000 K is most likely to be $\sqrt{3}$ x 6000 K/1.04 = 9992.6 K.

Velocity of light is the rate of pulses arriving at the receiver generated at the 10000 K

$$\text{Source} = c_s \equiv \sigma_s T^4/\sqrt{3}\ x1.04/1.04^{1/53} \equiv 3.000\ 000\ 53\ x\ 10^8\,m/s \tag{5.35}$$

The chosen quantum corrector has a high chance of being correct, if the accuracy increases suddenly by about two orders of magnitude. Here 53 is taken as being 40×1.154700^2. The variance is 5 quantum units of radiation in one orientation.

Sun attracts earth with mass 1×10^{30} kg with two of its 32 beams added at 90°, representing radiation from 1×10^{30} kg/32 = 3.125×10^{28} kg. Each beam is sourced from three orthogonal orientations in the atoms generating a probability $1/10^3$ for each beam and $1/10^6$ for two, which generate 1 N/kg time relativity over quantum corrector deriving the exact reference:

$$3.125 \times 10^{28} \text{ kg} \times 1/10^6 \times 1 \text{ N/kg} \times 1.1547 / 1.04^{23/50} = 3.5439 \times 10^{22} \qquad (5.37)$$

The mass of earth appears to generate the same force. The force 2 N/kg of mass direction is needed to keep the fields around earth "immobile" and keep earth stability with respect to sun:

Earth gravitation force =
$$2 \text{ N} \times 10^{24} \text{ kg} \times 1/10 \times 1/10 \times 1.1547^3 \times 1.111\ldots^{3/2} / 1.04^{9/20} = 3.5433 \times 10^{22} \text{ N} \qquad (5.38)$$

Sun diameter expressed in terms of velocity of light

Sun diameter = $2[A_{cirsun}/\pi]^{1/2} = 13.819766 \times 10^8$ m $\rightarrow x1/1.333333 \times 1.04^{9/10} = 1.00052 \times 10^9$ m
$$\equiv 3 \times 10^8 \text{ m/s} \times 3 \times 1.111\ldots \qquad (5.39)$$

If orientation is not specified the top and the bottom quanta energies have to be included and there are 32 quanta positions on the sphere of the sun which at sun temperature of 6000 K delivers per quantum energy:

$$1 \text{ kg at } (6000 \text{ K})^{1/32}/1.15470 \times 1.045/12 \equiv 1.00052 \text{ J or eV} \qquad (5.40)$$

The variance 0.00052 eV is taken to be 6 k_s.

The Newton's formula was adopted nine years ago as the earth gravitation reference. In integer cubic units the force of gravitation between the earth and sun is $3.600\ldots \times 10^{22}$ N, a value which is larger by the arbitrary chosen quantum corrector of $(1+1/63)$ deriving the value of Equation 6.1:

$$3.600\ldots \times 10^{22} \text{ N}/(1+1/63) = 3.543855 \times 10^{22} \text{ N} \qquad (5.41)$$

6. Earth gravitation balance in macroscopic terms

The sun rays are taken to travel in straight line towards each planet. Presently gravity will be defined by the classical SI units with gravitation constant G having units N-m^2/kg^2. The sun to planet Newtonian gravitation force of attraction is expressed by the product of their two masses over the square of the separation time the International Gravitation Constant G. For the sun to earth system in classical SI units the present reference is expressed by:

$$F_G = G x m_{earth} \text{ x } m_{sun}/L_{earth}^2 = 3.5438(2) \text{ x} 10^{22} \text{ N} \tag{6.1}$$

The uncertainty is the author assessment based on Equations 17.50-51. The integer model finding is that the above force is the total force between the sun and the earth, both bodies contributing ½ of the value directed toward them. Using the sun or the earth as the reference attributes all the gravity force to one of them.

Converting above value to atomic dimensions requires division by size converter $1 \text{x} 10^{10}$. Expressing force per cycle requires division by the frequency in one pathway $1 \text{x} 10^{12}$ Hz of the pair (proton+electron), expressing force per plane and requiring division by 3. Relativity and quantum corrections of 1.1547 and $1.04^{3/5}$ derive $3.5438 \text{x} 10^{22}/(1 \text{x} 10^{10} \text{x} 1 \text{x} 10^{12} \text{x} 3 \text{x} 1.1547 \text{x} 1.04^{3/5}) = 0.99921$ N/cycle $\equiv 0.99921$ eVat displaying variance of ~2 k_s. This is k_s for each body generating two c_s pulses. Energy of k_s at atomic level is subject to linear conversion by $1 \text{ x } 10^{-10}$ generating $h_s = 1 \text{ x } 10^{-33}$ J-s.

In integer cubic units the sun mass creates a force, which applied to the mass of earth creates the total force of attraction. The force of attraction has been for long time somewhat incorrectly considered to be balanced by the acceleration acting on the rotating bodies having particles moving at velocity 1/3 c_s induced by the centrifugal force as explained in detail Chapter 18. The numbers are equal with earth rotating:

$$F_{gs} = G_s x m_{earths} \text{ x } m_{suns}/L_{earths}^2 \text{ x } 1.1547^2$$
$$= 1 \text{x} 10^{-10} \text{x} 2 \text{x} 10^{30} \text{x} 6 \text{x} 10^{24}/(1.5 \text{ x} 10^{11})^2 \text{ x } 1.1547^2 = 4.00000 \text{x} 10^{22} \text{ N} \tag{6.2}$$

The reference force value requires two more corrections:

$$4.00000 \text{x} 10^{22} \text{ N} /(1.11111 \text{ x } 1.04^{2/5}) = 3.5439 \text{ x} 10^{22} \text{ N} \tag{6.3}$$

The accuracy displayed is the best indication that the nature is hiding its true values by not displaying relativity and quantum corrections. Special relativity has to be introduced to consider dilated values induced by different observers.

Using A) integer values for the masses of the sun and of the earth and their separation or
   B) 2/3 factor with gravity constant 10^{-10} expressing absence of pure electron radiation from earth and quantum corrector, derives the same numerical result, both within the limits of accuracy of the reference:

$$F_{Gs} = 2/3 G_s x m_{earths} \text{ x } m_{suns}/L_{earths}^2 \text{ x } 1.04^{1/12}$$
$$= 2/3 \text{ x } 10^{-10} \text{x} 2 \text{ x} 10^{30} \text{x } 6 \text{ x} 10^{24}/(1.5 \text{ x} 10^{11})^2/1.04^{1/12} = 3.54395 \text{ x} 10^{22} \text{ N} \tag{6.4}$$

The same accuracy is obtained by using SI value of gravity constants integer model values of sun and of earth masses, of their spacing, but with different quantum correction:

$$F_{Gs} = 0.66742 \text{ x}10^{-10}\text{x}2 \text{ x}10^{30}\text{x } 6 \text{ x}10^{24}/(1.5 \text{ x}10^{11})^2/1.04^{1/9} = 3.5441\text{x}10^{22} \text{ N} \qquad (6.5)$$

in the infrared frequency creating gravitation balance. These results are consistent with the earlier hypotheses that Bohr atom comprises three oscillators and only two interact

The Newton gravitation force of attraction is except for number of dimensions essentially **equal and opposite** to the centrifugal force F_C expressed by:

$$F_R = -F_G = m_{earth}\text{x}v_{earth}^2/L_{earth} = 3.5430 \text{ x}10^{22} \text{ N} \qquad (6.6)$$

The centrifugal force F_C becomes inward radially, if the center of mass is used as reference and radially outwards if the rotating body is used as the reference. At the end of the 11th year of investigation it was revealed that Equations 6.6 and 6.1 are accompanied by the conversion of protons to radiation. The above calculation repeated using integer cubic space constants with quantum correction derives the exact reference:

$$F_{Rs} = -F_{Gs} = 6\text{x}10^{24}\text{x}30000^2/1.5\text{x}10^{11}\text{x}1/1.04^{3/8} = 3.5436 \text{ x}10^{22} \text{ N} \qquad (6.7)$$

In Equations 6.6 and 6.7 the reference is the earth and all the force is attributed to it and not divided. Integer values and quantum correction in Equation 6.7 derives the F_{Rs} reference value.

The attractive force for satellites, as described by Kittel ET ales [46], is balanced by the centrifugal force acting in the outward direction. The fact that centrifugal force acts in one plane and gravitation forces may be formed in three planes does not create conflict because only 1D pulses interconnect. Figure 8 shows the two forces as function of L_{earth}. The intersection of two functions establishes the semi-stable balance between these two forces. With the sun and the earth shown on the same figure, the attractive force of gravitation has the opposite direction for each body. Since $dF_G/dL > dF_R/dL$ changes of total energy produce minute displacement $\pm dX$ moving the circulating body to a position, where F_G is greater than F_R. It is suggested now that this slight instability is the cause of celestial bodies moving very slowly towards the center of galaxy, or a planet moving very slowly toward the sun. The mass gradient and the temperature gradient are the two main sources of gravitation. Both gradients decrease radially outwards in galaxies and are the contributing factors to the overall centrally directed movement of masses leading to the formation of black holes.

The centrifugal force is balanced by unwinding in one plane equivalent (reduced) electrons from protons as explained in Chapter 5. The roots of gravity are based on discharge of energy radiated due to the end of the natural live of protons explained subsequently.

When the earth is the reference the sun rotates around the earth and its rotation around the earth balances the force of attraction induced by the earth. In each case the over the long distance the force of 1 N/kg mass times the gravity constant over the square of the separation creates a force, which is applied to the other mass producing the total force of attraction. That force is balanced by the acceleration acting on the rotating bodies induced by the centrifugal force as explained in greater detail in Chapter 18.

The temperature of the sun of 6000 K ≡ 6000 eV per pulse over a period of one of earth year of $31.557x10^6$ s produces with K_{rel} =1.1547, quantum correction and the power gradient of 1 [W/m], referred to 300 K, the amount of energy and of force of gravitation within the limits of the accuracy of the reference:

$$F_{Gs} \equiv 1[W/m] \; x \; 31.557 \; x10^6 \; x \; 6000^4/1.1547 \; x \; (1 +11/20000) \equiv \mathbf{3.5438 \; x \; 10^{22}} \; \mathbf{N} \qquad (6.8)$$

The number of seconds per year generates interesting results. During a quarter of the year the earth in one dimension (1D) is subjected to linear acceleration increasing its velocity from zero to 29786 m/s over the distance L_{earth} = $1.469x10^{11}$ m and in the period $1/4x31.557x10^6$ s. The resultant repulsive force, the $\sqrt{2}$ product of acceleration and of earth mass in kilogram with relativity correction 1.111 . . . is:

$$F_R = m_{earth} \; x \; acceleration \; x \; corrections \equiv 5.9742x10^{24}x29786 \; x\sqrt{2}/(1/4x31.557x10^6)$$
$$x1.111 \ldots$$
$$= \mathbf{3.5442 \; x \; 10^{22} \; N} \qquad (6.9)$$

The above expression recalculated using cubic integer physical constants with quantum correction (1+1/86) derives also value equal the reference:

$$F_{Rs} = m_{earth} \; x \; accel. \; x \; corrections \equiv 6x10^{24}x30000 \; x\sqrt{2}/(1/4x31.557x10^6)x1.111 \; \ldots \; x$$
$$(1+1/86)$$
$$= \mathbf{3.5440x10^{22} \; N} \qquad (6.10)$$

The average acceleration of earth measured toward sun due to the rotational velocity 29786 m/s is minute, 3.7755 and -3.7755 mm/s, but the force per kg of $0.0037755x\sqrt{3}/K_{rel}^{1/2}$ = 0.006086 [N-s/(kg-K^2)] balances the gravitation force of the sun F_R/m_{earth} = 3.5438 $x10^{22}$ N/5.9742x10^{24} kg = 0.006012 [N-s/(kg-K^2)]. When referred to 300 K and corrected by relativity the resultant becomes 1.2 *x* 0.006 *x* 300/1.04^2 ≡ 2 eV or 2 N-s impulse per kg expressing proton and (proton+electron) moments from earth balancing similar moments from the sun. This is the centrifugal force recalculated in a linear manner. It justifies the validity of using the number of seconds in an earth year in the previous equations. Acceleration requires energy source to balance the force of attraction. Maxwell, Larmor and Janes Jeans implied presence of radiation with acceleration and the above calculation confirms their views.

Earth circulates round a spherical volume of:

$$V=4/3 \; \mathbf{x} \; \pi \; \mathbf{x} \; L_{earth}^3 =4/3\pi(1.496 \, x10^{11})^3=1.402 \, x10^{34} \; m^3 \qquad (6.12)$$

because earth eccentricity is not considered. The quantum energy value for V becomes $\varepsilon_{os}V=1 \; x \; 10^{-11} \; x \; 1.402 \; x \; 10^{34}$ = 1.402 $x10^{23}$ J. The quantum energy value at atomic level with relativity correction ε_{os} in 1D has been derived in Equation 9.19 as $\varepsilon_{os}^{1/3}$. That value time relativity correction with energy per particle $2x10^{-23}$ J becomes equal $1.402x10^{23}x1.15470^{1/2}x2x10^{-23}$ = 3 J or eV. Multiplication by volume probability derives surprisingly 3/2 sun power of $3x10^{30}$ J. This is previous value, 3.86×10^{29} in 6 directions, with relativity corrections factor 2.3 for two oscillators:

$$6x3.86x10^{29}x1.1547^2/1.0328=2.99x10^{30} \text{ J} \tag{6.13}$$

For the progressing wave there are 5 pathways:

$$F_G x c_s x 1/5 =3.5439 x10^{22}x3 x10^8 x1/10 x 1/1.111 \ldots^{1/2} x1.04^{1/5} = 2.0007 x 10^{30} \text{ W} \tag{6.14}$$

Thus at earth distance from the sun, the power displayed by sun hydrogen components oscillators is 1 W per kilogram of its mass $\equiv$ 1 N, or 1 N-s/(kg-K²) (6.15) The above force of 1 N and Equation 6.13 indicate that oscillators of the sun mass 2.000 $x10^{30}$ kg, relativity corrected, at the squared distance L_{earth} apply pressure 1 N/m² per unit of local velocity $1/3 \times c_s$. This value, converted by linear probability to atomic units, becomes the gravity constant $G_s =1x10^{-10}$.

The local light velocity has to be used to derive from the sun mass the force 1 N/m²:
$m_{suns}/L_{earths}^2 x 1/(1/3 xc_s)x$ corrections =
2.000 $x10^{30}$ kg/(1.5 x 10^{11})² x 1/(1/3 xc_s) x1.111 \ldots x 1.04^{7/20} $\equiv$ 1.0013 N/m². (6.16)

Earth generates gravity force of 10 [N-s/(kg-K²)]. The force of gravitation of earth is given by the product: its mass times ten, the probability of forward vector formation, 1/1000, times relativity correction deriving accurately 5 x 10^{22} N, which divided by $\sqrt{2} x$ $1.04^{1/16}$ derive total force very close to the reference: (6.17)

$\sqrt{2} F_{Gs}= 6 \times 10^{24}$ [kg]x 10[N-s/(kg-K)] x 1/(3 x 300)x 1/1.154700² $\equiv$ 5.0000004663 $x10^{22}$ N. $F_{Gs} =5.000005 \times 10^{22}$ N/$\sqrt{2} \times 1.04^{1/16}=3.5442 \times 10^{22}$ N is within the reference accuracy.

The above calculation derived 10 [N-s/kg] $\equiv$ 10/(3 x 300 x 1.111 \ldots) = 0.01 N per plane per K equal in value to the value obtained in the Planck Radiation law. Using the SI value 5.9742 $x10^{24}$ kg for the earth mass derives 0.02% accuracy with corrector 1.111 \ldots$^{1/2}$ /1.04$^{7/24}$.
The Special Rel. sun gravity force = Geometric mass (Equation 17.24) $x c_s x1/10x1/10$
$$= 1 \times 10^{16}x3 x10^8 x1/10x1/10=3.0000 \times 10^{22} \text{ N} \tag{6.18}$$

Although most of the relations indicate that sun reacts only with part of its mass or temperature the following derives the integer cubic model gravitation value of

$$6 x 10^{24} \text{ kg x 0.006 N/kg} = 3.60000 \times 10^{22} \text{ N} \tag{6.19}$$

The above is derivative of classic reference $3.5438x10^{22}$ N=3.60 $x10^{22}/1.0328^{1/2}$ $x1.04^{1/100}$. The force generated per electron in six orientation is **0.006 N/1$x10^{-30}$ kg=6x10^{-33} $\equiv$ 6 h_s** expressing h_s per orientation requires further clarification and (6.20)

$1/3x\sigma_s xT^4x1/10^3xA$cirsun $x1.111 \ldots =1/3x5x10^{-8}x6000^4x10^{-3}x1.5 x10^{18}x1.111 \ldots =3.6000 x10^{22}$ N.

In the classical form the integer cubic model sun gravitation attraction with correction is

$$F_{Gs} = 2/3 \; G_s \; x \; m_{earths} \; x m_{suns} / L_{earths}{}^2 = 2/3 x 10^{-10} x 6 x 10^{24} \; x 2 \; x 10^{30} / (1.5 \; x 10^{11})^2 x (1 + 1/80) = 3.6000 \; x 10^{22} \; N$$
(6.21)

The geometric mass of the sun-earth couple is $1 x 10^{27}$ kg as in Equation 17.14. That mass interacts in gravitation. There are two bodies involved and probably for that reason the factor 0.006 is squared now: (6.22)
Gravitation force using above gravitation mass $= 1 x 10^{27}$ kg $x (0.006)^2 N/kg = 3.6000 \; x 10^{22}$ N.

Two sets of three orthogonal components of light velocity of the sun carrying energy gradient 1 J/m generate resultants in planes displaced by 90°. Corrected for quantum and added in two planes each subjected to probability 1/10 increase velocity by factor of four and derive the reference gravitation force of the sun:

$$F_{Gs} = 2 \; x \; \sqrt{3} (1 \; J/m \; x \; 1 \; x \; 10^8 \; m/s)^3 \; x \; 1/10 \; x \; 1/10 \; x \; 1.04^{29/50} = 3.5438 \; x \; 10^{22} \; N \quad (6.23)$$

The earth vectorized thermal flux of earth radiations $1 x 10^{-10}$ N surrounds the mass objects around the earth in the same way as the sun radiation surrounds earth a shown on Figure 7. Actually two graviton vector forces of 1 N/kg form wave function format generate without reflection $\sqrt{2}$ larger pulse further augmented by another $\sqrt{2}$ to $2 x 10^{-10}$ N. Multiplied by frequency $1 \; x \; 10^{-12}$ Hz generates 0.02 W, twice the Planck Radiation parameter. Corrected by temperature, relativity and quantum derive exact 10 W:

$$2 x 10^{-10} \, N \; x \; 1 \; x \; 10^{-12} \; Hz \; x \; 300 \; x 1.154700538^2 \; x \; 1.1111 \ldots^2 \; x \; (1 + 1/80) = 9.999 \; 999 \; 999 \; W$$
(6.24)

Pulses have magnitude and rate, and their product derives power. Two earth pulses of 300 K with 1 W/m² per K in wave function format derive 600 W deriving with relativity and quantum corrections: $600 x 1.154700^2 x 1.111 \ldots^2 x (1 + 1/80) = 999.999$ W/m², or 600 $x 1.3333 \ldots x 1.111 \ldots = 999.9999995$ W/m² that is eV per pulse. Actually two such pulses in wave function format generate 1 MeV loop progressing at c_s from earth.

Velocity of light c_s from sun is generated by three local velocities in orthogonal orientations generating $\sqrt{3} \; x \; 10^8$ m/s. Three such values appearing in orthogonal orientations generate another increase of $\sqrt{3}$ making $c_s = 3 \; x \; 10^8$ m/s.

The earth mass derivers only one half of the Special Relativity gravitation force of $1.5 x 10^{22}$ N that is spreading along only one plane due the acceleration in the plane of earth rotation. The earth mass over that force $6 \; x 10^{24}$ kg/$1.5 \; x \; 10^{22}$ N $\equiv$ 400 K $\equiv$ 400 W $\equiv$ 500 K/$[(1.111 \ldots^2 + (1 + 1/80)]$. The expression states that due to equipartition of energy only 1/2 of the 1000 W energy above is directed in balancing, the sun gravitation making the other half. The initially non functional 500 eV pulses from the sun have their balance! A spherical radiating structure has 30 quantum orientations and energies with two remaining invisible makes a total of 32. The orientation probability for the earth mass of $6 x 10^{24}$ kg derives mass of $(6 x 10^{24})^{1/32} x 1.04^{9/40} = 5.99998 \approx 6.0000$ kg per direction. That is 6 eV divided by above 1000 J pulse energy deriving 0.006 N/kg.

Can similar calculation generate reasonable results for proton of $2 x 10^{-27}$ kg mass or sun mass? The orientation probability mass $(2 x 10^{-27}$ kg$)^{1/32} / [\sqrt{2} x 10 (1 + 1/80)] = 0.010007$ eV express 1st Planck radiation term. For the sun $(2 \times 10^{30} \text{ kg})^{1/32} \times 1.111 \ldots \times 1.04^{17/40} = 9.9978$ ≈ 10 kg or eV. It appears that quantization of space orientation is size independent.

It would follow that Hartree energy might be 32 eV. This is true because earth from the whole surface generates $(6 \times 10^{24} \text{ kg})^{1/32} \times 1.04^{11/50} = (6 - 0.0022)$ eV per charge. The variance for 64 DOF is $22/64 \approx 0.344 \rightarrow \times 1.1547^3/1.0328 \approx 1/2 \ k_s$ (6.24a)

Energy 10 eV multiplied by rotational converter 10^{20} expresses its size in SI units at 1×10^{21} J or 100 eV per pulse appearing in some expressions.

The mass of sun is 1/3 of the mass of earth converted to energy by the square of its linear velocity time relativity corrected with linear probability cubed:

Mass of sun $= 2 \times 10^{30} \text{ kg} = 1/3 \times 6 \times 10^{24} \text{ kg} \times 30000^2 \times 1/10^3 \times 1.111 \ldots$ (6.25)
where $2 \times 10^{24} \text{ kg} \equiv (\sqrt{2} \times 1 \text{ J} \times 1 \times 10^{12} \text{ Hz})^2$ and $\sqrt{2}$ is the increment by one wave function.

Sun surface temperature is quoted on WEB pages at 5778 K. It is suggested that it is 6000 K relativity and quantum corrected with variance of 1 K per direction plus 2/3 K:

$$6000/1.111 \ldots^{1/2} \times 1.0328^{1/2} = 5784.7 \text{ K}$$ (6.26)

Gravitation force of above geometric mass $= 1 \times 10^{27}$ Kg $\times (0.006)^2$ N/kg $= 3.6000 \times 10^{22}$ N
Energy pulse $E_{searthSI3}$ in 3 planes is the earth mass per proton mass, converted to atomic units

$$E_{searthSI3} \equiv 6 \times 10^{24}/2 \times 10^{-27} = 3 \times 10^{51} \text{ protons} \rightarrow \times 1 \times 10^{-30} = 3 \times 10^{21} \text{ J}$$ (6.27)

Mass of earth over the life of proton time minimum energy pulse in SI joules derives the total pulse energy in SI joules in six space directions:

$$6 \times 10^{24} \text{ kg}/10^{10} \text{ per proton } 1 \times 10^{-14} \text{ J} = 6 \text{ J per atomic event}$$

(6.28)

7. Gravity balance of rotation between the space and the time field gradients of the sun

Proton mass converted to energy, spread in two opposite directions due to conservation of momentum, expresses energy in one direction:

$$m_{ps} \times c_s^2 \equiv 10^{-27} \times 9 \times 10^{16} = 9 \times 10^{-11} \text{ J per proton} = 9 \, \mathcal{E}_{os} \qquad (7.1)$$

Multiplied by the relativity corrector 1.111 . . . the value becomes numerically equal to linear probability 1×10^{-10} and the integer gravity constant G_s and c_s^2 $(\sqrt{3} \times 1/2c_s \times 1.15470)^2$.

The value 10^{-27} expresses equally (a) the cube of probability energy of $(1 \times 10^{-9})^3$, (b) the mass 10^{-30} kg of 1000 electrons, (c) the product $G_{suos} \times u_{os} \times \mathcal{E}_o = 1 \times 10^{-10} \times 1 \times 10^{-17} \equiv 1 \times 10^{-27}$ J.

Assuming that each kilogram of earth converts to energy E_n the mass of one proton per year, the power in watts delivered by earth is:

$$\partial E_n / \partial t = m_{earth} \times 10^{-27} \times 9 \times 10^{16} /(\text{seconds in one year}) \text{ W}$$
$$= m_{earth} \times (10^{-30} \, probability) \times 1/10^3 \times 9 \times 10^{16}/(\text{seconds in one year})$$
$$= 6.000 \times 10^{24} \times 9 \times 10^{-11} \text{ J}/31.557 \times 10^6 \equiv (17 + 1/9) \times 10^6 \text{ J for earth per year} \qquad (7.2)$$

Above power of earth repulsion per kg is produced with subjecting the force produced by accelerating 1 kg at 3.7755×10^{-3} m/s with velocity and with force of 1 N-s per kg at velocity c_s. Fitting these values with relativity corrector derives essentially the same power:

$$\partial E_n / \partial t \equiv (1 \text{ kg} \times 3.7755 \times 10^{-3}) \times c_s \times \sqrt{2} \times 1.1547^{1/2} = 17.2125554 \approx 17(1 + 1/80) \times 10^6 \text{ W} \qquad (7.3)$$

The variance is 0.54 W is taken to represent 0.54 $\times 1.111/(2 \times 300K) = 0.01$ W, a perfect result. Converting to energy, power integrated over a period of one year expresses the value of energy of repulsion given on left side of Equation 7.4 below. The right side adds atomic energy of the attraction involving its gradient $\partial E_n / \partial x = 6000$ J/m over the distance $L_{earth} = 1.4960 \times 10^{11}$ m with $\partial E_n / \partial x = 1$ J/m. The gradient of energy $\partial E_n / \partial x$ is a partial differential, because being in one dimension it does not interact with forces from other directions or with 1D gravitation force acting in a specific direction. For 1D or central system $\partial E_n / \partial t = dE_n / dx$.

For atomic particles at $c_s = 1.00$, $\partial x = 1.0 \, x \, \partial t$, $\partial x / \partial \partial = 1.0$, $\partial E_n = \partial E_n x \, \partial x / \partial t$ and
synchronized during rotation $\quad \partial E_n / \partial t = \partial E_n / \partial x$

Force of repulsion, accelerating earth $\quad$ Force of attraction
from converted mass $\quad \partial E_n / \partial t = \partial E_n / \partial x$, energy gradient of space

with corrections

Integrating total energy over one year and over the distance L sun to earth

$$\int_{year} \partial E_n / \partial t \equiv 1.04 / \sqrt{3} \, x \int_{L earth} \partial E_n / \partial x$$

$$17.038 \, x 10^6 x \, 31.557 \, x \, 10^6 \text{ s} \equiv 1.04 / \sqrt{3} \, x \, 6000 \text{ J/m} \, x \, 1.4960 \, x \, 10^{11} \text{ m}$$

$$5.3767 \, x \, 10^{14} \text{ J} \equiv 5.3896 \, x \, 10^{14} \text{ J} \qquad (7.4)$$

With two celestial bodies, one has to be taken as the reference; the inertial force on the other body has opposite sign to the gravitational force, the force inducing repulsion. This view does not conflict with the basic Einstein premises. The model adds information about the reality to current physics. The word observer should be changed in several occasions to an interpreter.

The data in Equation 7.4 is that in one dimension the two balanced annual energies attributed to the earth and to the sun are acting in opposite directions. Multiplied by relativity correction 1.1111 values in Equation 7.4 are $\sim 6 \, x 10^{14}$ J or $1 \, x 10^{14}$ J per direction, expressing the gravitation force of $3.0 x 10^{22}$ N of Equation 16.37 based on Special Relativity.

The above energy $5.37 x 10^{14}$ J$/1 x 10^{-19}$ represents $5.37 x 10^{33}$ eV, which times $1.111 x h_s$ $=1.111 \, x 1 x 10^{-33}$ derives 6 eVat, or 1 eVat per orientation. Using relativity and quantum corrections, $5.38 x 10^{14} x 1.04^{1/3} x 1.07457 x 1.0328 \approx 6.0008 x 10^{14}$ J, derives the value of the minimum pulse energy of $1 x 10^{-14}$ J in one of 6 orientations. Converting to atomic level by multiplying by volume probability 10^{-30} derives $B_{pulse1} = 1 \, x 10^{-44}$ Jat.

It is suggested that there are two types differently specified gravitation forces between the sun and the earth in two areas representing essentially the same forces. One type is three dimensional with forces surrounding the earth surface from all directions. The forces are vectotized thermal radiations from earth and the sun. Temperature was identified later in the research as pulsed rate. Pulses were finally identified as produced by the release of electron or proton charge caused by the collapse of the proton structure, a common cause of numerous effects. In thermal terms this will be explained by Equations 17.50-51. The author attributes Newton force spacifying sun gravitation in Equation 6.1 to be equally generated by sun and by earth.

The forces over the long distance 1.5×10^{11} m sun-earth separation are one dimensional in the local beams. They are described in Equation 7.4 and treat earth mass as a point at path termination. The earth macroscopic balancing force is the centrifugal force caused by the acceleration of the earth mass. Above equalities apply to vectorized thermal flux representing 1/1000 of the incoherent thermal flux reaching earth and of $1.111 x 10^{-7}$ total sun flux in one direction.

Number N_e of electron contained in earth mass/kg in its protons is evaluated in Equation 13.2 as $6\ x10^{54}$. The force of attraction, called above repulsion, by earth could be attributed to the energy released by the spin of these electrons in 6 orientations every spin discharging the force of 1 N on full rotation around the sun lasting one year. The result with relative correction is close to the SI reference:

$$\text{Total } F_R = N_e\ x\ 6h_s\ x1[\text{N}]/1.0329^{1/2} = 6\ x10^{54}\ x\ 6\ x10^{-33}/1.0329^{1/2} \equiv 3.5409x10^{22}\ \text{N} \quad (7.6)$$

Without correction the force value is exactly the integer relativity cubic SI model value $3.6000\ x10^{22}$ N. Thus each kilogram of $6\ x\ 10^{24}$ kg of earth sends out the force of 0.01 N as predicted by 1st term of Planck Radiation law generating the force $3.6000\ x10^{22}$ N.

The relativity correction 1.1547^2 applied to earth protons moving at velocity $1/2$ $x\ c_s$ adds to the mass of earth and produces two vectors of total velocity c_s. The resultant repulsive gravity force is $F_R = m_{earth}\ v_{earth}^2/x_c = M\ v_{earth}/x\ x\ c = M\ x_c/(x_c t) = \partial M/\partial t$
per K per kg or per proton along the earth sun axes. The incremental momentum or the impulse:

Force (or field) x time = mass x acceleration x time = Momentum M applied (7.7)

In author's experiments this simple relation was used to expand the dipole of Bohr atomic structure to dissociation length deriving several physical constants [2, 3, 4].
The energy and power related to 1 eVat derive also the gravitational force over (product of circular earth area and velocity of light) multiplied by relativity and quantum corrector, $1.111\ldots/1.04^{2/3}$:

$$F_G/(A_{cirearth}xc_s)xcorrectors=3.5438x10^{22}/(1.278x10^{14}x3x10^8)x1.111\ldots/1.04^{2/3}$$
$$=\textbf{1.0005}\ \text{W/m}^2 \quad (7.8)$$

With proton velocity $\frac{1}{2}\ c_s$ the result would be 2 W/m² derived earlier.

The power of sun radiation in 3D, derived on earth $P_{delivered}$, from sun constant and the circular earth areas using quantum and relativity correction for 1/3cs, is:

$$P_{delivereds} = 1.278\ \textbf{x}\ 10^{14}\ \textbf{x}\ 1372\ \textbf{x}\ 1.1111\ \textbf{x}\ 1.04 = 2.026\ \textbf{x}\ 10^{17}\ \text{W} \quad (7.9)$$

The power $P_{radiateds}$ radiated from earth surface in all directions using Stephan-Einstein constant and 300 K and the circular earth area:

$$P_{radiateds}=4\pi\ r_{earth}^2\sigma_s T_s^4=4\pi\ (6.3781x\ 10^6)^2\ \textbf{x}\ 300^4\ \textbf{x}\ 5\ \textbf{x}\ 10^{-8}= 2.070\ \textbf{x}\ 10^{17}\ \text{W} \quad (7.10)$$

Two oscillators radiate:

$$P_{radiateds}\ \textbf{x}\ (u_{os}\varepsilon_{os}) = 2\ \text{J} \equiv 2\ \text{eVSI derived also in Equations 16.7 and 16.9} \quad (7.11)$$

The above agreement shows that the stable earth temperature is obtained by both formulas, although source of heat is not indicated.

Using SI value for σ and Radkiewicz's temperature for earth 288.5 K obtains

$$P_{earth\ radiated} = 4\pi\ r_{earth}^2\ \sigma T^4 = 4\pi\ (6.3781\ x10^6)^2 x\ 288.5^4 x5.6704 x10^{-8} = 2.008 x10^{17}\ W \quad (7.12)$$

derives better results, because it is closer to an integer value, the concept postulated now. Surprisingly the power $2x10^{17}$ W, subjected to space quantization of $1/10^3$ converting it into a vector force, divided by relativity and quantum factors and multiplied by 3 planes derives

$$2\ x\ 10^{17} x\ 3/(10^3 x\ 1.1547 x1.04^{7/8}) = 5.377 x10^{14}\ W,\ \text{the value of Equation 7.4} \quad (7.13)$$

Equation 7.4 implies the proton to be a conglomerate of 2000 electrons having nulled neutralized momenta.

It should be noted that the repulsive force of Equation 7.4 derives with corrections the force of attraction of the sun to earth:

$$\sqrt{3}\ x\ 5.38\ x10^{14}\ Jx1x10^8\ m/s\ x\ 1.1547/[3(1 + 1/80)] = 3.5424\ x10^{22} \quad (7.14)$$

The mass represented by the number of electrons moving with the velocity of mass derives the current displaying Ampere's Law effects. Mass was found to generate the force of 10 N/(kg-K^1) reduced to 1 N/(kg-K^1) in linear interactions when expressed in atomic units. Billion years ago the force of 1 N/(kg-K^1) lead to the formation of the sun and its thermo-nuclear mass to energy conversion, producing thermal flux. The results presented up to now find that thermal flux, vectorized by space quantization, generated gravitation force, which is then balanced by the acceleration of the centrifugal force.

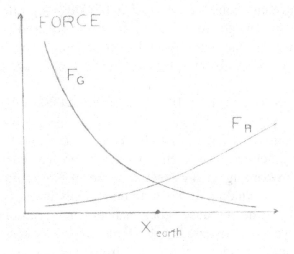

Fig. 8 Balancing the sun gravitational force acting in one plane with the centrifugal force

The mass of earth 6×10^{24} kg over electron mass of 10^{-30} kg displayed presence of 6×10^{54} electrons on earth with each generating per revolution per second energy (momentum) of $h_s/(1s) = 1 \times 10^{-33}$ J deriving for earth the force of 6×10^{21} N. During the period of one year earth covers the sun-to-earth celestial space four times increasing the above value to 2.4×10^{22} N. The force of repulsion is that value corrected by three relativity and one quantum correctors:

The force of earth repulsion induced by unwinding of its angular reduced electron spins on rotation per second:

$$= 2.4 \times 10^{22} \text{ N} \times 1.1547^{3/2} \times 1.0328/1.04^{1/12} = 3.5428 \times 10^{22} \text{ N} \qquad (7.14a)$$

Above value deviates from the reference sun gravitation by 0.001×10^{22} N, which time 1×10^{-19} derives 1 eV, a quantum in SI units per orientation at 300 K. Being able to express accuracy by natural limit of a quantum is a very important argument for accepting main findings of the model postulated.

It is suggested that both the force of attraction and the force of repulsion (momenta in atomic terms) require energy sources to display forces or propagate momenta. **The force of sun attraction is not an independent gravitation force**, but it may be expressed by wave converted to inverted pulse by reflection from the mass of the back potential barrier of the atom (see page 27 Num. 20). **The force earth repulsion of 1 N is attributed to the releases of energy of angular momentum of reduced electron spin on each annual rotation expressed by Planck constant of h_s in 6 space orientations. Thus 6×10^{54} electrons on earth generate an equal and opposite force to that of sun attraction. The apparent independent forces are considered to act in three dimensions and can be better expressed as generated by radiation of mass, by vectorized thermal flux, by current and by temperature. Acceleration and the centrifugal force act in one plane and express the repulsive force of Equation 7.6.**

In one of ten orientations the energy **5.3767×10^{14} J/kg**, after relativity and quantum corrections, becomes an integer 6×10^{13} J/kg. Its value squared (expressing wave format) times proton frequency in one pathway and over square of the light velocity derives the Special Relativity force of repulsion of Equations 16.53 and 16.55:

$$6 \times 10^{13}) \text{ X } (6 \times 10^{13})(1 \times 10^{12})/(3 \times 10^8)^2 \equiv 4.0000 \times 10^{22} \text{ N} \qquad (7.15)$$

The use the squared velocity of light seems to indicate that unwinding of angular electron momentum generates mass to energy conversion. The energy manifesting the force of repulsion **is the unwinding of the electron spin of the reduced electrons** expressing also protons and hence unwinding of the earth mass. Only masses of four planets have been found to act that way either because other planets cannot sustain stable spin orientation for a period of one year or they are too far from sun to express any mutual inductance. Each earth rotation unwinds some spins of electrons with the energy 6×10^{14} J/year being released from the mass of earth converted to energy and it is expressed in the mass of earth comprising 6×10^{54} reduced electrons expressed by the oscillatory probability of power 1/4 with 2 expressing doubling by pulse reflection:

$$2 \text{ x}(6 \text{ x}10^{54})^{1/4} \text{x}1.04^{1/4} = 0.9996 \text{ x}10^{14} \text{ J} \approx 1\text{x}10^{14} \text{ J} \qquad (7.16)$$

Calculations of Chapter 7 suggested presence of components of velocity of light 10^8 m/s in orthogonal orientations. The repulsion force in Special Relativity value created by earth acceleration of Equation 7.4 using relativity and quantum corrections requires velocity in each of three space directions velocity 1/3 c_s:

$$1/2 \text{ x } 5.38 \text{ x } 10^{16} \text{ x}1/3 \text{ x}10^8 \text{ x } 1.1111 \text{ x}1.04^{1/12} = 3.00065 \text{ x } 10^{22} \text{ N} \qquad (7.17)$$

It is suggested that both velocity of light and temperature express for the receiver the rate of the arriving pulses.

The same amount of momenta and energy formed by the sun is flowing in the opposite direction from the earth. Both sources are generating equal opposing energies and opposing momenta in SI units. Both in SI terms and in the model integer terms, radially directed gravitation from the earth and from the sun dissipates vectorized energies in one plane of rotating directions. Energies are spent both by earth and by sun with no data being available on the product of their interaction. The opposing momenta are likely nulled at the atomic level.

Above explanation applies to the earth-to-sun gravitation and with minor corrections shown subsequently applies also to other planets. Neptune and Jupiter generate some forces unrelated to sun radiation making the gravitation formula less applicable.

The common Equation 7.1 is true, but misleading because mass cannot move with velocity of light. It creates for some researchers doubt about validity of the Special Relativity Theory. Equation 7.1 in modified format has two pulses, in one plane in wave function format, generating "mass to energy conversion". The wave function formation expresses process sustaining velocity of light of radiation loop while converting one loop into another displaced by 90°:

$$m_{ps} \text{ x } c_s^2 \equiv [(1\text{x}10^{-27} \text{ kg})^{1/2}\text{x } c_s]\text{X}[(1\text{x}10^{-27} \text{ kg})^{1/2}\text{x } c_s], \text{ where } c_s = \sqrt{3} \text{ x}1.1547005\text{x}1/2 \text{ } c_s \qquad (7.18)$$

In three orthogonal orientations probability is 1/3 generating $(1\text{x}10^{-27} \text{ kg})^{1/3} = 1\text{x}10^{-9}$ which times $1 \text{ x } 10^{-10}$ obtains the integer model electron charge $e = 1\text{x}10^{-9}\text{x}1\text{x}10^{-10} = 1 \text{ x}10^{-19}$ C.

Temperature effect

The sun temperature of 6000 K is reduced by a factor of 1/6 expressing interacting fraction of the sun radiation with earth. The interacting temperature of 5000 K expresses the sum of five space quantization power components of 1000 W each discharging in the direction of earth as shown on Figure 1 and 1000 W expressed by the sum of three orthogonal 500 W pulses multiplied by relativity corrector for 1/2 c_s:

$$5000 \text{ x } 1/5 \text{ W} = 1000 \text{ W} \equiv 500 \text{ W x } \sqrt{3} \text{ x } 1.154700538 = 999.9995 \text{ W} \qquad (7.19)$$

The power 500 W is balanced by three 300 K ≡ 300 W pulses in orthogonal orientations with uncertainty $1/2\ kT$ for each of the 17 degrees of freedom deriving 0.2 J, which with 1.04 correction applied to 500 J derives the balance with earth pulses:

$$\sqrt{3}(300\ J + 8.5\ kT) = 500\ J \times 1.04$$
$$519.96 \equiv 520 \qquad (7.20)$$

The temperature of 5000 K or 6000 K expresses also the reference gravitation force of the sun and of the earth derived otherwise in Equations: 1.28, 6.8 and others (7.21)

Gravitation of sun to earth=Sun circular area x 5000 x 3 x $\sqrt{2} \times 1.04^{3/4} \times 1.1111 = 3.5439 \times 10^{22}$ N and 1000 W x $1/3\ c_s$ x $\varepsilon_{os} = 10^3 \times 10^8 \times 10^{-11} = 1$ W the sun pulse or unity event probability.

Numerical value of a radiation pulse does not disclose features of the force and of gravitation radiations. They may express: sum of ten or five pulses radiating together; sum of ten pulses radiating in sequence; pulse rate expressed by pure number; total input energy size; partial energy input per reference pulse; pulse per one of four space orientations; sum of energy or rates of 12 or 34 pulses; sum of two pulses displaced by 90°; sum of pulses from three orientations; electron, proton or equivalent particle pulse; pulse per orientation; pulse energy at 1K, 300 K, 500 K, 1000 K, 3000 K, 6000 K and 10000 K; energies at atomic level, and SI level in joules or electron volts or expressing 10:1 energy ratio between SI and atomic levels before size conversion.

For all planets with minor corrections sun gravitation force is counter balanced by the force of acceleration *F=ma* induced by mass *m* of planet rotational acceleration. **The force is kinetic and it has no units of time or length.** Mass of sun generates 1 N/kg and its temperature increases the force to power eight. Two force pulses in wave function formation with corrections are needed to express the mass of sun:

Mass of sun=$[(6000K)^4 \times 1N/kg \times (6000K)^4 \times 1N/kg] \times 1.154700^{3/2}/1.045^{1.05} = 1.99999424 \times 10^{30}$ kg =$(2-0.0000058) \times 10^{30}$kg, **the variance is expressing approximately the mass of earth 6 $\times 10^{24}$ kg** (7.22).

For earth the identity derived is $[(900K)^4 \times 34)]^2 \times 1$ N/kg $\approx 6 \times 10^{26}$ kg/1.2 with three planes 3 x 300 K and 900 x1.1111 ≡ 1000 J being the pulse balancing the sun and the decrement 1.2 expressing not interacting radiation. Also $600^4 \times \varepsilon_{os}/(1.1547^{3/2} \times 1.04^{1.1})$ ≡ 1.00038 W or eV (7.23)

And 1 kg at $300^4 = 8.1 \times 10^9 \rightarrow$ x probability 10^{-9} derives 8.1 J which converts to 10 J.

8. Frequencies and radiation parameters of cubic geometry of Bohr atom and more attributes of temperature

The model proposes that the information transmitted in space by the elementary particles is most often based on the structure of a single hydrogen atom, the Bohr atom, or its molecule. The atom has two components, the electron and the proton formed from a congregate of electrons discussed subsequently. These two components vibrate with their natural resonate frequencies and according to the author form three oscillators: electron, proton and the pair, (electron+proton). The electron oscillates at 5×10^{15} Hz, the proton at 5 $\times 10^{12}$ and the pair at some 0.1% lower values than the proton. I. Turcu, R. Allot, S. Huy and N. List announced in WEB under Tera frequencies that doublet values measured for 0-0H coupling is 2.510 and 2.514 $\times 10^{15}$ Hz fitting well the earlier data quoted. However they attribute them to reflection, while the integer model attributes these frequencies to different components of Bohr atom generating them.

The distribution of energy of particles in translation motion is expressed by Boltzmann factor $e^{-w/kT}$ so that, if the only variable is temperature, the number of particle at the energy state w is greater by a factor of $e^{6000/300} \approx 5 \times 10^6 \times 10 \times 10/1.04^{4/5}$ for temperature of 6000 K as compared to 300 K.

Effectively the same above value 5×10^6 is derived from ratios of areas $A_{cirsuns}/A_{cirearths}$ times the square of temperatures $(T_{suns}/T_{earths})^2$ and a guessed quantum corrector:

The number of particle $=1.5 \times 10^{18}/1.2780 \times 10^{14} \times (6000/300)^2 \times 1.1547 \times 1.04^{8/5} = 4.996 \times 10^6$

$$(8.1)$$

The sun signal of 50 J times $\mu_{os}/\varepsilon_{os}$ is also 5×10^6 J. In terms of ε_{os}:

$$\sqrt{2} \times 3\varepsilon_{os} \times c_s^2 \times 1.11111^{5/2} \times 1.04^{1/6} = 5.0016 \times 10^6 \text{ J} \equiv 5.0016 \times 10^6 \text{ eV} \qquad (8.2)$$

And
$$5.0 \times 10^6 \text{ eV}/(1/3 \ c_s) \times 2/3 \times 1/1.111\ldots = 0.03 \text{ eVat} \qquad (8.3)$$

Five 1 MeV pathways create 5×10^6 J. Now 1 MeVSI is becoming 10^{14} J at $1/3 \ c_s$:

$$1 \text{ MeVSI} \times 1/3 \ c_s \equiv 1/(\text{minimum energy}) = 1/(1 \times 10^{-14} \text{ J}) = 1 \times 10^{14} \text{ J and for 3}$$

planes at 6000 K, 10^{14} J expresses the sun power:

$$3 \times 1 \times 10^{14} \text{ J} \times 6000^4 \text{ K} \equiv 3.8880 \times 10^{29} \text{ W} \qquad (8.4)$$

Since at 1 K quantum value $k_s = 1 \times 10^{-23}$ J energy 1×10^{-19} J represents 10000 K $\equiv$ three orthogonal 6000 K vectors on the surface of the sun displaced by 90° generate:

$$\sqrt{3} \times 6000/1.04^{0.98} = 10000.4 \text{ K} \qquad (8.5)$$

deriving in Conclussion 44 the atomic constant of gravity expressed by $T_{sun} = 10000$ K.

The electron mass energy generated by the infrared frequency of 5×10^{12} Hz:

$1x10^{-30}$ $x5x10^{12}/1x10^{-19} \equiv 50$ eVat making the value 50 eV rediations energy uncorrected for relativity and quantum:

The product 50 eVat $x1/(1.15470^{3.5})$ $/1.04^{1/5}$ derives $29.985 \approx 30.0$ eVat $\qquad$ (8.6)

The more detailed derivation of Hartley energy of hydrogen atom is the product of atomic force $F_{at} = 1x10^{-7}$ N and the radius of atom $a_{os} = 1/2 \times 10^{-10}$ m, corrected for relativity and quantum:

$$(1 \times 10^{-7} \text{ N } x \ 1/2 \times 10^{-10} \text{ m })/1x10^{-19} = 5 \times 10^{-18}/1x10^{-19} = 50 \text{ eV} \qquad (8.6a)$$
$$50 \text{ eV}/(1.154700538^2 \ x \ 1.111 \ldots^2) \ x \ (1+1/80) = 30.000 \ 000 \ 00 \text{ eV} \qquad (8.7)$$

Electron equivalent energy $= h_s v_{Lms} = 1x10^{-33} \ x \ 5 \ x \ 10^{15} = 5 \ x10^{-18}$ J $= 50$ eV $\qquad$ (8.8)

All result indicate that the electron moves with proton at frequency 5×10^{12} Hz, any minor mismaches being hidden in the arbitrary choice of quantunm corrector or quantum uncertainty. This indicates the need to propose presence in physics of **two more physical constants, the Boltzmann constant for electron $k_{selec} =1 \ x \ 10^{-26}$ J and the Planck constant for electron $h_{selec} = 1x10^{-36}$ J-s** deriving Planck Radiation energy value:

Electron energy $= h_{selec} v_{se} = 1x10^{-36} x \ 1x10^{15} = 1x10^{-21}$ J$\rightarrow x \ 1/e_s = 0.01$ eV $\qquad$ (8.9)

Proton energy requires introducing $= h_s v_{Lms} = 1x10^{-33} \ x \ 1 \ x \ 10^{12} = 1 \ x10^{-21}$ J $\equiv 0.01$ eV, which has the same value as reduced electron due to the equipartion energy rule.

The product of proton mass in one direction $1x10^{-27}$ kg and at local velocity $1/3 \ c_s$ derives energy, and not momentum, because the velocity term in atomic units is actually squared and divided by velocity on conversion to SI units:

Local pulse energy $= 1x10^{-27}$ kg $x \ 1 \ x \ 10^8 = 1 \ x \ 10^{-19}$ J $\equiv e_s$ $\qquad$ (8.10)

Since integer cubic muo10$_s$'s mass in SI units is $1 \ x \ 10^{-28}$ kg it would indicate presence of ten muo10$_s$ orientation masses and ten mu10s orientation forces within the atom.

The radiation energies which express clearly infrared frequencies in atomic structures are often incorrectly attributted to electrons generating radiation energy creating misleading presentation of events. The oscillator in classical SI units is the reduced electron mass (equivalent electron) $m_{es}m_{ps}/(m_{es}+m_{ps}) = 1x10^{-27} \ x \ 1 \ x \ 10^{-30}/(1.001x10^{-27}) = 0.999x10^{-30}$ kg generated by masses of the electron mass and of the proton. The application of the geometric mass value of the sun-earth couple and of the electron-proton pair is a different issue and it is discussed in Section 17.

The above important statement claims that Bohr atom SI energies attributed in typical textbbooks of physics to energies of electrons are really due to reduced masses of electron and of proton. Consequently the frequency $5x10^{12}$ Hz generating the correct energy should be attributed to the mass of mesmps/(mes+mps). The atomic energy derived from the electron mass and electron frequemcy is $1 \times 10^{-30} \times 5 \times 10^{15}$ Hz/$1x10^{-19}$ = 50000 eVSI $\equiv 5x10^{-15}$ J is clearly incorrect. Applying 10^{-10} and multipliers $3x10^8$, 300 K and 1.11 . . . derives 10 J expressing a correct energy value.

Distribution of values of various parameters is not considered, **but their integrals over time in the integer model are unity** justifying accurate results.

The two frequencies are defined by $3k_sT_{300}$ *ln2*/(h_s x 1.111 . . .^{2} x 1.04$^{1/4}$) deriving the infrared value at 300 K:

$$v_{Lms} = 3 \ x1 \ x \ 10^{-23}x300 \ xln2/(1 \ x \ 10^{-33} \ x \ 1.111 \ . . .^2 \ x \ 1.04^{1/4})=5.0038 \ x10^{12} \ Hz \qquad (8.11)$$

Above frequency has been derived from earlier experiments of the author and it has been derived from extreme accurate measurements of Weitz and ales [10], for which these researchers were awarded Nobel Price.

The frequency 5×10^{15} Hz replaces with accuracy 2 in 10^5 the constant C*=3.29 $\times 10^{15}$ Hz discovered in the year 1885. The constant C* defines the frequency emitted by hydrogen in the following expression expressing C* in terms of 5×10^{15} Hz and the proposed corrections:

$$v = C^*[1/2^2-1/n^2] = 5 \ x10^{15}x1.04^{1/3}/(1.1547^2x1.07456)x[1/2^2-1/n^2] \qquad (8.12)$$

The research presented indicates at the very common presence of the hydrogen structure in the Universe expressed by the value of 1 eVat or 1 J derived also from the Planck's Law in Equations 22.51 and 22.54. Milliken voltage of 1.82 eV is that of 1/2(13.62 - 10), that is 1/2 of n = 2 to n = ∞ radiation. The energy of 10 eVat is believed to represent the input of ten 1 eVat radiations in ten space quantization orientations, producing one 1 eVat vector force in one direction. The ratio of the product of the mean infrared proton and (proton+electron) frequency times linear probability 10^{10} and quantum corrector over $\sqrt{2}$ derives exactly the reference gravitation force of sun:

$$F_{Gs} = 4.9975 \ x \ 10^{12} \ x \ 1.04^{1/14} x10^{10}/\sqrt{2} = 3.5437 \ x10^{22} \ N \qquad (8.13)$$

Thus on earth at 300 K infrared frequency delivers $10^{10}/\sqrt{2}$ N per cycle in the integer cubic model SI units. At atomic level this is reduced to $1/\sqrt{2}$ N per cycle, so that two pulses displaced by 90^0 deliver 1 N.

The option that electron frequency generates the sun constant using the expression

$$5.0 \ x \ 10^{15} \ x \ 10^{10}/3.5438 \ x10^{22} = 1410.9 \ eV \qquad (8.14)$$

is not acceptable because the electron frequencies effectively do not interact with the infrared and hence they are not diverted in the magnetic lens, but move in the straight line. Hence the total force induced by electron frequency is related only to the earth circular surface

$$5.0 \times 10^{15} \times 10^{-10} \times 1.278 \times 10^{14} \, m^2 = 6.39 \times 10^{19} \, N \tag{8.15}$$

representing a small fraction 0.18% of F_{Gs} force. It reduces the accuracy of calculations of gravitation although the physical constants are integer values.

Temperature acts along 5 pathways of space quantization to generate 5.0×10^{15} Hz oscillations

$$5 \times (6000-300)^4 / 1.04^{1.4} \equiv 4.996 \times 10^{15} \, Hz \tag{8.16}$$

The electron frequency with 50 radiation directions, 6x8 cube corners = 48 + 2, derived with two degrees of freedom from a surface and losing 1/6 on reflection is given by:

$$\nu_{Hms} \equiv (1-1/6) \times 50 \times 2 \times k_s T_{sun} / h_s = 5/6 \times 50 \times 2 \times 10^{-23} \times 6000 / (1 \times 10^{-33}) = 5 \times 10^{15} \, Hz.$$

In this Equation k_s may be replaced by k_{selec} and h_s by h_{selec} \hfill (8.17)

The author uses the mass centre as an independent reference, because in his NaCl experiments such treatment produced correct dissociation energy values. Similarly Einstein treats the center of mass as the absolute reference.

The two radiating particles of opposing momenta move relatively to each other with velocity c, with proviso that relatively to the center of mass and of energy they move with velocity 1/2xc. Their velocity of one half of the velocity of light, 1/2x c, requires vector and relativity correction of:

$$\text{Total correction for 3D display is } \sqrt{3} \times 1.154700 = 1.999999 \approx 2.0 \tag{8.18}$$

$$K_{rel} = 1.154\,700\,538 = 1.333333333^{1/2} \tag{8.19}$$

Such K_{rel} correction, applicable to energy, to mass and to momentum is incorporated in numerous discovered identities of the proposed gravitation theory. In cubic geometry shown in Fig. 4 and 5, the integer model light velocity c_s is the sum of three 1/2xc_s vectors displaced from each other by 90° producing resultant velocity c_s:

$$\sqrt{3} \times \tfrac{1}{2} \, c_s \times K_{rel} = 299\,999\,999.9 \, m/s \tag{8.20}$$

Properties of protons, that is their masses, energies, powers or momenta are carried in space in relation to the center of energy at velocity $\frac{1}{2} c$ and so require suitable relativity corrections K_{rel}. The measured classical value of c varies from the integer cubic value only by quantum corrector:

$$c_s = c \times 1.04^{1/56} = 3.000020388 \times 10^8 \, m/s \tag{8.21}$$

And $\qquad c_s = c \times 1.0328^{1/56} = 2.99965 \times 10^8 \, m/s \approx 3 \times 10^8 \, m/s$ suggesting that the variance between c_s and c is due relativity correction of the proton moving at ¼ c and there being 6×10^{54} protons/kg with the probability being $(1/4)^2 = 1/16$ generating 10^{56} events. Such non meaningful guesses created this book.

Converting to atomic units by 1×10^{-10} derives 0.03 eV, which is the energy quantum for the three oscillators.

The radiation may appear in any six orientations, which decreases to three if one broad direction is chosen. It seems illogical to attribute velocity of light to radiation energy or mass, because according to the theory of relativity that mass would become infinity. Representing light by three vectors at velocity $1/2xc_s$ or $1/4xc_s$ eliminates the problem. It is mainly the relativity factor $(1.04^{1/2}) \times K_{rel} \cong 1.04^{1/2}x1.1547$, which to sustain the energy value converts the measured proton frequency $v_{Lm} = 4.23392x10^{12}$ Hz into $v_{Hms}=5x10^{12}$ Hz and electron frequency $v_{Hms}=4.23392x10^{15}$ Hz into $v_{Hms}=5 \times 10^{15}$ Hz.

Depending on the number of components required in the pulse formation process the various proposed identities may require relativity correction: $K_{rel}=1.1547$, $K_{rel}{}^2=1.333$ . . ., $K_{rel}{}^3=1.5396$, $K_{rel}{}^4=1.778$, $K_{rel}{}^{1/2}=1.07457$, $(K_{rel}{}^{1/3})^2=1.234567901$ and $K_{rel}{}^{1/4}=1.0328\approx1.04^{4/5}$. The product of two vectors moving at velocity of proton $1/4xc_s$ requires the correction of 1.0328. For $1/3$ c_s the corrector is 1.111. . . . Since $1.1547/1.0328 = 1.11803$ it is difficult to differentiate this value from the mean velocity relativity corrector creating possible errors. Application of Special Relativity corrections generate several values of the gravitation force between the sun and the earth.

The quantum correction, [3, 30 p.40], $1.04068449 = 1/(2xln2^2) \approx 1.04$; the use of this expresses uncertainty of four directions of Planck radiation terms 0.01 J and it generates often values within H. u. limits. Alternative value is $ln2/(2/3)=1.03972$, because $2/3$ of atomic energy interacts with probability factors adding power indexes. The quantum correctors used were $1.04^{1/12}=1.00327$, $1.04^{1/6}=1.00655$, $1.04^{1/3}=1.01316$, $1.04^{1/2}=1.019804\approx 1.02$ and others. Please note that $1.333 . . ./1.111 . . . = 1.2$. The numerical equality $(1.2x1.111 . . .)^{1/2} = 1.154700$ is deceptive because two terms describe different formative process. The other factor affecting values numerically is temperature.

Conversion from natural atomic units expressed in SI units has problems, because the classical SI list of physical constants has values, which express atomic parameters. Linear dimension vary by a factor of 10^{10}, while volumetric properties including mass by factor of 10^{30}. Rotational interaction include two linear events require factor of 10^{20}. Conversion process from sun may differ from that from earth.

The atomic force, which is 1D converts by multiplying by 10^{10} from atomic value $F_{sat} = 1 \times 10^{-7}$ N to $F_{satSI} = 1000$ N identified with the 1000 eV, or 1000 K pulse generated by the sun and by the earth. If the pulse of 1000 N is generated by components displayed in three orthogonal orientations the resultant 1000 N expresses product of three probabilities requiring input energy of 1 G eV.

The apparent conflict between values 1 eV and 10 J needs an explanation. The atomic (equivalent electron) velocity of 1×10^7 m/s times the atomic force 1×10^{-7} N generate energy 1 J or 1 eV at atomic level, representing energy of one oscillator formed from two pulses and so it is subjected to multiplication by 10^{20} on conversion to true SI units. Converting those units to electron volt requires multiplication by 1×10^{-19} generating a multiplier of 10. Thus while 10 J obtained in several expressions derives 13.62 eV ionization energy per oscillator of Bohr atom it is actually 1 J at the atomic term discussed above. At times the energy has to be doubled to allow for radiation in the other direction. This is a research book so all possible alternatives have to be considered.

Temperature according to the integer cubic model expresses the pulse rate of the source. Three pulses of 6000 K in orthogonal orientation generate pulse rate, which time ε_{os} over quantum corrector derive sun pulse per change event: $6000^3 \equiv 2.16 \times 10^{11} \rightarrow$ x $\varepsilon_{os}/1.04^2 \equiv$ $1.9970 \approx 2$ J or eV. The variance is $30k_s$ each k_s comprising 3 quanta $1/3\ k_s$. Spherical shape has 30 interacting quantum orientations; two pulses, with resultant pulses joining, have 60 orientations and one plane has 10 orientations. Their inverts become probabilities deriving the following energies of pulses:

$$(6000^3)^{1/30} \text{ x } (1.15473^{3/2} \text{ x } 1.04^{1/3})] \equiv 3.000527 \text{ eV or J} \tag{8.22}$$

$$(6000^3)^{1/60}/1.04^{3/4} \equiv 1.500142 \text{ eV or J} \tag{8.23}$$

Multiplying by 2/3 eliminates electron components and derives 2 J and 1 J. With 10 orientations the resultant is the familiar 13.62 eV:

$$(6000^3)^{1/10} \text{ x } 1.04^{1/25} \equiv 13.618 \text{ eV} \tag{8.24}$$

The identity displaying 10 J shows that the sun radiates in two opposite directions and derives very accurately variance of $5k_s$ because 0.0005 eV $= 5k_s$:

$$1/2 \text{ x } (6007)^{1/3} \text{ x } 1.111\ldots/1.04^{1/4} \equiv 10.0004998 \text{ J or eV} \tag{8.25}$$

Six kelvin K for three planes is an acceptable uncertainty of 1 K per orientation and 1 K uncertainty. For both directions with correction the energy is 20 eV as quoted in some textbooks for energy of Bohr atom.

Author's books and papers do not deal with waves or photons, but with chains of particles, from interacting waves. Individual particles appear on the second differentiation of the recorded current signal. A wave would remain a wave. Waves may be excited in the receivers. Such pulses arise mainly as the result of hydrogen molecule or equivalent energy converting into vectorized radiation initially by moving with local velocity of light and moving at velocity of light consistent with the Special Relativity.

Protons are much heavier than electrons and their mean velocity was found to be $1/4xc_s$. This value was chosen from the consideration of energy equalization. Three $1/4c_s$ components along X, Y and Z co-ordinates of space with 1.1547 correction add vectorially to $1/2\ c_s$. Thus in space units proton radiation mass m_{ps}, times velocity expressing momentum, times c_s converter to SI units and times $\sqrt{3}$ multiplier derives ε_{os} expressing energy density of space converted from proton mass:

$$\varepsilon_{os} \equiv m_{ps} \text{ x } 1/\sqrt{(1-1/16)} \text{ x } 1/4^2 x\ c_s^2 \text{ x } \sqrt{3} = 1.00623\ x10^{-11} \text{ J/m}^3 \tag{8.26}$$

where $c_s^2 = (\sqrt{3}x1/2c_s x1.15470)^2$.

Dividing above by $1.04^{1/6}$ reduces the error from 0.6% to 0.3%.

The natural frequency of oscillations of the proton, 5×10^{12} Hz, and of the electron, 5×10^{15} Hz, was considered earlier. The natural frequency of the oscillations of the pair (proton+electron) is attributed to their reduced mass. For linear velocity it is proposed now that for a coupled movement the two particle momenta have to be equal. The above value, usually in the form $1/(1-1/10)=1.111\ldots$ was considered initially to be a reflection coefficient. It appears very often in Tables 4-12 evaluating gravitation forces both for the earth and for the planets. In Chapter 23 it was realized that local velocities $1/4 \times c_s$ and $1/2\ c_s$ applied to the oscillations of formative stages of vector formation of the radiation pulse; particles were found to move with local velocity $1/3\ c_s$ requiring relativity correction $1.111\ldots^{1/2}$. Two applications derive $1.111\ldots$

Mean velocity of radiation particles is 1/3 of the integer cubic model velocity of light c_s $=3 \times 10^8$ m/s. With proton velocity being decreased by 1/1000 by electron moving with it, is the geometric mean value with relativity and quantum corrections:

$$(8.27)$$

Geometric mean velocity $=[\sqrt(1/2 \times 1/4 \times 0.999\ c_s)/(1.111\ldots^{1/2} \times 1.04^{1/7}=0.333369\ c_s \equiv 1/3\ c_s.$

Late finding indicated that $1.1547/1.111\ldots \approx 1.04$ with expressions unchanged. The finding has not been introduced through the book.

Because the corrector $1.111\ldots$ produced best accuracy, it was used intuitively also to represent $K_{rel}^{1/2}K_{rel}^{1/4}=1.1098$ all through the text. At the final review more accurate relativity corrections were calculated:

$$K_{rel1/3cs}=1/[1-(1/3c_s/c_s)^2]^{1/2}=1/(1-1/9)^{1/2}=(1.0/0.9)^{1/2}=1.1111\ldots^{1/2}=1/1.054092553.$$

And $K_{rel1/2cs}=1/[1-(1/2c_s/c_s)^2]^{1/2}=1/(1-1/2^2)^{1/2}=1.154700538.$

And $K_{rel1/4cs}=1/[1-(1/4c_s/c_s)^2]^{1/2}=1/(1-1/4^2)^{1/2}=1.03279559 \qquad (8.28)$

The derivation of the force $1/6\ \mu_{os}\ I_{earth}$ incorporating earth current was obtained from $F_R=I_{earth} \times B$ balancing the sun gravitation force F_{Gs} on page 68 of book 1 [45]. The first expression using classical SI values for the mass of earth and its linear velocity derived accuracy of 8 parts per 10^4:

$$F_{Gs} \equiv 1/6\ \mu_{os}\ I_{earth}\ K_{rel}^{1/2}=$$
$$=1/6 \times 1/9 \times 10^{-5} \times 5.9742 \times 10^{24}/1 \times 10^{-30} \times 1 \times 10^{-30} \times 29786 \times 1.07457= \mathbf{3.5411 \times 10^{22}\ N}$$
$$(8.29)$$

When SI mass and linear velocity is now explained by rounded values with quantum and relativity correction for $1/3\ c_s$ the accuracy increases to 5 parts per 10^5:
$$F_{Gs} \equiv 1/6\ \mu_{os}\ I_{earth}\ K_{rel}^{1/2}=$$
$$=1/6 \times 10^{-6} \times 6 \times 10^{24}/1 \times 10^{-30} \times 1 \times 10^{-30} \times 30000 \times 1.039742^{1/5}= 3.5437 \times 10^{22}\ N \qquad (8.30)$$

Above accurate derivation of the gravitation force confirms validity of treating protons as congregations of electrons and of expressing mass of 1 kg by 6×10^{54} electrons.

The above identity in classical SI units added with $K_{rel}=1.1547$ derives also an accurate value of F_G:

$$F_G \equiv 1/6 \, \mu_o \, I_{earth} \, K_{rel} = 3.5432 \times 10^{22} \, \text{N}$$
$$= 1/6 \times 4\pi \times 10^{-7} \times 5.9742 \times 10^{24}/9.10938 \times 10^{-31} \times 1 \times 10^{30} \times 29786 \times 1.1547 \, \text{N} \quad (8.31)$$

Equation 8.30 has highest accuracy in deriving the force in SI newtons. The current I_{earth} replaces I^2 of the classical expression of the formula expressed in electron-volts per unit velocity of atomic terms.

In atomic units, the unit of charge is one electron. Although the body of earth may have large charge inducing current and momentum, the structure of mantel of a planet may be such that circulating currents are produced decreasing the external effect of the original charge. The Equation 8.30 has been applied to other planets in Chapter 20.

The classical Balmer formula is the procedure of deriving spectral wavelength, λ, from two radiation wavelengths for Balmer series. With Rydberg constant $R=1.097 \times 10^7$ expressing the infrared radiation for $n = \infty$:

$$1/\lambda = R \times (1/2^2 - 1/\infty) = 1.097 \times 10^7 \times 1/4 = 2.5 \times 10^6 \, \text{m}^{-1}, \text{ and } \lambda = 0.4 \times 10^{-6} \, \text{m} \quad (8.32)$$

In the integer model R is expressed by the atomic force $F_H = 1 \times 10^{-7} \, \text{N}$, so that the total ionization energy of hydrogen atom of 13.62 eV is using a term $\sqrt{3}$, quantum corrector 1.04, $h_s \, c \, E_n / \lambda$ and kT factor

$$\text{Energy} = \sqrt{3} \, h_s \, c_s / \lambda \times 1.04 + k_{s300} T_{300} ln2/e_s = 13.62 \, \text{eVat} \quad (8.33)$$

Using F_H derives the Hartree energy and its component 10 eV, expressing 1/3 of the atomic momentum carried at atomic level by the electron proton and the proton+electron pair:

$$\text{Hartree energy} = F_H \, h_s \, c_s/e_s = 1 \times 10^{-7} \times 10^{-33} \times 3 \times 10^8/10^{-19} = 30 \, \text{eVat} \quad (8.34)$$

becoming 10 eV for $1/3 \, c_s = 10^8 \, \text{m/s}$ for each of three orthogonal orientations.

To obtain radiation energy of 1 eV delivered by proton radiation in Pfunt series with n=5 and n=∞ requires using relativity K_{rel}^2 and quantum correctors $1.04^{1/3}$:

$$1/\lambda = R \times (1/5^2 - 1/\infty) = 1/(22.789 \times 10^{-7}) \, \text{m}^{-1} \quad (8.35)$$

Above generating energy = $h_s \, c_s / \lambda \, K_{rel}^2 \times 1.04^{1/3} = 1.00031 \times 10^{-19} \, \text{J} \equiv 1.00031 \, \text{eV} \quad (8.36)$

If energy is lost from the Bohr atom it must appear in the radiation beam. That energy is not evaluated in classical physics because so far science does not recognize presence of local velocities requiring relativity corrections.

While $1/\lambda_{2-3} = 1.097 \times 10^7 \times (1/2^2 - 1/9) = 0.138888 \ldots \times 10^7$ derives $\lambda_{2-3} = 720$ nm it is the product of the measured value of 656.3 nm in the infrared times the Rydberg factor 1.097.

Textbooks such as A. Beiser in his Concepts of Modern Physics 1973 derive using his Equation 4.20:

$$\lambda = h/e \text{ x } \sqrt{(4\pi\varepsilon_o r_d/m_e)} \tag{8.37}$$

the wavelength of de Broglie Bohr electron to have the circumference 3.3 $\times 10^{-10}$ m of the electron orbit in Bohr atom. Applying the same equation using integer space cubic constants with corrections 1/6, relativity and quantum, but replacing π by 3 derives

$$\tag{8.38}$$
$$\lambda/6 = 1\times10^{-33}/1\times10^{-19}\times\sqrt{(4\times3\times1\times10^{-11}\times \tfrac{1}{2}\times10^{-10}/1\times10^{-30})} \times 1/1.111^2\times1.04 = 1.005 \times 10^{-10} \text{ m.}$$

This represents Bohr atom dimension and λ wavelength as the sum of half-wavelengths fitting the plane sides of the cubic atom in six orientations. The term wavelength describes inaccurately the parameter, because in 1D it represents the amplitude of the electron vibration in the direction of the propagation. To produce a gravity vector signal requires several radiations of the Bohr atom combining together and forming a wave from three dimensions.

Energies used rarely in classical physics:

For electrons $E = h_s v_{Hms}/e_s = 1 \times 10^{-33} \times 5 \times 10^{15}/1 \times 10^{-19} = 50$ eVat $\tag{8.39}$

For protons $E = h_s v_{Lms}/e_s = 1 \times 10^{-33} \times 5 \times 10^{12}/1 \times 10^{-19} = 0.05$ eVat $\tag{8.40}$

For electrons in SI, but using correctors

$$E = (h_s v_{Hms} \text{ x } 1.04^{11/5} - kT_s)/e_s = 27.253 - kT \approx \text{Hartee energy} \tag{8.41}$$

Investigating energies comprising temperature term

Temperature in physics is badly defined. The power density per electron cycle is the thermal field gradient with its square value $(\varepsilon_{os}T_{6000})^2/1\times10^{15}$ per cycle and volume and quantum correction multipliers deriving energy in SI units in each of 4 directions

1/4 $\times(6000$ K $\times1\times10^{-11}$m$)^2\times10^{30}\times1.111 \ldots/1\times10^{15}$ Hz$=1.0000000000$ W/m^2 per electron cycle $\tag{8.42}$

The thermal potential is the temperature in degrees kelvin. Thermal power surface density P_s converted to integer SI units are expressed by the product of space energy density per cubic meter, ε_{os}, the kelvin temperature, the velocity c_s and the relativity corrector for the average local velocity of 1/3 cs:

Thermal power surface pulse density $P_s = \varepsilon_{os}$ [J/m^3] $T_{Ks} c_s$ [m/s] $\times1.111 \ldots$ W/m^2 $\tag{8.43}$

From earth $P_s = 1 \times 10^{-11} \times 300 \times 3 \times 10^8 \times1.111 \ldots = 0.999\ 999\ 999$ J $\equiv 1$ W $\tag{8.44}$

becoming with surface probability at the atomic level 1 W **x 1 x** $10^{-20} \equiv 1/10$ e_s with two pulses required to generate velocity of light deriving 0.01 J main term of Planck radiation and expressing energy per plane at 1 K:

$$E_{1K} = 1 \text{ x } 10^{-11} \text{ x } 1 \text{ K x } 3 \text{ x } 10^8 \text{ x} 1.111 \ldots = 1/3 \text{ x } 0.01 \text{ J} \tag{8.45}$$

From sun $P_s = 1\text{x}10^{-11}\text{x } 6000 \text{ K x } 3 \text{ x } 10^8 \text{ m/s x}1.111 \ldots = 19.\,999\,999\,999 \text{ J} \equiv 20 \text{ W} \tag{8.46}$

expressing 20 eV from $\varepsilon_{os} = 10^{-11}$; similarly (vector sum of $\sqrt{3}$ **x** three directed energies 10 eV in orthogonal directions) times 1.154700538 equals 19.99 999 999 eV.
The temperature 6000 K is well defined from the pulse of 50 eV, derived from 30 eV:

$$\text{At atomic level } 6000.016 \text{ K} \equiv \{[\sqrt{3} \text{ x } 50/1.111 \ldots \text{ x}1.0]^2/1.04^{19/60}\}. \tag{8.48}$$

Quantum corrector looks odd, but the index 19/60 fits the temperature 6000 K very well.

Equation 8.48 clarifies definitely that the energy of 50 eVat expresses the sun pulse and it defines very accurately the gravitation force of the sun:

$$\text{Sun gravitation force } 1/3\text{x}1/10 \text{ x}[50/(0.5\text{x}10^{-10})]^2\text{x}1.111 \ldots^{1/2} \text{ x}1.04^{13/60} = 3.5438 \text{ x } 10^{22} \text{ N} \tag{8.49}$$

Force F_{Gs} varies in value as predicted by the Special Theory of Relativity discussed in Chapters 16-18:

$$F_{Gs} = 1/3\text{x}1/10 \text{ x}[50/(0.5\text{x}10^{-10})]^2/1.111 \ldots = 3.000\,000\,000\text{x}10^{22} \text{ N} \tag{8.50}$$

The large gravitation force of attraction is all along the sun-to-earth distance and it is expressed all the way through in the sun beam by the square of the atomic space gradient of the electrical kelvin field E_K, i.e. 6000 K, with proper conversion and corrections:

$$F_{Gs} = 6000 \text{ K x}1.04^{1/18}/(\sqrt{2} \text{ x}1 \text{ x}10^{-19}) \text{ x}1.111 \ldots/1.1547^2 = 3.5432\text{x}10^{22} \text{ N} \tag{8.51}$$

Or in Special Relativity terms $F_{Gs} = \frac{1}{2} \text{ x } 6000 \text{ K}/1\text{x}10^{-19} = 3.0000000000\text{x}10^{22} \text{ N} \tag{8.52}$

With 60 DOF the probability is 1/60 and the sun temperature of 6000 K derives with relativity and quantum corrections the sun pulse of 1 J

$$6000^{1/60} \text{ x } 1/1.1547 \text{ x } 1/1.04^{1/40} \equiv 1.00017 \text{ J or eVat} \tag{8.53}$$

The variance is 2 k_s the uncertainty in SI units. Thirty DOF is due to angular momentum of electrons. Although in discussing spin, term electron is used in all textbook, the integer space model considers them to be reduced electrons. Equivalent mass of electron and proton move with infrared frequency.
  With 17 degrees of freedom in space each degree requiring energy of 1 J the probability is 1/17 with the sun temperature delivering:

$$6000^{1/17} \times 1/1.1547^{3/2} \times 1.04^{1/5} \equiv 1.00045 \text{ J or eVat.}$$

The variance 4.5 k_s is 1/4 k_s per degree of freedom $\hspace{2cm}$ (8.54)

Assuming that the pulse is generated in one earth direction and three pulses are needed to generate velocity of light the probability becomes $(1/6)^3 = 1/216$ and 6000 K delivers

$$6000^{1/216} /1.04^{1.03} \equiv (1 - 0.00012) \text{ J or eV expressing uncertainty of } 1.3 \ k_s \hspace{1cm} (8.55)$$

One dimensional frequency of the light velocity is $(3 \times 10^8)^{1/2} = 17320.50808$ Hz.

Dividing by $\sqrt{3}$ and 1.04 derives the sun temperature:

$$\text{The sun temperature} = (3 \times 10^8)^{1/2}/\sqrt{3} \times 1.04 = 6004.4428 \text{ K} \hspace{1cm} (8.56)$$

The value 4.44/1.111 ≈ 4 K or 1 DOF per orientation each of four expresses H. u.
Proton radiates in three orthogonal directions with 10 quantum positions in each direction. The probability of radiation taking place in one chosen direction is 1×10^{-10} making total probability 10^{-30}. This is shown in $1 \times 10^{-27} \times 1/10 \times 1/10 \times 1/10 = 10^{-30}$.

Generated power density with corrections is reduced by 1.2^2, per local unit velocity $1/3 \ c_s$:

$$\text{Atomic power density} = \sigma_s T_{6000}{}^4 \times 1.2^2 \times 1.1547/(1 \times 10^8 \times 1.04^{1/15}) = 1.000245 \text{ units.} \hspace{0.5cm} (8.57)$$

Expressed in three orthogonal orientations generates $\sqrt{3} \times 1.1547 = 1.999999 \approx 2$ units

Referred to 6000 K over sun power density in two orientations (Equation 1.4) derives power density in each of three local orthogonal orientations expressed as product of probabilities:

$$1 \times 6000 \text{ K} /6 \times 10^9 \text{ m} = 1 \times 10^{-6} \text{ units} \equiv 0.01 \text{ J} \times 0.01 \text{ J} \times 0.01 \text{ J} \hspace{1cm} (8.58)$$

Power density at 6000 K has to use index 4; converting to SI with 1×10^{-6} eV/m² $= \mu_{os}$:

$$1 \times 10^{-6} \text{ eV/m}^2 \times 6000^4 \text{ K} \times 1 \times 10^{-19}/(1.1547 \times 1.04^{1/4}) = 1.00028 \times 10^{-10} \text{ W/m}^2 \equiv G_s \hspace{0.3cm} (8.59)$$

So total temperature index is 5; this is confirmed using energy 1×10^{-15} W in evaluating the gravity force pulled by sun equal and opposite to the gravity force pulled by earth:

$$\sqrt{2} \times 1 \times 10^{-15} \times T_{6000}{}^5 \times 1.1547^{1/2} = 1.7725 \times 10^{22} \text{ N} \hspace{1cm} (8.60)$$

$\hspace{1cm}$ Conclusion 34 deduces that sun reacts with earth with one of its six beams, so that although $T_{6000} = 6000$ K, the pulse rate is 1000 in each circular orientation once the source receiver axes has been set. Thus in 2D presentation the related energy radiated corrected for relativity is:

$$1000^{1/2} /1.111 \ldots^{1/2} = 30 \text{ J or eV} \hspace{1cm} (8.61)$$

exactly with four 30 eV components in wave format generating the sun pulse of 1 MeV as given in Equations 10.16-19 and 10.21-2:

$$(30 \times 30)X(30 \times 30) \times 1/11111\ldots^2 = \mu_o/1.111\ldots \times 10^{12}\ \text{Hz} = 1 \times 10^6\ \text{J exactly} \qquad (8.62)$$

Multiplied by 10^{-30} for atomic level derives: 10^{-24} J or eV $\equiv 1/10^{12}$ Hz $\times 1/10^{12}$ Hz $= 1/10\ k_s$.

In 1D format the 1000 J energy is 10 eV or J $1000^{1/3} = 10$ J or W (8.63)

Sun spectrum on the top of the atmosphere is shown on WEB pages to fit well the black body radiation at 5250 K. Temperature expresses energy and it is suggested that it is subject to relativity and quantum corrections. Applying correction for the velocity of protons 1/2 c_s derives the sun temperature at:

$$5250\ \text{K} \times 1.1547/1.04^{1/4} = 6003\ \text{K} \qquad (8.64)$$

Energy radiated E_n is proportional to frequency, $E_n = h_s\ v_T$, and to temperature, $E_n = k_s T$. Hence the product of two pulses displaced by 90° is needed to generate velocity of ½ c_s and the energy generated is proportional to T^4 as required by Stephan-Boltzmann Law. Variance is ½ K per space direction.

The integer muo10$_s$ particle responsible for inner atomic force has been now found to derive the international gravity constant G_s. Muo10$_s$ is a short lived pulse, most probably because its energy is quickly diverted to other degrees of freedom, a typical characteristic of all high energy pulses. At atomic level the pulse of G_s is also short lived, and it expresses its force at its maximum value. While radiation pulse of muo10$_s$ or of G_s is activated, the electron requires some 1000 steps to transfer its energy to proton.

The muon energy 106 MeV measured in cosmos experiments by Carl Anderson is considered in the integer cubic model to be closely related to 1000 eV or J representing vectorized pulse energy of earth and of sun. Incoherent energy 106 MeV requires vectorization and correction for relativity and quantum:

$$106 \times 10^6 \times 1/10 \times 1/10/(1.1111\ldots \times 1.04^{1/7}) = (1 - 0.0000143)\ \text{MeV vector} \qquad (8.65)$$

Anderson's muon is related, but not the muo10$_s$ of the book. It is suggested that hydrogen atom has two sets of ten quantum oriented energies 10^{-28} J of muo10$_s$, or simply energies (1×10^{-28} J), spread in two planes of the total mass 2×10^{-27} J representing the proton. At the atomic level in the final conclusions there is no distinction between neutrons, kilograms, joules or watt. **All these units represent probability of occurrence of the reference unity event in six orientations with the source subjected to quantum orientation probability, relativity and quantum corrections:**

$6 \times (2 \times 10^{-27}) \times 1.1547/(1+1/70) = 1.000277$ eV with uncertainty $3\ k_s/1.04^2 = 1.00 \times 10^{-23}$ J (8.66)
Mass of sun in 2D without electrons (factor 2/3) relativity and quantum corrected:

$(2 \times 10^{30}\,\text{kg})^{1/2} \times 2/3 \times 1.1111\ldots^{1/2} \times 1.04^{1/2} = 1.0003255 \times 10^{15}\,\text{kg} \equiv 1.0003255 \times 10^{15}$ protons.

Number of protons times power per proton as per Equation 2.36

$$1.0003255 \times 10^{15} \times 1.0 \times 10^{-15}\,\text{W} = 1.0003255\,\text{W or J} \qquad (8.67)$$

expressing variance of $3k_s$. The mass of 1 kg in SI units expresses 1 proton at atomic level. The earth temperature of 300 K expresses value above the absolute zero and

$$273.15\,\text{K} \times 1.1547^2 = 300.64 \approx 300 + 2/3\,k_s T \qquad (8.68)$$

expressing variance of 2 DOF.

9. Source, observer and signal geometry

The reality of the world is presented in the front of our eyes, as shown in Figure 9. From the observer's view point the world is represented by straight lines of rays entering the eye from all the directions generated by protons and electrons oscillations at different temperatures producing a range of frequencies. At 300 K the gravity model proposed considers only the mean frequency of the (proton+electron) pair oscillating at 5×10^{12} Hz. Two eyes of animals contribute to three dimensional visions. Taking measurements over time extracts some 3D experimental data. The latter procedure has been used by the author in his relaxation and in high field conduction experiments.

Single measurements miss energy or momentum components along the propagation axes leading to misjudgments of radiation data. In electrical engineering the invisible component is described in the imaginary $j = \sqrt{-1}$ algebra, expressing virtual numbers and completed by a sinusoid. Radiation generated in one plane and one direction has equal probability of being discharged in all orientations of a semicircle. Hence the average amplitude of the unity source force vector (or momentum) moving along the pathway source-receiver is $cos\ 45^0$. The imaginary j components of energy and of momentum are real values, which are not carried in radiation rays and so they may cause an under-assessment of radiation energy and momentum.

The other cause of incorrect estimates is absence of electron vibrations at low temperatures because the physical constants used assumes their presence. Hence the true energy evaluated next is only 1 J:

$$h_s/(k_s T_{300})xv_{Lms}/1.111\ldots \equiv 1 \times 10^{-33}/(1 \times 10^{-23} \times 300) \times 5 \times 10^{12}/1.111\ldots = 1.5 \text{ J} \qquad (9.1)$$

Both thermal and gravity rays move along straight paths and are one dimensional (1D). The basic ID interaction is common to all string theories; this means that the only variable is along well defined propagation axis; the interacting thermal flux is restricted to the beam covering the earth area, while the interacting gravity flux being vectors is deflected from a wide volume around the earth as shown in Figure 7.

The probability to generate a rotating square in one plane with four available orientations Y, X, -Y, -X, is $\frac{1}{4} \times \frac{1}{4} \times \frac{1}{4} \times \frac{1}{4} = 1/256 = 1/4^4$. Millikan electron charge identified now with value 1×10^{-19} C derives the 3 W/m^2 pulse density with sun pulse on earth at 300 K. The square and cubic structures are independent of space orientations in a chosen source-receiver axis. Each pixel of the recorded world around us is formed from the probability of rotating squares and cubes of original source directions. Generation of 1 J or 1 eVat pulses from five available local pathways shown on Figure 1 in a specific direction requires input 4^4 per pathway and a total input $4^4 \times 5 \times 1.111\ldots = 1422.22$ eV or J per plane. The temperature input generating 1 J/K value corrected for quantum derives for three planes the temperature of the sun:

$$1422.222 \times 3 \times \sqrt{2}/1.04^{1/7} = 6000.263 \text{ K} \qquad (9.2)$$

With $k_s \times 6000/e_s = 0.6$ eV, 6000.263 K ≈ 6000 K $+ 1/2 x k_s T$ with $k_s T/10$ taken to be H. u.

The same value derives the sun constant:

$$1422.222 \text{ eV}/1.04 = 1367.5 \text{ eV} \rightarrow +4 \text{ eV} = 1371.5 \text{ eVsi} \qquad (9.3)$$

At the atomic level an elementary energy Universe pulse will be called B_{pulse}. It has three values, which are derived from atoms of H, H_2 or He. Elementary input $B_{pulse1} = \varepsilon_{os} h_s$ or $B_{pulse1} = \varepsilon_{os}$ x volume probability x space quantization of $1/10^3$:

$$B_{pulse1} \equiv 1\text{x}10^{-11} \text{ x}1\text{x}10^{-30} \text{ x}1/10 \text{ x}1/10 \text{ x}1/10 = 1\text{x}10^{-44} \text{ J/K} \qquad (9.4)$$

The universal atomic B_{pulse} is derived in expressions as a product of sequential probability of variety of vectors. In SI space units B_{pulse1} is derived by square of the proton mass time velocity of light multiplied by two invert space quantization of 10 over $\sqrt{3}$:

$$B_{pulse1} \equiv [(m_{pds} \text{ x}c_s^{1/2}) \text{ x } 10/\sqrt{3}] \text{ X } [(m_{pds} \text{ x}c_s^{1/2}) \text{ x}10/\sqrt{3}] = 1\text{x}10^{-44} \text{ kg}^2\text{-m/s} \qquad (9.5)$$

It appears that

$$c_s^{1/2} = (3\text{x}10^8)^{1/2} = 17320.050808 \equiv \sqrt{3} \text{ x } 10000 \text{ K} \approx \sqrt{3} \text{ x } 1.666 \ldots \text{ x } 6000\text{K} \qquad (9.6)$$

establishes an **exact** relation between the velocity of light in two dimensions and the sun temperature. Another relation between c_s and T^2 is expressed in Equation 18.17.

Thus to derive the minimum pulse energy, e.g. $(1 \text{ kg x } 1 \text{ x}10^{-21} \text{ Hz x}1/10)^2 \equiv 1\text{x}10^{-44}$ kg²-m/s two pulses (e.g. *ExE*, *ExB*, *VxI*), each expressed by their own probability are needed to be displaced by 90° angle.

Corrected for relativity and quantum 1D display of velocity of light $3 \text{ x}10^8$ m/s:

$$c_s^{1/3}/(1.15470^2 \text{ x } 1.04^{1/10}) \equiv 500.11 \text{ eV} \qquad (9.7)$$

With $k_s = 1\text{x}10^{-23}$ J space pulse is per kelvin and the waveform pulse formation is the product $(k_s\text{x}10)$ squared or h_s x ε_{os}:

$$\text{Energy of } B_{pulse} \equiv (k_s\text{x}10)\text{X}(k_s\text{x}10) = 1 \text{ x}10^{-44} \text{ J/K} \qquad (9.8)$$
$$\equiv h_s \text{ x } \varepsilon_{os} = 10^{-33} \text{ x } 1 \text{ x } 10^{-11} = 1 \text{ x } 10^{-44} \text{ J/K}.$$

Above energies per kelvin express values produced by repulsive forces generated by earth acting in the plane of earth rotation generating with reflected pulse and a pulse of light velocity: $\sqrt{2}$ x $[\sqrt{2} \text{ x}1/4 \text{ } c_s + \sqrt{2} \text{ x } 1/4 \text{ } c_s] = c_s$.

The proton radiated mass equivalent $m_{ps1D} = 1 \text{ x}10^{-44}c_s^2 \text{ x}1.111 = 1\text{x}10^{-27}$ J, while the pulse 10^{-44} J/K derives the space energy density, the muo10$_s$ mass density also per K:

$$1\text{x}10^{-44} \text{ x } (1/3c_s)^2 \equiv 1\text{x}10^{-28} \text{ J/m}^3. \qquad (9.9)$$

Since $h/2\pi = 1 \backslash \text{x} 10^{-27}$ erg x 10^{-7} (erg/J) $\equiv$ 1 x 10^{-34} J replacing π by 3 for cubic system and applying relativity correction derives the classical crossed h; e.g. $h/2\pi$, divided by relativity 1.111 . . .$^{1/2}$ is $h_s/10$ and $h_s/10 \equiv 6.626$ x$10^{-34}/(2\text{x}\pi\text{x}1.111$ . . .$^{1/2}) = 1.0004$ x10^{-34} J $\approx 1\text{x}10^{-34}$ J and so **the classical crossed h corrected by relativity is identified with integer $h_s/10$** and pulse B_{pulse1} is crossed h_s time linear probability:

$$B_{pulse}/\text{energy} \equiv h_s/10 \text{ x } 10^{-10} \text{ J} \qquad (9.10)$$

With energy radiation mass $m_{ps} = 1\text{x}10^{-27}$ kg, relativity corrections $K_{rel} = 1.15457$ and $K_{rel}{}^{1/4}{}_{cs} = 1.0328$, velocity of light $c_s = 3\text{x}10^8$ m/s, $1/10^3$ vector creator, each momentum of 3x1.111 . . . x1/10^3 x 1.0328xm_{ps} x ¼ c_s x $K_{rel}{}^2$ x 1.03281= 0.02977 J $\approx$ 0.03 eV = 1/2 x $k_s T_{300}/e_s$.

Squared value derives $(1/2 \; k_s T_{300}/e_s.)^2$ x 1.111 = 0.001 eV, while the smallest energy decrement measured by the author [2, 3, 45] has nearly 10 time larger value 10 [0.001/ (1.1547 x 1.0328)] = 0.00833 eV.

Figure 9. Our eyes see millions of pixels (points) in space all generating space particles of all temperatures, colors and frequencies with eccentricities of the sources.

Energy of 0.01 eV has been derived from Planck Radiation law and 0.01/0.00833 = 1.2 is the radiation energy decrement per plane.

The total radiation strength decreases with the square of the distance from the sun while individual signals keep their energy, momentum and frequency unchanged. The ray movement in one direction variations take place only along the axis of propagation interlinked with rotations comparable with the size of the source. The proper number of dimensions is related to the number of types of independent oscillators, which under semi-stable conditions share equal amounts of energy with other oscillators. That energy is expressed by a quantum value or/and its integer multiplier.

The radiation pulse 1×10^{-27} J moving at c_s^2 expresses with linear probability at 300 K at atomic level energy 30 eVat: (9.11)

$1 \times 10^{-27} \times 9 \times 10^{16}/1 \times 10^{-19} \times 10^{-10} \times 300$ K $\times 1.111 \ldots = 900$ MeVx10^{-10}x300 Kx1.111 $\ldots = 30$ eVat.

For the first stage of excitation the smallest signal is 0.01 J Planck Radiation and the signal has specific orientations. In the model the distributions of signal value are generally not considered. Thus in reality there is a cloud of values around each amplitude line. The value often quoted 0.05 J = 3 oscillators x 0.01 J x$\sqrt{3}/1.04$.

For some radiation space pulses quantization is applied to two planes, for other radiation quantization is applied to three planes of the orthogonal orientation of space. Present interpretation of Bohr atom is that $l = 2$ represent excitation of the atom in one orientation having three derivations of the 10 eVat value. With three orthogonal directions corrected for relativity the resultant becomes the Hartree energy of 30 eVat exactly exceeding slightly the SI value of $27.2 \ldots$ eVat.

The probability of generating one 1 eVat signal along the specified source-receiver axes, from discharges in three planes each due to conservation of momentum and space quantization having ten h_s value pathways, is $1/10 \times 1/10 \times 1/10 = 1/1000$. Consequently with $v_{es} = 1 \times 10^7$ m/s, the total discharged electron energy in thirty pathways:

Electron energy $= \frac{1}{2} m_{es} v_{es}^2/e_s = \frac{1}{2} \times 1 \times 10^{-30} \times (1 \times 10^7)^2/1 \times 10^{-19} = 500$ eV (9.12)

The value of $v_{es} = 1 \times 10^7$ m/s in above statement requires justification.

According to textbooks the circular SI electron velocity v_e is the velocity of light over the fine structure constant

$$v_e = c/\alpha = 2.99792 \times 10^8/137.03599 = 2.187688 \times 10^6 \text{ m/s} \qquad (9.13)$$

In the space model the circular SI electron velocity $v_{es} = k_s/m_{es} = 1 \times 10^{-23}/1 \times 10^{-30} = 1 \times 10^7$ m/s, and the previous value represents component local velocity in five space pathways

$$\text{Calculated } v_e = 2.187688 \times 10^6 \text{ m/s} \times 5 \times /1.0328^3 \times 1.04^{1/5} = 1.0007 \times 10^7 \text{ m/s} \quad (9.14)$$

The above result confirms the earlier finding that local velocities in quantum pathways add to make the resultant. **In atomic terms velocities express pulse rate, which adds like vectors; since the signal is 1D adding rates involves arithmetical summing.**

It is suggested that several parameters of the integer cubic space constants display in space different numerical values related to their one dimensional (1D), two dimensional (2D) or three dimensional (3D) display. Dimension displays convert by the power indexes of the first value, which might be 1, 2/3, 1/3, or 3, 2, 1 or 3/2. Energy oscillating in three dimensions is large, it is reduced by the probability requirements in two dimensions and in one dimension it converts into a vector oscillating in the direction of propagation alternating with rotation. The power index 1/3 applied to the total probability of sequential displays of three orientations become display in one orientation. Applying it to H atom the proton component velocities from three orthogonal orientation increases to $\sqrt{3} \times \frac{1}{4} c_s \times$ 1.1547 ≈ 1/2 c_s. This is repeated from three atoms generating velocity c_s or from three pairs of atoms each pair generating each $\sqrt{2} \times 1/2 \times c_s \times \sqrt{2} \approx 1 c_s$. Final postulates suggest that in a hydrogen atom there are two sets of 10 orientation muo10$_s$ masses ten times larger than muo10$_s$ mass 1 x 10^{-28} kg in integer SI units making the proton mass 2×10^{-27} kg. In this case each set increases in three orthogonal orientations (and also in three planes) proton velocity to 1/2 c_s. Two such values at right angle increase the value by $\sqrt{2}$ and with relativity and quantum correction generate light velocity to c_s value with very acceptable value 1.5 k_s after application of 1.04 quantum correction:

Pulse from one H atom having proton mass comprising two muo10$_s$ masses, each of 10 masses of 10^{-28} kg:

$$= \sqrt{2} \times 1.1547 \times 1/2 \, c_s \times 1.111\ldots^2/1.04^{1/5} = 1.0001438 \, c \qquad (9.14a)$$

Conversions of 10 eVat signal into 1 eVat and 1 eVat into a 10 eVat energy given below help to comprehend the process.

$$(10 \text{ eVat})^{2/3} \equiv 4.64155888 \qquad\qquad (9.15)$$
$$4.64155888^{1/3}=1.6681 \text{ eV}; \qquad 1.6681 \text{ eV}/(1.2 \times 1.111\ldots^{4/3} \times 1.04^{1/3})=1.0002 \text{ eVat}$$

The variance from 1 eVat expresses $2k_s$ accuracy, while $4.6415 \times 1.1111^{1/2} \times 1.04^{1/3} \approx 5$ eVat, and $5/\sqrt{3} \times 1.04 = 3.0022$ eVat.
In the inverse direction
$$(1.66666\ldots \text{ eV})^3 = 4.63923 \text{ eVat}; \qquad (4.63923 \text{ eVat})^{3/2} =9.961377\text{eV};$$
$$9.961377\text{eV} \times 1.04^{1/10} = 10.0005 \text{ eVat} \qquad\qquad (9.16)$$
The variance from 10 eVat expressing $1/2k_s$ accuracy at Boltzmann level, while

$$4.6391 \times (1.1111^{1/2} \times 1.04^{1/3}) \approx 5 \text{ eVat} \qquad\qquad (9.16a)$$

The sun mass displays two dimensions in gravitation relation to earth; sun mass radiating with temperature 6000 K and electron frequency from two planes displays one step of space quantization:

$$(2 \times 10^{30} \text{ kg})^{1/2} \equiv 6000^4 \times 1.111\ldots/1.04^{9/20} \text{ K}^4 \equiv \sqrt{2/5} \times 5 \times 10^{15} \qquad (9.17)$$

$$c_s \equiv 300^3 \times 3^{1/9}/1.0328^{1/2} \equiv F_{Gs}^{1/3}/1.111\ldots \times 1.04^{1/3} = 30017127 \text{ m/s}. \qquad (9.18)$$

Permittivity of vacuum in 1D, $\varepsilon_{os}^{1/3}$, converted to SI by multiplying by e_s and corrected for relativity and quantum generates in two orientations Boltzmann constant, k_s, which is essentially H. u. at 1 K: $\qquad\qquad (9.19)$

$$k_s = \frac{1}{2} \varepsilon_{os}^{1/3} e_s \times \text{corrections} = \frac{1}{2} \times (10^{-11})^{1/3} \times 10^{-19}/(1.1547^{1/2} \times 1.04^{1/16}) = 1.00001 \times 10^{-23} \text{ J}.$$

Boltzmann constant also derived from electron mass in rotational (2D) display per meter of local light velocity:
$$k_s = (10^{-30})^{1/2}/(1.0 \times 10^8) = 1.0000 \times 10^{-23} \text{ J} \qquad\qquad (9.20)$$

Radiation mass 10^{-27} kg in 2D rotation displays energy per plane corrected for relativity:

$$(1 \times 10^{-27})^{1/2}/(3 \times 1.111\ldots^{1/2}) = 1.000\,000\,000\,0 \times 10^{-14} \text{ kg} \qquad (9.21)$$

Squared with inverted linear probability derives back $(1.0 \times 10^{-14} \text{ kg})^2 \times 1/10 = 1 \times 10^{-27}$ kg. **Thus 1.0×10^{-28} energy density J/m^3 is not energy of space, but energy density of a progressing pulse expressing mass and force of muo10$_s$ force in integer cubic SI units.**

Over volume 10^{-30} m^3 and time $1/10$

$$1.0 \times 10^{-28}/10^{-30} \times 1/10 = 10 \text{ J or eVat} \qquad (9.22)$$

Energy density 1.0×10^{-28} J/m^3 per unit Bohr atom volume of 5 local pathways, three planes and 6 directions expresses sun radiation temperature

$$6 \times 3 \times 5 \times 1.0 \times 10^{-28}/10^{-30} \equiv 6000 \text{ K} \qquad (9.23)$$

Minimum pulse mass time local light velocity derives μ_{os}:

$$1 \times 10^{-14} \text{ kg} \times 1.0 \times 10^{8} \text{ m/s} \equiv \mu_{os} \qquad (9.24)$$

And the product of 2 eV x 6000 K decreased by 1.2 factor and converted to classical ST is

The 10 times the minimum SI energy = 2 eV x 6000/1.2 x1 $10^{-19} \equiv 1 \times 10^{-15}$ J, the energy expressed in Equations 11.10-11.12.

During above analyses the velocity of light to power 3/2 has been derived as the infrared frequency:

$$(3 \times 10^{8})^{3/2}/1.04^{29/30} \equiv 5.0028 \times 10^{12} \text{ Hz} \qquad (9.25)$$

Dividing k_s by electron charge converts the integer SI value to per electron (or per event value since e_s may be the equivalent electron) $10^{-23}/1 \times 10^{-19} = 0.00001$ eV, which is the product (or 2D presentation) of two pulses each of ten quanta in two planes perpendicular to each other the resultant pulse in one direction occurring with probability of 10^{-20} and 10 being the converter.

Energy in two degrees of freedom is $k_s T$. Boltzmann constant k_s in one orientation and in 1 D is expressed by $1/3 \times (1 \times 10^{-23})^{1/2}/1.111\ldots^{1/2} = 1.000\ldots \times 10^{-12}$ J. Multiplied by 1.00×10^{12} Hz derives 1 J. It is to be noted that 1.00×10^{12} Hz frequency has not been considered to comprise temperature component and the result obtained refers to 300 K. Dividing by 300 K/1.11 ... delivers $10c_s$ value and dividing by 3000 K/1.11 ... delivers = c_s value.

In c.g.s atomic units current is 10 times larger in abampers. This situation arises at times in relations proposed needing a factor of 10 or 1/10. For example Boltzmann constant k_s converted to atomic units k_s/e_s x10 =0.001 eV = $\mu_{os}^{1/2}$. Note $\mu_{os}^{1/3}$= 0.01 J =$(\varepsilon_{os}/10)^{1/6}$.

In c.g.s atomic units current is 10 times larger in abampers. This situation arises at times in relations proposed needing a factor of 10 or 1/10. For example Boltzmann constant k_s converted to atomic units k_s/e_s x10 =0.001 eV = μ_{os}^{2}.

Multiple information in sun gravitation expression

Sun gravitation expressing both the integer SI model and SI reference value is formed by two pulses vectorized 1000 K = (1/6 x 6000 K) thermal pulses to power four each formed by vectorial additions of vectors in three orthogonal orientation expressed by $\sqrt{3}$, the SI value relativity and two quantum corrections:

Sun force: $[\sqrt{3}(1000)^4 \times 1/10] \, X[\sqrt{3}(1000)^4 \times 1/10] = 3 \times 10^{22}$ N $\rightarrow$ x1.15470 x1.04$^{3/5}$/1.04$^{1/6}$
$$= 3.5444 \times 10^{22} \text{ N} \qquad (9.26)$$

Inserting 1/10 into the brackets converts it to $1/10^{1/4} = 0.562341 = (1/1.2)^3/1.04^{3/4}$. Pulses of 10 J decremented by factor 1.2 and relativity and quantum corrections display energy 7.9 J or eV. The value 7.9 has been experimentally derived by the author and as the average [12] of 40 researchers and in classical physical textbooks as the activity energy of the typical Bohr atom structure of the single valence NaCl crystal defect:

$$10 \text{ J} \equiv 7.9 \times 1.15470 \times 1.111 \ldots /1.04^{1/2} = 10.004 = 10 \text{ eV} + 4 \, k_s \qquad (9.27)$$

The pair of terms $(1000 \text{ K})^4 \equiv (10^3)^4 \equiv 10^{12}$ Hz generate 1 J/wave in wave function format. For three planes time 10^{10} power produced of Special Relativity is 3×10^{22} W of sun gravitation 3×10^{22} N. Sun gravitation force can be expressed in terms of squared invert permittivity of space ε_{os}: 3×10^{22} N $\equiv 1/(\varepsilon_{os}/\sqrt{3})^2 = 1/(10^{-11}/\sqrt{3})^2$ m^6/J^2. Dividing by velocity of light 3×10^{22} W/3 $\times 10^8$ m/s $= 1 \times 10^{14}$ J, which time the energy 1×10^{-14} J (of Equation 2.54-5, which is also the square of the atomic force 10^{-7} J) derives 1 J per pulse. Dividing now 3×10^{22} J by square velocity of light derives radiation mass 3×10^{22} J/(9 $\times 10^{16}$ m^2/s^2)=3.3333 $\times 10^6$ kg, which is 1/15 of the 5×10^8 values in Eq.3.17-19, expressing mass in one of five quantum orientations of 3 planes.

10. *Integer values of physical constants, oscillators of Bohr atom and radiations from space and from planets*

The model uses extensively the old rule of energy equalization. The equalization is now attributed to occur between all degrees of freedom requiring a stable Bohr atom to share its energy equally between electron, proton and (electron+proton) pair. The importance of energy equalization was realized from the experiments in NaCl crystals [3, 4] in which double valence impurity created energy of Bohr atom structure from charged crystal vacancies. Velocity of light in Equation 3.12 is derived from the frequency frequency 5×10^{15} Hz considered to be electron in origin. The model postulates that this value should be replaced by $\sim 10^3 \times 5 \times 10^{12}$ Hz frequency expressing the developed forces attributed to vectorized infrared oscillations of (electron+proton) pair.

The proposed cubic constants often generate integer values of radiation parameters and derive a large number of numerically correct data. The cubic geometry generates equal probability of independent radiations in its six orthogonal orientations. Relaxation and high field conduction studies were first to show that the energy of single hydrogen structure has three equal energies attributed to proton, electron and (proton+electron) pair. At its relatively low temperature of 300 K earth generates, difficult to differentiate from a normal distribution, two base infrared frequencies varying in value by only 0.1%.

Correct value of the repulsive force was calculated by assuming that each kilogram of earth mass generates the model force of 10 newton subjected to three steps of quantization of 1/10 and division by $\sqrt{3}$. The application of a quantum correction derives the force value equal the reference: (10.1)

Force of earth repulsion $= 6 \times 10^{24} [\text{kg}] \times 10 [\text{N/kg}] \times 1/10 \times 1/10 \times 1/10/\sqrt{3} \times 1.04^{29/50} = 3.5438 \times 10^{22}$ N.

Using the classical earth mass derives a less accurate force:

$$5.9742 \times 10^{24} [\text{kg}] \times 10 [\text{N/k}] 1/10^3 \times 1/\sqrt{3} 1.04^{7/10} = 3.5452 \times 10^{22} \text{ N} \qquad (10.2)$$

Energy of repulsion attributed to relativity correction 1.111 . . . applied to the integer cubic gravitation $3.6000 \times 10^{22} \times 1.111 \ldots = 4.0000 \times 10^{22}$ N. Repulsion is also expressed by 2/3 N mutual interaction by 1N/kg oscillations of two oscillators reduced by $1/10^2$ on conversion to vectors:

$$= 6 \times 10^{24} [\text{kg}] \times 1 [\text{N/kg}] \times 2/3 \times 1/10 \times 1/10 = 4.0000 \times 10^{22} \text{ N} \qquad (10.3)$$

And since vectorized 10 [N/kg] $\times$ 2/3 $\times 1/10 \times 1/10 \times 1/10 /1.11111 \ldots = 0.006$ N [N/kg] exact cubic integer force of earth repulsion $= 6 \times 10^{24}$ [kg] $\times 0.006$ [N/kg] $= 3.6000 \times 10^{22}$ N.

Using integer for the earth mass derives integer force. Several calculations producing now the correct numerical results lead to the deduction that both sun and earth act as ideal black surfaces radiating power given by the product of Stephan-Boltzmann constant σ_s and kelvin temperature to power 4.

The sun quantum pulse is 6000/300 = 20 times larger than the quantum 0.0174 eV measured in authors experiments making it 0.348 eV. Converting for two degrees of freedom using relativity corrections derives 0.348 x2x1.15470$^{3/2}$ = 0.997 eV ≈ 1 eV. In the region close to earth, discussed above, the power density of the pulses from the sun and of the pulses generated by the earth become equal in terms of (quantum intensity x rate of appearance) producing maximum interactions. The pulses of sun attraction are cancelled by the pulses of earth repulsion, or earth attraction with reference changed.

The assumption that Bohr atom has under dynamic equilibrium three equal energies requires justification. This is obtained by additional interpretation of Frank-Hertz experiment.

__Additional interpretation of Franck-Hertz experiment.__ **The present proposition is that total quantum energy represents the number of all independent oscillators plus one is illustrated in Franck-Hertz experiment, in which energy of electrons decreases in steps of 4.88 eV on exciting mercury vapor. Atomic number of Hg is 80 with 14 groups of electrons and one nucleus making a total of 95. The vapor is diatomic increasing by two the total energy: 2 x 95 x kT = 4.807 eV. An extra kT ≈ 0.025 eV is required to free the atoms and $2kT$ to balance the created momenta, making a total of 4.882 eV; this value compares well with the original 4.88 eV.**

The last stage of sun life involves mass to energy conversion requiring the formation of iron. The number of oscillators in atomic iron, has been found **NOW** to be the sum of the atomic number of iron, 26, plus one for the nucleus and plus 26 electron groups forming individual oscillators: 1s -2, 2s - 2, 2p - 6, 3s - 2, 3p - 6, 3d - 6 and 4s - 2. Thus the total number of oscillators is 26 + 1 + 26 = 53. Squared value of 53, in a manner similar to Moses' rule, expresses 2809 degrees of freedom. The total energy, with each degree requiring 1/2 kT, amounts to 1405.5 kT ≡ 1405.5 eV ≡ 1405.5 W/m^2. The latter value, reduced by 53 uncertainties of 1/2 kT x $ln2$, derives within 0.5% the accuracy of the `sun constant:

Sun constant P_{sun} =1405.5/1.04$^{3/5}$=5 x 300/(1.1547$^{1/2}$x1.04$^{5/11}$)=1372.8 W/m^2 (10.4)

It could be argued that 1405.5 W/m^2 is the sum of earth radiation 701.78 W/m^2 and of the sun radiation 701.78 W/m^2 of Equation 5.15. Both bodies pull in the opposite directions with vectorized power density of:

$$1405.5 \text{ W/m}^2/\sqrt{2} \; x \; 1/10^3 \; x1.04^{1/6} = 1.00036 \text{ J/m}^2 \qquad (10.5)$$

The sun pulse, the reference of 1 J, is derived below using the classical SI units:

$$2xkxT_{sunmean}/e=2x1.38065 \; x10^{-23}x5800/1.602176 \; x10^{-19}=0.9996 \text{ J}≈1.000 \text{ J} \qquad (10.6)$$

The expression deriving an identical value using integer cubic space constants requires the factor 1.2 neglecting flux radiated at 90° to the propagation. orientation, 20% of the total value:

$$2xk_sxT_{ssunmean}/e_s \; x \text{ cor.} = 2x1x10^{-23}x \; 6000/1x10^{-19} \; x \; 1/1.2 = 1.000 \text{ J} \qquad (10.7)$$

It should be noted that $2k_sT$ is the vector sum of three k_sT vectors in orthogonal orientation times the relativity correction $2xk_sT = \sqrt{3}\ xk_sTx1.154700538 = 1.999999999kT$.

At 300 K the energy of pulses from Bohr atom radiations expressed both in integer model and in the classical SI units take the same 20 times smaller values 0.05 J deriving 0.01 J per orientation, which is the 1st Planck Radiation term.

Unity values at the atomic level such as above or in Equations 8.42, 8.44, 14.7, 15.1 and 15.15 are not approximations. Unity values are exact statistical average values of measured pulse of t duration, which within the interval -½ t to +½ t is displaying quantum variations in amplitude with normal or other distribution produced by statistic of non-events and quantum events.

Equipartition of energy occurs not only in objects of atomic size, but also in large areas of space, because time to reach equilibrium may take many millions of years. Consequently microwave frequency measurements located thousands of miles away on earth are in phase and display quantum properties. It is now proposed that a group of atomic moments sustains for a period of time the same phase positions in widely different environments. These environments are: in a small crystal, in space extending over light-years and in the body of the sun. Careful measurements display quantum energy momenta having steps unaffected by neglecting natural time distributions in the model. The radiations from the sun are the result of space quantization, size conversion, unit conversion and the sequential probability formation of radiations in two orientations subject to the event probability of their joined appearance. Omitting the probability requirement reduces on earth the vector power of sun to 1 W/m² per oscillator, the quantum reality event, a Bohr atom event.

Magnetic domains generate also similar display. The same phase position of atomic force vectors was believed to be present in 0.1 mm thick layers of NaCl crystal in author's experiments [3] producing quantum display. Quantum display was also shown to be derivative of potential barriers formed by common flow rates.

Two dimensional (2D) comparable oscillations of surface water volumes in a water pool have been observed on midday in June by the author. They were formed by reflections from pool sides, being created rapidly by changes of sequences of maximized sun-light intensity shapes covering large areas. They were areas of about 1/2 foot squares, parallelograms, octagonal, pentagonal and near-circles displayed in 2D fitting various turbulences. In three dimensions the similar shapes, differing obviously by not more than Boltzmann constant energy limits, are the sphere and the cube of the atomic space energy model. Such display of several identical energy configurations contributes to simultaneous exhibition of energy equipartition and to energy minimization rules. Daniel SHECHTMAN was awarded Nobel Price discovery of quasi-crystals having pentagonal symmetry.

The proposed frequency of 5×10^{12} Hz (also a derivative of the Nobelist WEITZ, M. famous measurement [10]) represents the radiation of space measured at wavelength of 0.21 m. Frequency of 5×10^{12} Hz allows for a different interpretation of that space wavelength from that expressed on WEB pages. Wavelength of 0.21 m represents no doubt the space temperature of 2.725 K. Converting from 300 K to 2.725 K and multiplying by $\sqrt{3}$ and 2 since H2 is twice as heavy as Bohr atom derives:

Infrared space frequency = 5 x 10^{12} Hz x $(2.725/300)^2$ x2 x$\sqrt{3}$= 1.429 x 10^9 Hz   (10.8)

**Above space frequency fits the well known and
measured space wavelength of $3 \times 10^8 / 1.429 \times 10^9 = 0.2099$ m** $\qquad$ (10.9)

Only one of five quantization components has the correct radiation orientation. Both electron and proton frequencies display 1/5 components along the quantization orientations. The sun pulse frequency $1/5 \times 5 \times 10^{12}$ Hz, squared to express magnitude, is decreased in probability by two step quantization of 1/10 and increased by $\sqrt{2}$ by the presence of two components in progression along one plane. It equals the value of one orthogonal direction of the force of reflection generated by the earth in three directions. Please note the exactness of the force of earth gravitation with the reference in Equation 6.1:

$$F_{Gs} \equiv \sqrt{3} \times (\sqrt{2} \times 1/5 \times 5 \times 10^{12})^2 \equiv 3.5438 \times 10^{22} \text{ N} \qquad (10.10)$$

Due to equipartition of energy the proton frequency signal from the sun generates also gravitation force value after one step quantization, quantum correction and 3D display having index 3/2. The following identity indicates that 5×10^{15} Hz is a 2D display:

$$F_{Gs} \equiv (5 \times 10^{15})^{3/2} \times 1/10 \times 1.04^{1/17} \equiv 3.5436 \times 10^{22} \text{ N} \qquad (10.11)$$

Electron frequency expressed in integer space value corresponding to SI infrared 1.429×10^9 Hz is 1000 time higher, amounting with the relativity and quantum corrections to:

Integer electron frequency/K $= 1.429 \times 10^9 \times 1000 \times 1.1111^{1/2} / 1.04^{1/10} \approx 1.500 \times 10^{12}$ Hz $\quad$ (10.12)

For hydrogen molecule with two atoms the integer electron frequency/K is 0.75×10^{12} Hz. For radiation group of four electrons the integer electron frequency/K is 0.375×10^{12} Hz.

First tests of the APEX Alacama Telescope mounted at 5100 m in Chile with range 1.5 mm to 0.2 mm wavelength gave publicly quoted values 1.300×10^{12} Hz for the SI frequency converting now for model velocity $\frac{1}{2} c_s$ to the integer electron frequency/K of

$$1.300 \times 10^{12} \text{ Hz} \times 1.1547 = 1.500 \times 10^{12} \text{ Hz} \qquad (10.13)$$

For Jupiter the frequency value was eight hundred and twelve THz and for Venus four hundred and sixty two THz. Converting these frequencies to integer electron frequency/K using relativity and quantum corrections to obtain sub-harmonic values the following are derived

For Jupiter $0.812 \times 10^{12} / (1.1547^{1/2} \times 1.04^{1/5}) = 0.74975 \times 10^{12}$ Hz
For Venus $462 \times 10^{12} / 1.1111^2 = 0.3742 \times 10^{12}$ Hz $\qquad$ (10.14)

These values are taken to indicate that radiation is coming from molecules having two electrons and four electrons oscillating as a group.

For the first excitation step of Bohr atom WEB pages quote frequency 2.46×10^{12} Hz. In the model this converts to 2.46×10^{12} Hz$/(300K \times 1.1111) \times 1.04^{5/12} = 0.748 \times 10^{12}$ Hz in good agreement with previously quoted experimentally values.

At 3000 K per one side of the sun the Bohr integer space electron frequency becomes $v_{es} = 1.500 \times 10^{12}$ x (3000 K) x 1.111 . . . = 5×10^{15} Hz and the energy of electron oscillation **in integer space units converted to SI units:**

Bohr atom dissociation energy
$= \frac{1}{2} m_{es} v_{es}^2$ x cor.$= \frac{1}{2} \times 1 \times 10^{-30}$ x$(5 \times 10^{15})^2$x$1.1547^{1/2}$x$1.04^{1/3}=13.61$ J $\equiv 13.61$ eV (10.15)

is the value, which was earlier expressed as 10 eV converting to 30 eV for three planes.

The electron frequency of 1.5×10^{12} Hz/K can be easily mistaken for an infrared value. Proton frequency at 300 K of 10 orientations per direction is $10/6 \times k_s T_{300}/h_s = 5 \times 10^{12}$ Hz. Energy due to electron oscillations expressed per K, $\frac{1}{2} m_{es} v_{es}^2 = \frac{1}{2} \times 1 \times 10^{-30}$x $(1.5 \times 10^{12})^2 = 1.125 \times 10^{-6}$ J, converting now to 300 K requires multiplying by 300^2 while applying 1 step of space quantization, 1/10, derives energy 0.01 eV. With $(e^{hv/kT} - 1)$ being unity, the result obtained indicates that Planck Radiation energy refers equally to electron radiation at 300 K. Atomic radiation energies are reduced mass equivalent frequencies, while one of gravitation forces is the vectorized thermal radiation.

Radiation expressing apparent mass converted to energy

The energy in three planes at 50 eV/K referred to 6000K corrected by relativity factor of 1.111 . . . derives:

Sun pulse energy $= 1/\mu_{os} = 10 \, \mu_{os}/\varepsilon_{os} \equiv 3$ x 50 eVx 6000 x 1.111 . . . = 1 MeV
(10.16)

Above is related to 1 J or eV of Equations 10.6 and 10.7 because 1000 K per orientation generate 1 J/K and two pulses in two orientations produce:
$$\sqrt{3}(1000 \text{ eV}) \times 1.1547 = 2 \text{MeV}.$$
In terms of 3000 K sun pulse is

$$(3000 \text{ K} \times 1.1547^2/\sqrt{2}) \times (3000 \text{ K} \times 1.1547^2/\sqrt{2}) = 1 \text{ MeV} \quad (10.17)$$

The same value of pulse energy is expressed by

$$\text{Energy of sun pulse} = \varepsilon_{os} (\sqrt{3} \times 1/2 \, c_s \times 1.1547)^2 \times 1.111 \ldots = 1 \text{ MeV} \quad (10.18)$$

Replacing $(\sqrt{3} \times 1/2 \, c_s \times 1.1547) = c_s$ by c_s is not allowed because ε_{os} is relativity sensitiove becoming infinite at c_s. In the model proposed particles move in 10 quantum pathways with velocity $1/3 \, c_s$ generating the Poynting vector of Equation 2.14a by:

$$\text{Sum of energy in 10 quantum pathways} = 10 \, \varepsilon_{os} \times (1/3 \, c_s)^2 = 1 \text{ MeV} \quad (10.19)$$

With the term c_s^2 the relation appears to look like mass (energy) to energy conversion because of the presence of the squared velocity term. It is proposed that it displays presence of **two pulses** in wave function form, each being a square root of the above: (10.20)

ε_{os} x 1/3 $(\sqrt{10}\sqrt{c_s})$X$(\sqrt{10}\sqrt{\varepsilon_{os}}$ x 1/3 $c_s)\equiv1000^2$ eV$\equiv$(300 K)X(300 K) x1.111 . . . x 10 $\equiv$ 1 MeV.

Or $\qquad\qquad$ $(3\mu_{os}$ x 1/3 $c_s)$X$(3\mu_{os}$ x 1/3 $c_s)$ x 1.1111 . . . x 10 = 1 MeV $\qquad\qquad$ (10.21)

Please note that $\sqrt{10}$ x $\sqrt{\varepsilon_{os}}$ = 1 x 10^{-5}.

Discharge progress, which is the velocity of light, is the consequence of two events of radiation occurring at the right angle to each other generating the resultant at 45°, like in a directional antenna made of two loops. Their resultant, square area of flux, equals the rate of change of flux, displaced forward by ½ [m] in integer model SI units and ½ x 10^{-10} [m] in atomic terms; its delay of formation derives the velocity of light c_s. The delay is the time interval needed for the reflected to join the forward pulse. The resultant pulse is at 90° generating rotation. The time constant τ of the wave packet has been derived in Equation 15.25 as 1.666 . . . x 10^{-8} s. Hence since τ x c_s = distance:

Velocity of light c_s = 5 m/1.666 . . . 10^{-8} s = 3.000 . . . 10^8 m/s $\qquad\qquad$ (10.22)

The linear probability has been derived as

$$1 \times 10^{-10} = 10\,\varepsilon_{os} = 82\,\mu^2_{os}/1.012345679 \qquad\qquad (10.23)$$

Referring the pulse to 1000 K requires multiplication by the square of temperature that is 1000^2 converting 1 eV to 1 MeV, which can be alternatively expressed also by

$$3k_sT_{6000} \text{ x 5 x 1.111 . . . x } (1/3\ c_s)^3 \equiv 1.0 \text{ MeV} \equiv 1.0 \text{ MW per oscillator} \qquad (10.24)$$

The square of two pulses in wave function format $(3\text{x}10^{-14}\text{x}1.1111$ . . . x 3 x $10^8)^2$ x 3 x 10^8 = 1 MW generate a third pulse moving at c_s with double the energy making 1 MW values 2 MW.

The physics would be more simply expressed, if the magnetic permeability were expressed by its invert: $\qquad$ **$1/\mu_{os} \equiv 1.0$ MeV per oscillator** $\qquad\qquad$ (10.25)

Energy of space ε_{os} created by mass conversion of protons or Bohr atom conversion into electrons

The conversion of electrons into proton was shown on Fig. 6 in Section 2. Equations suggest that the process is reversible. Two thousands and one electrons forming the proton represent in classical SI units 2001 electron mass = 2001 x 9.109 x10^{-31} kg =1.8277 x10^{-27} kg. The classical mass of proton is 1.6727x10^{-27} kg. Hence the conversion of proton to electron requires an additional mass of 1.55x10^{-28} kg, which times squared velocity of light c_s^2 = 9 x10^{16} derives energy 1.395x10^{-11} J. That value divided by relativity 1.15470^2 and quantum corrections $1.04^{1.154}$ obtains the net energy needed for conversion:

Energy needed=1.395x10^{-11}J/(1.1547^2x$1.04^{1.15/4}$)=1.000108x10^{-11} J is contained in a unit of space

$$\equiv \varepsilon_{os} = \tfrac{1}{2} \, m_{ps} \times (1/3 \times c_s)^2 = 1 \times 10^{-27} \times (1 \times 10^8)^2. \; \text{J/m}^3 \qquad (10.26)$$

Thus conversion of proton or of a hydrogen atom to electrons consumes the mass of proton and converts the mass of proton into the energy, the two generating the sum of masses of the electrons. The converted mass of proton into energy is ε_{os}, the energy density of the radiation pulse. Energy released from the collapsed space $\varepsilon_{os}/(2\pi \times 10^{15})$ per cycle of electron frequency times correction $1.1111 \times 1.04^{4/5}$, is sufficient:

$$1.0001 \times 10^{-11}/(2\pi 10^{15}) \times 1.1111 \times 1.04^{4/5} \; \text{J} = 1.825 \times 10^{-27} \; \text{kg} \qquad (10.27)$$

to create 2001 masses of electrons.

Two thousands and one electrons in the integer cubic SI units each of mass 1×10^{-30} kg produce a total mass of 2.001×10^{-27} kg. This is one half of hydrogen molecule H2 being split in two on radiation. With proton moving with velocity $\tfrac{1}{4} \, c_s$ and electron circulating around proton with velocity $v_e = 1 \times 10^7 = 1/30 \times c_s = 1 \times 10^7$ m/s the energy of proton converted by square of its velocity is

$$2 \times 10^{-27} \times (3/4 \times 10^8)^2 \, /(1.1111 \times 1.04^{1/3}) = 0.999 \times 10^{-11} \; \text{J} \equiv \varepsilon_{os}. \qquad (10.28)$$

On the other hand the expression $1 \times 10^{-30} \, (3/2 \times 10^8)^2 \approx 2 \times 10^{-14}$ J derives a total energy of 1.0×10^{-11} J $= \varepsilon_{os}$, with the same space energy density as before $\qquad (10.29)$

Radiation mass 1×10^{-27} kg displayed in 1D, $(1 \times 10^{-27} \text{ kg})^{1/3}$, expresses orientation probability 10^{-9}. Two such 1D probabilities are needed to produce the power, energy or momentum output:

$$10^{-9} \times 10^{-9}/1 \times 10^{-19} \equiv 10 \; \text{eV per plane.} \qquad (10.30)$$

Above are important findings. Energy of nulled moments of electrons may form protons and thus mass of planets and of stars may be generated in this process. And again in the opposite process, an instability of Bohr atom energy or momentum may lead to electron and thermal radiations.

Possible structures of hydrogen and its conversion to radiation

Sun temperature of 6000 K in one of four orientations, per one plane of rotation and in one of ten quantum pathways derives difficult to justify value of 50 J or eV

$$6000 \; \text{K} \times \tfrac{1}{4} \times 1/3 \times 1/10 \equiv 50 \; \text{eV} \qquad (10.31)$$

Energy of 50 eV is also expressed in Equations 8.6a and 8.8. The structure of the radiation pulse having five 10 eV carriers justifies now 50 eV value. The sun pulse arriving on earth at 300 K is

$$50 \times 300 \times 1/10 = 1500 \; \text{W, contributed by a quarter of 6000 K,} \qquad (10.32)$$

and $\qquad$ 50 x1.15470^{2}1/111111 = 60.00 eV = 2 x 30 eV

1500 W deriving the sun constant: $1500/1.11111$x$1.04^{5/12}$ = 1372.2 W/m^2 $\qquad$ (10.33)

Thus hydrogen molecule has energy 60 eV or 50 eV and six or five main energy and momentum carriers carrying each 10 eV. Two moments carried by oscillations of m_p are anti-phase so that net energy carried is close to 10 eV. Two masses of protons moving at light velocity carry energy of linear momentum reduced by the factor 1.2:

$$2 \text{ x } 2 \text{ x } 10^{-27} \text{ kg x } 3 \text{ x } 10^8/(1 \text{ x } 10^{-19} \text{ x } 1.2) = 10 \text{ eV} \qquad (10.34)$$

The energy of the infrared oscillator of proton mass m_{ps} has the same value: ε_{os} x 1 x 10^{12} = 10 eV. The fourth 10 eV energy component is the infrared oscillator of the pair (proton+electron) mass marginally different from 10 eV, but only electron energy can to considered. The fifth and the sixth energy carriers are the angular magnetic momenta of two electrons moving with velocity v_e=10^7 m/s displaying each:

$$\boldsymbol{\mu_{os}v_e\text{=10}^{-6}\text{x10}^7 \equiv \text{10 eV}} \qquad (10.35)$$

There might be also an energy carrier due to electron mass oscillating in local pathways with frequency 10^{15} Hz generating

Electron H.F. energy = m_{es} x $(1/5v_{Hms})^2$ = 1 x10^{-30} x (1 x 10^8)^2 = 1 eV or J $\qquad$ (10.36)
The above energy in integer SI units corrected for relativity:

m_{es} xc_s^2x1.1111 = 1x10^{-30} x 9x1x10^{16} x1.1111 =1x10^{-23} J $\qquad$ (10.37)
which is Boltzmann const. = minimum input energy pulse x 10^{-10}=10^{-13}x10^{-10}= 1x10^{-23} J.

In the integer cubic model of radiation an atom or molecule is represented by a spherical structure on earth or a cubic structure on sun, comprises radially positioned chains of interlocked oscillations with four sequential elements between the atom center and the outside surface, where reflections take place. The chain interlinked loops rotate in four planes in quantum displacement steps of 90°. The centre flux of the first loop can be represented by a tail of an arrow *, centre flux of the 2nd loop by downwards arrow ↓, the 3rd loop by the tip of the arrow ·, the 4th loop by the upwards arrow ↑. Thus one cycle is *, ↓, ·, ↑. Arrows of flux can be more comprehensively represented by the spin of the flux inducing them and generating the cycle □, ↓, □, ↑. These elements form the 32 quantum orientations of four units in the hydrogen atom and 128 energy components. This number is 1000 K/(6 x 1.1111^2) ≡135 J less 7 J uncertainty. In earth gravitation all 32 orientations are interacting, but in linear relations discharges at right angle to the receiver are not received so only 28 are interacting. Various probabilities were applied to the mass of sun, of proton and of electron in Chapter 22. It was found that most correct energies were displayed by assuming that mass reacted with one side the signal generated with probability 1/14 as evaluated in Equations 22.123a.

Discovery of the relation 'mass = f (current)' explained in Chapter 22 suggests H$_2$ radiation expresses linear cycle unwinding the rotation momenta of particles contained in the hydrogen molecules. The cycle is formed by continuous changes of moments $M = -dM/dt$ according to the old hand's rules with vectors changing direction in each step inducing quantum step rotation. Three such pulses in orthogonal directions generate 1D pulse of radiation incorporating all 64 (32 **x2**)components:

$$(64/[1.111 \ldots^{1/2} \times (1+1/80)] = (60 - 3/2 \, k_{s300}T_{300}) \text{ eV}; \lambda=60 \times 0.5 \times 10^{-10} \text{m}= 3 \text{ nm} \qquad (10.37a)$$

deriving:

three Planck radiation terms 3 nm $\times$ 1 $\times$ 10^{15} Hz $/1 \times 10^{10}$=0.03 J $\rightarrow$ x 300 K = 10 J (10.37b)

Single structures have no magnetic or electric features. For the spherical hydrogen on earth individual elements are interlinked squared or rounded loops. Three, four element components, occurring in orthogonal orientation increases velocity by two = $\sqrt{3}$ x 1.1547005 = 2 and sums energy of sources. The cycle may be formed by four sequentially interlinked elements changing planes and direction of oscillations, so that one energy element oscillates sequentially in eight positions.

In the cubic structure the loop is replaced by the square of four independent components in orthogonal orientations three of which are shown with directions forming a square; the next quantum is in a plane perpendicular to this surface of the paper and the arrow indicates the direction of the flux created by the square; the directions change in third square and the flux orientation in the fourth square: $\uparrow^{\rightarrow}\downarrow, \downarrow, \downarrow\rightarrow\uparrow, \uparrow$.

The whole mass of the atom converts energy to two directions the signal in one direction is the graviton energy:

Converted mass = 1/2 x (2 x 10^{-27}) x(3 x 10^8)^2 x 1.1111= 1 x 10^{-10} J; This is gravitation constant including electron oscillation G_s = 1 J converted to SI by multiplying by 10^{10} = 3 J for three oscillators. On division by 300 K this value becomes 1st Planck Radiation constant = 0.01 J and gravity energy 0.01666 J after corrections

$$0.01666 = 0.01 \times 1.2^2 \times 1.1547 \text{ J} \qquad (10.38)$$

Alternatively (1 x 10^{-27}) X (3 x 10^8)^3 x 1.1111=0.03 J = Planck Radiation for 3 oscillators. **Power c_s^3 indicates wave function producing velocity of light and energy through mass to energy conversion that energy being injected at velocity of light obtaining 0.03 W. Corrected 2D proton mass displayed in 1D becomes minimum mass and μ_{os} proton momentum of such mass:** **(10.39)**

(1x10^{-27}) $^{1/2}$/1.1547$^{1/2}$x1.04$^{1/2}$≡3.00111x10^{-14}J and 3.0x10^{-14}Jx3 x10^8 x 1.111=10x10^{-6}≡10μ_{os}.

For both directions energy per each of 64 above mentioned quanta expresses also power:

2 x10^{-10} W/64 x 1/1.04$^{1.04}$=3.00009736 x 10^{-12} J, which at 1x10^{12} Hz delivers 3 W (10.40)

Numerically the product 3.0 x 10^{-12} J x 1 x 10^{-19} C $\equiv$ 3 x 10^{-31} $\equiv$ 300 h_s represents 300 K.

One direction in a cubic structure involves four corners each discharging radiation energy $G_s/4 = 2.5 \times 10^{-11}$ J from three adjoining planes. That energy is dielectric permittivity ε_{os} corrected for relativity:

$$1.1547005^{7/2} \times 1.1111^3 \times 1.0238^3/[1+(1/2.5 \times 1/800) \times 10^{-11} \text{ J}] = 2.500012\ \varepsilon_{os} \text{ J} \qquad (10.41)$$

There are 32 loops of quantum energy in the spherical structure. Since the radiation energy after mass conversion is 2×10^{-10} J the energy per square shaped loop is:

$$2 \times 10^{-10} \text{ J} /32 \times (1/1.04^{1.04}) = 3.00009736 \times 10^{-12} \text{ J} \qquad (10.42)$$

which generates power of 3 W at frequency 1×10^{12} Hz. Thus all 1, 2 and 3 values are power in watts although they might have been called joules in the text.

The frequency doublet in the hydrogen oscillation indicates the presence of two energy levels demonstrated by the Nobelist W. Lamb and confirmed by others at $\sim 4.5 \times 10^{-5}$ eV. Presently correction of $1.2 \times 1.111\ldots^3$ has to be used to derive this value from the increment of the oscillating proton mass at 5×10^{12} Hz by one electron mass:

$$[(m_p v_{Lms}^2) - (m_{ps} + m_{es})v_{Lms}^2] \times \text{corrections} = \qquad (10.43)$$
$$= \{[(2.000 \times 10^{-27} \times (5 \times 10^{12})^2] - [2.001 \times 10^{-27} \times (4.995 \times 10^{12})^2]\}/(1.2 \times 1.1111^3) = 4.456 \times 10^{-5} \text{ eV}.$$

Formation of c_s from three 1/3 c_s components at orthogonal orientations from two hydrogen atoms, the two forming 1 J x pulse per oscillator

In a cubic integer geometry event occur in total independence in six space directions Probability of radiation taking place between the location 1 and the location 2 along the longest cube axis is calculated as follows. The probability of the first radiation P1=1/4 is with 4 optional directions. For the pulse to recur from the same of 8 corners, P2 =1/8. The probability for the pulse to be directed at right angle to the first pulse towards node 3, P3=1/4. The last two probabilities are repeated for the third radiation event. The total probability of this occurrence is Ptotal = P1xP2xP4xP5 = ¼ x 1/8 x 1/4 x 1/8 x 1/4 = 1/4⁵ = 1/1024. For ½ of the mass of sun interacting with earth, with mass identified as force:

$$(1 \times 10^{30} \text{ kg})^{1/1024} \times 1.1547^{1/2} = 1.00455 \text{ J or eV with variance } 45.5\ k_s \qquad (10.44)$$

Variance applies to 7x4=28 events with uncertainty per event value $1.5\ k_s = 3 \times \frac{1}{2}\ k_s$, may be formed and total variance becomes $42\ k_s$.

A second hydrogen atom is taken to be positioned below the first one. It is generating a similar pulse from the same location and across the long cubic structure that is displaced by 90° from the first one. The pulse would be located along the long axes of the lower cube in the same plane with the pulse in the upper structure. Two such pulses with wave function format create one resultant pulse, having early components which started moving at $1/3\ c_s$ doubled c_s by $\sqrt{3} \times 1.111\ldots$ to $2/3\ c_s$ as shown on Fig. 4. The velocity is further increased by $\sqrt{2} \times 1.111\ldots \times 1.04^{0.15} = (1-0.00033)$ eV, expressing correct variance of $\sim 3\ k_s$.

Above calculation, an example of options, omitted reflection and also did not consider that moments at right angle to the propagation direction are not transmitted to the receiver, so that the energy received, the output pulse is really

$2/3$ x $\sqrt{3}$ x$1.111 \ldots$ x$\sqrt{2}$ x $1.111 \ldots$ x $1/ 1.04^{0.15} = 2.0042$ 15eV, requires replacing correcting Equation (10.44) by 2 x $(1 \text{x} 10^{30} \text{kg}) 1/^{1024} /1.1547^{1[4} = (2 -0.0045)$ J or eV

(10.45)

11. Coalessence of hydrogen into cubic domains, the space pulse from the crandle of the Universe and Coulomb Law

Spherical structure represents the most efficient method of containing energy of a single particle and such geometry applies probably to hydrogen on the earth surface. At the surface of the sun the gravitation pressure is increased $1/3 \times 10^6$ times in proportion of mass of sun to the mass of earth. The spherical structure of atoms converts into a cubic shape maximizing energy density of the hydrogen. The sizes of atoms of Na of 1 A and Cl of 1.8 A are different in NaCl crystal. All hydrogen atoms are of the same size. They are attracted to each other and they minimize their energy. It is suggested that relatively close to the sun surface hydrogen atoms form transient clusters of regular cubes from two units to 12 units of atoms. The decrement of energy caused by coalescence is likely to be the same as for the NaCl crystal. This value has been calculated at 8.1 eV and confirmed by chemicals data [see Dekker 53]. The 8.1 eV value subjected to two relativity corrections derives the ionization potential of the hydrogen atom of 10 eV:

$$8.1 \times 1.1545 \times 1.07547 = 10.06 \text{ eV} \qquad (11.1)$$

The cubic domains comprising 4 to 2197 atoms are of particular interest because they generate energies fitting the experimental data, or their components.

$2^2 = 4$ atoms; $2^3 = 8$ atoms $3^3 = 27$ atoms; $4^3 x\ 1.1111^2 x\ 1.04 = 49.85$ (i.e. **50 W/m²**);
$5^3 = 125$, $6^3 = 216$; $7^3/1.1547 x 1/1.04^{1/4} \equiv \textbf{300.1 K}$; $8^3 = 572$, $9^3 = 729$; $\textbf{10}^3 \equiv \textbf{1000 W/m}^2 =$ power per orientation; $11^3 \equiv 1331$ atoms; and $12^3/(1.111\ x\ 1.04^{1/2}) \equiv \textbf{1372.5 W/m}^2 =$ sun constant; and $13^3/(1.1111 x 1.33331) x\ 1.04^{1/3} \equiv 1502.5$ W W/m².

A cube of $27 = (3^3)$ quantum pulses of hydrogen atoms excited by the sun in six octagonal orientations generate $3^3 x\ 1.1111/1.2^3 x\ 1.04\ x\ 1.1111 = 20$ J $\equiv$ 20 photons ($18 x 1.1111 = 20$) counted per K: $20\ \boldsymbol{x}\ 10^{-27} = 1/5\ h_s x 1/3\ c_s = 1/5 x 10^{-25}$ J. At 3000 K the pulse becomes $6\ k_s$. Sun pulse of 50 eV expresses a total dissociation of hydrogen molecule with 30 such pulses expressing 1500 W per orientation, which times $1.04^{13/30}/1.1111 = 1373.1$ eV derives the value of the sun constant. The energy of 50 eV is released, when in sun plasma near sun surface, two electrons and two protons form hydrogen molecule H2.

The other likely options to generate the measurable sun constant of 1372 W/m $\equiv$ 1372 eV:
The cube of 11^3 of atoms generating $(11\ \boldsymbol{x}\ 11\ \boldsymbol{x}\ 11)/1.04^{3/4} \equiv 1371$ eV $\qquad (11.2)$

Alternatively cubes of 12^3 quantum steps generating $(12\ \boldsymbol{x} 12\ \boldsymbol{x} 12) \equiv 1728$ eV appear also likely with two possible estimates: $\qquad (11.3)$

$$12^3/(1.111^2 \boldsymbol{x} 1.04^{1/2}) \equiv 1372.5 \text{ eV or } 12^3/(1.1547\ \boldsymbol{x} 1.07457\ \boldsymbol{x} 1.04^{2/5}) \equiv 1369.8 \text{ eV}.$$

The input energy/power density energy of value $12^3 \equiv 1728$ eV $\equiv 1728$ W/m², exceeds the sun constant by 356 W/m². That value expressed by $(\textbf{7} \boldsymbol{x} \textbf{7} \boldsymbol{x} \textbf{7})\ x 1.04 \equiv 356$ W/m² is postulated to be absorbed by the earth. Further interpretation attributes 1728 W/m² to be the resultant of the pulse generated from three lines of hydrogen atom along the cube

$(10)^3$ produce by vector addition of three pulses at right angles to each other. Due to space quantization 1000 eV $x \sqrt{3} \equiv 1372.05$ W/m^2:

$$(10 \, x \, 10 \, x \, 10) \, x\sqrt{3} \equiv 1372.05 \text{ W/m}^2 \qquad (11.4)$$

The above calculations show clearly that a single numerical agreement is insufficient to firmly demonstrate the presence of a postulated process.

Late assessment attributes sun energy of 50 eV to relativity corrected four orientations of energy of H atoms totaling 30 eV:

$$50/4 \, x \, (1.111111^{1/2}) = 13.176 \text{ eV} \qquad (11.5)$$

And $\qquad 50/[1.154700^2 \, x \, 1.1111111^2 \, x \, (1+1/80)] = 30.0000028$ eV.

Equation 17.50 favors strongly sun pulse of 100 J, which may be generated from 4H2 molecules or 1 J pulses from a cube of (5 atoms)^3/[1.11112)1+1/80)]=100 J (11.5a)

The pulse from the outskirts of the Universe

The logic for contributing the molecules H$_2$ being the source of space pulses of wavelength 0.21 m are given in Equation 10.9. The molecule H$_2$ radiates in 10 directions to conserve momentum, although it could be argued that it radiates in all 12 directions, with two not being observed. It is suggested that the reflections from backward direction cancell the forward pulse, and the wave being generated from the radiating mass converts into one pulse. Huge areas of space forming the outskirts of the Universe generate pulses of 0.21 m wavelength.

The pulses from other masses and the pulses from the sun, which according to Weidner and Sell [49] may contain 12 photons are not considered to be from the outskirts of the Universe. Elecromagnetic waves or the variety of pulses generated by human technology are also not considered. The wavelength of 0.21 m is the pulse of nature created by H$_2$ molecule converted to radiation.

Huge areas of space generate infrared radiation that radiation that can be attributed to (proton+electron) or proton suggesting presence of hydrogen in gaseous envelopes of planets both acting as black bodies. In the presence of purely elastic collisions resulting black body radiations are directed in six of the orthogonal orientations, each radiation discharges energy of the source completely in the direction, where the potential barrier is broken. The observer sees only the radiations taking place in his orientation.

Since radiation events are repeated the radiation at the receiver is not the result of a single event from a localized position. It may incorporate components of surrounding radiations reflections preceding the event. Thus even at the most elementary level past events are likely to affect the current event.

It is likely that the pulse is produced by the energy of mass of two protons, which radiate in two opposite directions to satisfy conservation of momentum rule. It appears that mass contained in an element of radiation moving at velocity of light and corrected for relativity is expressed by

$$\text{Proton discharge energy} = \varepsilon_{os} c_s^2 \times 1.111 \ldots = 1\times10^6 \text{ J} \qquad (11.6)$$

The above statement looks correct, but does it represent the events? Actually the proton discharges at local velocity $1/3\ c_s$ in 10 orientations of the rotational plane:

$$\text{Proton discharge energy} = 10\ \varepsilon_{os} \times (1/3\ c_s)^2 = 1\times10^6 \text{ J} \qquad (11.7)$$

Above SI energy discharged in three planes of two step space quantization delivers at the atomic level vectorized energy: 1×10^6 J x $\mu_{os} \equiv 1\times10^6$ J $\times 1\times10^{-6}$ = 1 J, W or eV. Expressed also by 1×10^6 J $\times 1/10^2\ \times 1/10^2 \times 1/10^2 = 1$ J or 1 eV/particle $\qquad (11.8)$

Allowing for linear probability 1/10 and conversion to electron volts the output is the minimum pulse energy:
Minimum pulse energy = 1×10^6 J $\times 1/10 \times 1 \times 10^{-19} = 1 \times 10^{-14}$ J $\qquad (11.9)$

The component of energy in SI integer units is the electon mass moving with radial velocity $v_e = 1 \times 10^7$ m/s converting mass to energy over inverse of linear probability:

$$\text{Minimum integer vectorized SI energy} = m_{es} x v_e^2 \times 1/(1/10) = 1\times10^{-30}\times(1\times10^7)^2\times10 = 1\times10^{-15} \text{ J}$$
$$\backslash(11.10)$$

Ten quantum components in two directiions 1×10^{-15} J make the minimu pulse 1×10^{-14} J
Above energy is expressed also in Equations 2.36-37 and it is expressed also by proton oscillation energy per second, that is power for each of ten quantum pathways:

$$1/5 \times 5 \times 10^{12} \times 10^{-27} = 1 \times 10^{-15} \text{ W} \qquad (11.11)$$

Due to equipartition of energy the electron oscillation energy has the same value:

$$1/5 \times 5 \times 10^{15} \times 10^{-30} = 1 \times 10^{-15} \text{ W}. \qquad (11.12)$$

Period of radiation is $1/1\times10^{15} = 1 \times 10^{-15}$ s generating power of 1 W/per plane.
Energy radiation in space (10^{-28} J/m^3) times linear length 1×10^{-10} over area 10^{-20} m^2 is the forward field of the elementary beam 10^{-18} J/m or force 10^{-18} N. The 10^{-18} N force is balanced by the force F_e produced by the magnetic field generated by the electron rotation around the area 1×10^{-20} m generating the electric current $e_s v_e$:

$$F_e = \mu_{os} e_s v_e = 1 \times 10^{-6} \times 1 \times 10^{-19} \times 1 \times 10^7 = 1\times10^{-18} \text{ N} \qquad (11.13)$$

The energy $[10^{-28}\text{J/m}^3]/1\times10^{-19} \equiv 10^{-9}$ J is the probability, which times invert of linear converter derives 10 eV. Muo10$_s$ mass is 10^{-28} kg.

The radiation energy of two oscillators 2×10^{-27} J has been identified with two vectors of minimum energy 1×10^{-14} J, displaced by $90°$ to generate between them a signal moving at the velocity of light energy requiring relativity correction 1.111 . . . of magnitude derived from the excess mass of 18 atoms of hydrogen converted to 5 atoms of helium:

$2 \times 10^{-27} = (1 \times 10^{-14} \text{J}) \times (1 \times 10^{-14} \text{J}) \times 1.111 \ldots \times 18 = (0.01 \text{J}/10^{12} \text{Hz}) \times (0.01 \text{J}/10^{12} \text{Hz}) \times 1.111 \ldots \times 18$

And $\qquad 0.01 = 1/6 \times 6000 \times 1/10^3 \times 1/10^2 \qquad\qquad$ (11.14)

These two relations display conditions and electromagnetic properties: two wave function formatted pulses; pulse produced is relativity corrected; residual mass of H atom is needed; pulse rate 6000 is in 1/6 th of orthogonal directions; probability factor $1/10^3$ of local discharges is three with vectorization factor that needs to be satisfied to generates a radiation pulse carrying force, graviton, momentum $1/10^2$. In conclussion the four step radiations of $2 \times 8 \times 1.1111 \times 1.0328^{1/2} = 18.0670$ protons of oxygen in one plane are shown to be able to balance two stepped 3 D radiations of $2 \times 9 = 18$ protons from the sun each step increasing local proton velocity by two. Variance is $0.067 \text{eV}/300 \text{K} = 2/3\ k_s$.

12. Atomic units, exactness of the gravitation force, and acceleration on earth and thermal balance

Sun-planet pairs have to be considered as structores of components each requiring balancing forces determined by local geometry. The forces have to be balanced in macroscopic terms, in atomic terms, in classical SI units and in integer cubic SI units. Some physical constants are superflous leading to further increase of forces to be considered. Equipartition of energies lacking directions display further equalities that have to be considered.

Electron charge e_s has been identified to express mommentum of proton mass increased by relativity corrector 1.1547^2 moving at velocity ½ x c_s. Radiation mass 1×10^{-27} kg = 1/2 x 2×10^{-27} kg is the mass of the proton momentum split in two opposite directions. A delema arises because of the 2/3 factor that appears at least in several gravitation expressions for earth and for planets, as well in Chapter 5:

$$1/2 \times 1.1547^2 = 2/3 = 0.66666 \ldots \tag{12.1}$$

The 2/3 factor has been interpretted as indication of the absence of electron frequency absorbsion by earth at 300 K.

The electron charge e_s has been identified to be the radiation mommentum of two oscillators absorbed on earth of 2/3 of the sun momentum of proton moving at **1/4xc_s**:

$$\textbf{2e}_s \equiv \textbf{1.1547}^2\textbf{x}\textit{\textbf{m}}_{\textbf{pds}}\textbf{x1/4x}\textit{\textbf{c}}_s = \textbf{1.1547}^2\textbf{x2x10}^{-27}\textbf{x1/4x3x10}^8 = \textbf{1.999998x10}^{-19}\textbf{ C} \equiv \textbf{2 eV} \equiv \textbf{2 J} \tag{12.2}$$

The 1 J value is the radiation momentum of proton (½ of $\textit{\textbf{m}}_{pds}$) and proton with electron moving at the velocity **1/3xc_s**:

$$\textit{\textbf{e}}_s \equiv \textbf{1x10}^{-27}\textbf{x1/3 x3x10}^8 = \textbf{1x10}^{-19}\textbf{ C} \tag{12.3}$$

The electron chage e_s is at the same time mass **2x10^{-27} kg** moving at **1/4xc_s** and relativity corrected for that speed. In atomic terms energy, charge or momentum e_s expresses vector sum of three orthogonal components $1/2 \times k_s T_{6000}$ in:

$$e_s \equiv \sqrt{3} \textbf{ x}1/2 \times k_s T_{6000} \textbf{x}1.111 \ldots \equiv \sqrt{3} \textbf{ x } 1/2 \textbf{ x}1\textbf{x}10^{-23}\textbf{x } 6000 \textbf{ x } 1.111 \ldots = 1\textbf{x}10^{-19}\textbf{ J} \tag{12.4}$$

The quantity e_s expresses one of 6 space orientations with discharges following sequentially from three planes. The relation states also that electron charge per degree corrected for relativity, non interacting flux and quantum is the Boltzmann constant:

$$e_s/6000 \textbf{ x } 2/3 = 10^{-19} \textbf{ x } 1/6 \textbf{ x } 1/10^3 \textbf{ x } 2/3 = 1.000 \textbf{ x}10^{-23} \textbf{ J} = 1.2\textbf{x}1.1547^2\textbf{x}2/3 \textbf{ x}10^{-23} \approx 1.000\textbf{x } k_s \tag{12.5}$$

In atomic natural units the momentum is e_s = 1.000 and light velocity c_s = 1.000. Then the product (momentum x c_s) = 1.000 represents energy (mass times c_s^2) of 1 eV at

300 K per oscilator. The corresponding energy of the earth pulse per kelvin is 20 times smaller at 0.05 eV.

Planck constant can be considered to be probability per unit particle or per proton.

The circular sun area A_{cisunr} generates in each of six directions power in three planes:

$$A_{cisunr} \, c_s \, x \, 6000 \, K_{rel1/3cs} = 1.5 \, x10^{18} \, x \, 3 \, x \, 10^8 \, x \, 6000 \, x \, 1.111 \ldots = 3 \, x \, 10^{30} \, \text{W} \tag{12.6}$$

The above value per plane time invert of volume probability derives above 1 eV $\equiv$ 1 W representing the unity natural atomic event. Total sun power $3x10^{30}$ W represents likelyhood of that event on the sun in three planes.

Special Relativity Theory has to be applied to explain variations of values of gravitation from the Reference Equation 6.1. Now the product of three integer circular areas of sun and its 6000 K temperature derives an exact value of the force of gravitation predicted by Special Relativity:

$$F_{Gs} = 3 \, A_{cirsuns} \, x \, 6000 = 3.000 \ldots 10^{22} \, \text{N} \tag{12.7}$$

The square of the distance to earth X_{earth} time $\sqrt{3}$ divided by the relativity corrector for $1/3 \, x \, c_s$ derives the same force with slightly different value:

$$F_{Gs} = 1 \, \text{N/m}^2 \, x \, \sqrt{3} \, (L_{earth})^2/K_{rel1/3cs} = \sqrt{3}(1.5x10^{11})^2/1.111 \ldots = 3.5074x10^{22} \, \text{N} \tag{12.8}$$

Above force value is also obtained from corrected Equation 19.28 [45]

$$F_{Gs} = A_{cirsun} \, c_s \, (1/10^2)^2/(1.2 \, x1.111 \ldots) \, x1.1547^{1/2}/1.04^{3/4} = 3.4983 \, x10^{22} = 3.500x10^{22} \, \text{N} \tag{12.9}$$

Integer force times $1.04^{1/4}$ is close to the reference

$$F_G = 3.500x10^{22} \, \text{N} \, x1.04^{1/4} = 3.5420x10^{22} \, \text{N} \tag{12.10}$$

Thus at the distance L_{earth} the density of the sun force is 1 N/m² and the density decreases with the square of the distance from earth with multiplier $\sqrt{3}$ and $K_{rel1/3cs}$ correctors confirming the average velocity $1/3xc_s$ and vectorial addition of three orthogonal velocity vectors. These very simple relations for three planes, with 2/3 absorption, relativity and quantum corrections demonstate otherwise the validity of the differential equation expressing $L_{earth} \cdot dF_{Gs}/dL_{earth} = 0$, which on integration derives

$$F_{Gs} = 1 \, \text{N/m}^2 \, x \, 3 \, x \, 2/3 \, x \, L_{earth}^2/2 \, xK_{rel}^3 = 3 \, x2/3 \, x \, (1.5x10^{11})^2/2 \, x1.1547^3 \, x \, 1.04^{1/4}$$
$$= 3.500009 \, x10^{22} \, \text{N} \tag{12.11}$$

Above equation shows that the energy contained in the whole length of the straight beam of sun contributtes to the earth gravitation force. The force contained in the beam is the same along the beam and so it creates energy in the product $F_{Gs}L_{earth}$. If the 300 K beam takes three planes of length of earth to sun distance, the energy gradient of 1 J/m is increased by 300 K and combines to 600 K; kelvin is treated as rate multiplier. Temperature squared, multiplied by the square of the velocity of light $3x10^8$ m/s and multiplied by relativity 1.1111 ... derives exactly the gravitation force in integer relativity model units:

$$\text{1 J/m } x \text{ } 600^2 \text{ K } x(3 \text{ } x \text{ } 10^8 \text{ m/s})^2 \text{ } x \text{ } 1.111 \ldots \equiv 3.600 \ldots x10^{22} \text{ W} \qquad (12.12)$$

It should be noted that change of the distance between the sun and the earth requires adding or subtracting energy to the beam.

Treating the Planck constant $1x10^{-33}$ as probability per unit energy with reflection of ½ and multiplication by $K_{rel}/1.04^{1/4}$ derives:

$$\tfrac{1}{2}F_{Gs} L_{earth} h_s K_{rel}/1.04^{1/4}=3.500 \text{ } x10^{22} x1.5x10^{11} \text{ } x1x10^{-33}x1.1547/1.04^{1/4}=3.0015 \text{ eV} \qquad (12.13)$$

The international gravitation constant G_s of the cubic space model is the gradient of the infrared radiation energy 0.05 eV within Bohr atom converted to SI units:

$$G_S= 3/2 \text{ } G = h_s v_{Lms}/a_{os} =1x10^{-33}x5x10^{12}/0.5 \text{ } x10^{-10} = 1x10^{-10} \text{ J/m} \qquad (12.14)$$

To express the value in atomic terms the above energy gradient has to be divided by $e_s = 1 \text{ } x \text{ } 10^{-19}$ C generating 1 000 000 000 = 1 GeV/atom. Multiplied by probability 10^{-9} derives 1 eV or J.

According to Newton's second law of motion the gravitation force F is also equal $m_{earth} \text{ } x \text{ } g$, where g is the earth acceleration due to gravity; therefore $g = G \text{ } x \text{ } m_{sun}/L_{earth}^2$ at the earth's surface in the classical SI units equals:

$$g = 0.66742 \text{ } x \text{ } 10^{-10} \text{ } x \text{ } 1.9891 \text{ } x \text{ } 10^{30} /(1.496 \text{ } x10^{11})^2 = 0.0059319 \text{ N/kg} \qquad (12.15)$$

Multiplying by the mass of earth $5.9742 \text{ } x \text{ } 10^{24}$ derives the reference $3.5438 \text{ } x \text{ } 10^{22}$ N.

In comparison the model integer values derives

$$g'=2/3 \text{ } x \text{ } 10^{-10} \text{ } x \text{ } 2 \text{ } x \text{ } 10^{30} /(1.5 \text{ } x10^{11})^2= 0.00592593 \text{ N/kg} \qquad (12.16)$$

Multiplying by the integer mass of earth $6 \text{ } x \text{ } 10^{24}$ over quantum correction $1.04^{17/200}$ obtains also the reference earth gravitation force $3.5437 \text{ } x \text{ } 10^{22}$ N. Again both classical SI units and integer space cubic SI units derive almost identical values.

Acceleration on earth

It is reconfirmed [45] that the gravity force on earth per kilogram is actually 10 N. The mass of 1 kg due to the velocity of protons of $1/4xc_s$ is relatively increased by $1/[1-(\tfrac{1}{2}c/c)^2]^{1/2}=1/[(1-1/4^2)]^{1/2}=1.0328$ or in 1D by $3.28\%/\sqrt{3} =1.894\%$. Thus 10 N produces acceleration of $10/1.01894= 9.8141$ m/s, which agrees well with the experimental value of 9.8178 m/s varying slightly with the position of earth. The earlier mentioned 0.006 N/kg is per K. Referred to 300 K over $\sqrt{3} \text{ } x \text{ } 1.04$ derives 9.9923 N/kg.

Correcting earth acceleration 9.81 m/s to 10 m/s by quantum factor is less accurate:

$$9.81 \, x \, 1.04^{\,7/10} \approx 10.00 \text{ m/s} + 3k_s T_{300}/e_s \qquad (12.17)$$

With earth mass $5.9742 \, x \, 10^{25}$ kg the attractive force generated towards earth should be

$$F_{Gs} = 5.9742 \, x10^{24} \text{ kg } x10 \text{ N/kg } x1/10^3 \, x1/1.2 \, x1/1.1547^2 x1/1.111\ldots^{1/2} = 3.5422 \, x10^{22} \text{ N} \quad (12.18)$$

Using integer earth mass and quantum corrector derives F_{Gs} within limits of its accuracy

$$F_{Gs} = 6.0000 \, x10^{24} \text{ kg } x10 \text{ N/kg} x1/10^3 x1/1.2 \, x1/1.1547^2 x1/1.11\ldots^{1/2} x1.041^{/10} = 3.5436 x10^{22}$$
$$\text{N} \qquad (12.18a)$$

Centrifugal force of repulsion of the sun

Viewed from earth the sun extends over 0.53 degree. Total outward spherical view covers $4\pi(180/\pi)^2 = 41253$ stearian degrees so that sun covers $1/[41253/(0.53)^2] = 1/146860$ fraction of the spherical view surface from earth. With the sun temperature of 6000 K the attractive force generated towards sun is $10x6000/300 = 200$ N/kg originated from one side of the mass $2x10^{30}$ kg

$$F_{Gs} = 1x10^{30} x \, 200 \text{ N/kg } x1/10^3 \, x \, 0.14686 \, x10^{-6} x1.1547 x1.04^{11/10} = 3.5411 \, x10^{22} \text{ N} \qquad (12.19)$$

The force 0.006 N/(kg-K^2) is per space quantization step relativity corrected for velocity $1/3 \, c_s$:

$$\text{The force 0.006 N/(kg-K)}^{-1} = (10 \text{ N/kg})/300 \text{ K } x \, 1/5 \, x1/1.111\ldots \qquad (12.20)$$

Moving around the earth, sun with mass of $1x10^{30}$ kg exerts a repulsive force of $1x10^{30} \, x \, 30000^2/1.5x10^{11} = 6 \, x10^{27}$ N reduced by the above fraction $1/146860$, by relativity 1.1547 and by quantum $1.04^{1/32}$ and derives the exact opposing value of the force of attraction, $6 \, x10^{27}/(1468601.1547) \, x1.04^{1/24} = 3.54395 \, x10^{22}$ N. This evaluation together with that of Section 7 generates an overwhelming evidence for the centrifugal force to be one of the forces balancing the sun gravitation. The force of gravity generated by static mass is spread by planets equally in all orientations and so while it cannot balance a large unidirectional force of the sun it can keep local forces acting on the whole earth or planets surfaces balanced. For an observer moving in the plane of the couple rotation at half the velocity of light, with current $I_{earth} = 1.8x10^{29}$ A and with the Paris reference force $2x10^{-7}$ N/A-m, the force observed is repulsion having the exact value of $3.6000\ldots x10^{22}$ N, expressing the integer cubic gravitation value.

Thermal balance between the sun, the earth and the Universe

Earth spherical surface rounded to an integer value is:

$$4 \, A_{cir \, earth} \text{ x quantum corrector} = 4 \, x \, 1.278005 \times 10^{14} \text{ m}^2/1.04^{1/6} = 4.993 \times 10^{14} \approx 5 \times 10^{14} \text{ m}^2 \qquad (12.21)$$

Thermal energy radiated by earth at 300 K= $\sqrt{3}\sigma_s T_{300}{}^4$=701.4 W/m² as in Equation 5.13 generating after corrections total power 5 x 10¹⁴ m² x 701.4 W/m²[1,15470(1+1/80)] = 3 x 10¹⁷ W of Equations 2.2-2.2a expressed per plane.

In integer relativity model atomic units the electric and the magnetic energies of space pulses are the same:

Electric space power delivers = $(\varepsilon_{os}/10)^{1/6}$ x 1/3 c_s=1 x 10⁵ W   (12.22)

Magnetic space power using permeability = $(\mu_{os})^{1/3}$ x 1/3 c_s ≡ 1 x 10⁵ W of area 1.278005 x 10¹⁴ m² and sun constant 1371.2 W/m² derives 1.7528 x 10¹⁷ W taken to be √3 x 10¹⁷ W. In integer SI units:

$$\varepsilon_{os} \text{ x } 1 \text{ x } 10^{12} \text{ Hz} = 10 \text{ W/m}^3.$$ (12.23)

This is derivative of 3 W/m³.

The wavelength of space radiation from the edge of the Universe measured by several astronomers is 0.21 m. The product of 0.21 m and of velocity of light corrected by relativity and by quantum factor derives:

Model energy of space radiation = 0.21 m x 3 x 10⁸ /1.11111¹ᐟ² x1.04¹ᐟ¹⁰ = 0.2 J

(12.24)

With five pulses per orientation the energy is 1 J or power 1 W. It is suggested that above

$$701.4 \text{ W/m}^3 \text{ x } \sqrt{2} \text{ x} 1.04^{1/40} =1001 \text{ W} = 1000 \text{ W (or eV)} + 33 \, k_s T_{300}$$ (12.25)

Equipartition of thermal powers besides gravitation forces indicates that the Universe obeys the laws of physics and is not in chaos. It is in long term stability. The observed explosions are extremely rare events in the huge Universe.

13. Proton as 2001 spinning electrons; electron content defined by body mass; Ampere's Law displayed in planet rotation; is it 1 J/kg, 1 W/kg-year or 1 N/kg?

In Equation 5.2 of Book 2 the electron minimum charge e_s expressed in SI units has been identified to represent 2/3 radiation momentum of proton moving at velocity $\frac{1}{4} c_s$, while in Equation 2.83 charge e_s has been derived from the probability of its occurrence. **It is now suggested that proton is just a set of 2001 electrons forming three sets of 667 units comprising four elements rotating in opposing phase orientations as shown in Figure 6 nullifying the total momentum.** While the proton mass is taken in integer space units as 2×10^{-27} kg, the elementary radiation mass progressing in one direction has a mass of 1×10^{-27} kg. Total number of electrons $667 \times 3 = 2001$ of zero momentum is accompanied by one electron rotating with radius of $1/2 \times 10^{-10}$ m. The momentum of the outside electron cancels the momentum of the proton so that the resultant mass of proton is $m_{dps} = 2000\ m_{des}$. The radius of the proton is assessed now at

$$r_p = 3/2 \times 1/10^3 \times [1 \times 10^{-30}]^{1/2} = 1.5 \times 10^{-18}\ \text{m} \tag{13.1}$$

According to CODATA 2010 the proton radius is about $0.877(5) \times 10^{-15}$ m. **The integer cubic model suggests that there are three orthogonally directed energies combining into one vector value $\sqrt{3} \times 0.877 \times 10^{-15}$ m $\times 1 \times 10 \times 1 \times 10 \times 1 \times 10 = 1.519 \times 10^{-18}$ m.**

Using the classical mass, the number of protons N_{pr} on earth is (SI earth mass)/ (SI proton mass) $= 5.9741 \times 10^{24}/1.6728 \times 10^{-27} = 3.5713 \times 10^{51}$ units. Because of volume probability the expression for the mass of a body in SI units (kg) expresses the number N of electrons in the rotating earth:

$$N = m_{earth}/m_{es} \times 1 \times 10^{-30} = 5.9742 \times 10^{24}\ \text{kg}/1 \times 10^{-30} = 5.9742 \times 10^{54}\ \text{units} \tag{13.2}$$

Since in integer cubic model units earth has 6×10^{24} kg, N is 6×10^{54} units.

With the later adoption of 6×10^{24} kg for mass of earth, the earth contains 6×10^{54} electrons. **Consequently these earth electron charges rotate around the sun at linear velocity of $v_{earth} = 29786$ m/s. The product of earth mass expressed in terms of the number of electrons, of e_{earth}, of unity charge and of SI to atomic level converter generates current at atomic natural level:**

$$I_{earth} = N_{earth}\, v_{earth} \times e_s \times 1 \times 10^{-30} = 1.7795 \times 10^{29}\ \text{A} \tag{13.3}$$

Actually the corrected current is an integer $I_{earth} = 6 \times 10^{54} \times 30000 \times 1 \times 10^{-19} \times 1 \times 10^{-30} \times 1.111 \ldots = 2 \times 10^{29}$ A. The earth rotating around the sun as shown on Fig.10 can be then represented by a ring of current of 2×10^{29} A, inducing by the Ampere law on the sun the flux $B = \mu_{os}\, I_{earth}/2\pi L_{sun}$. Equal opposing flux is created by the rotation of the mass of the sun. The repelling force using integer relativity space units converted to SI is obtained by the observers travelling at $1/2 \times v_{earth}$, causing two linear currents flowing in the opposite direction generating between them the force proportional to energy and so decreasing with the square of the velocity v_{earth}, being given by the gradient dB/dX_{sun} and the repelling force varying linearly with I_{earth}:

$$F_{GR} \equiv dB/dL_{sun} \equiv 1/6\mu_{os} I_{earth}$$
$$= 1/6 \times 1 \times 10^{-6} \times 1.8000 \times 10^{29} = 3 \times 10^{22} \text{ N, the Special Relativity term} \qquad (13.4)$$

where in integer cubic units $1/6 \equiv 1/2\pi$.

Since earth is rotating it is subjected to acceleration and deceleration causing mass to energy conversion requiring space mass multiplication by $10^{-10} c_s$ to convert it to SI units with each square meter of surface generating $\varepsilon_{os} \times c_s$ J/(K-m^2):

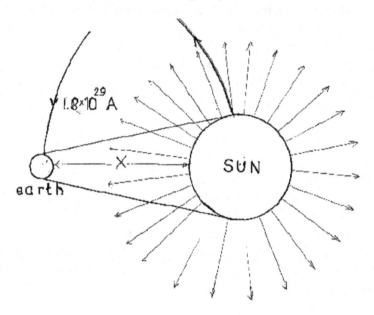

Figure 10. The cone of sun incoherent radiation interacting with earth, the radial sun radiation and the currents of the sun and of the earth causing repulsion; to satisfy momentum conservation rule the sun current is effectively that of earth

The earth repulsive force = Individual proton X-section x 1 J/(K-m^2) x Linear earth velocity x ε_{os} x c_s^2 x N_{pr} x 300 (K) x linear probability x $\sqrt{3}$ x $1.04^{1/5}/1.11111 = \pi$ x $(1.5 \times 10^{-18})^2$ x1 x29786 x1x10^{-11}x9x10^{16} x3.5713x10^{51}x300 x 1x10^{-10} x$\sqrt{3}$ x1.04$^{1/5}$/1.11111 $\equiv$ 3.5449x10^{22} N   (13.5)

The resultant force value is the effectively the reference Equation 6.1.

The proton cross-section area on earth is expressed by the product of the proton surface area and the number of protons on earth:

Proton-X-section area in 4 directions on earth=$4\pi(1.5 \times 10^{-18})^2$x3.57133x10^{51}
=1.01x10^{17} m^2 = $(\mu_{os}\varepsilon_{os})^{-1}$ (13.6)

This numerical value is expresses $E_{sspaceSI}$ = 1.01x10^{17} (J-s) of Equation 2.2.
The radiation from sun circular area of 1.5279 x10^{18} m^2 produces on earth vectors of 1 eV, 1 W and 1 N per squared meter. These values are reduced by 1/10 of space quantization and

further by 2 because only 1D component interacts with the proton reducing the effective sun surface area to $1.0733 \times 10^{17} \text{m}^2/1.1547^{1/2} = (1-0.0028) \times 10^{17} \text{m}^2$. It represents energy in eV, power in watts and force in newton of interacting photons with earth. Thus

$$\text{Proton X-section area on sun} \approx 1.000 \times 10^{17} \text{ m}^2 \qquad (13.7)$$

Proton X-section area on sun is close to the earth value suggesting that all earth photons generate pulses, which interact with sun proton pulses at distance around 15500 km derived in Equation 4.1. The pulses from sun concentrate on earth against the pulses from earth causing their neutralization with total momenta becoming zero. In SI language lines of forces from positive charges end on electrons. The concept of momentum nulling is used with reluctance because it seems only to be valid for atomic vectors.

The probability of three sequential vectors of proton charges $e_s = 1 \times 10^{-19}$ C forming pulse 1×10^{-27} J is shown in Figure 6 and 7. That process has to be repeated again to generate the minimum atomic energy pulse converted to SI:

$$B_{pulse} \text{ momentum} \equiv (e_s \times 1/10^3) \text{X} (e_s \times 1/10^3) = 1 \times 10^{-44} \text{ [kg-m/s]} \qquad (13.8)$$

It is suggested that the atomic unit force 1×10^{-7} N, which in atomic terms is also the energy gradient, transfers its value in the radiation process to the gravity force beam both for earth and for sun. Hence energy gradient, which is the gravitation force from earth, is assessed with 0.03% accuracy:

$$2 \times 10^{-7} \text{ N/m-kg} \times I_{earth} /\text{cor.} = 2 \times 1 \times 10^{-7} \times 1.7795 \times 10^{29} /1.04^{1/10} \equiv 3.5451 \times 10^{22} \text{N}. \qquad (13.9)$$

The current 1.8×10^{29} corrected for relativity 1.111111 times 2×10^{-7} N/m generates force derived by Special Relativity 4.00000×10^{22} N.

The 1×10^{-7} N force has the same value along all the sun to earth distance.

Sun mass reacts in rotation only along one plane expressed by radiation of 10^5 kg and 10^{10} converter, $(1 \times 10^{30} \text{ kg})^{1/2} = 10^5 \text{ kg} \times 10^{10}$, or $10^{30} \text{ kg} = 10^{10} \text{ kg} \times 10^{20}$ converter; this fits better the squared and cubed pulses of radiations. For sun the pulse $10^5 \text{ kg} \equiv (1000 \text{ eV})^2 \times 1/10$, for earth $10^5 \text{ kg} \equiv (300 \text{ K}^2 \times 1.111 \ldots \times 1/10)$ two sets of three pulses in X-Y-Z orientations and $1/1 \times 10^{-19}$ converter derive SI sun momentum:

$$(13.10)$$

Interacting with earth per plane: $[(10^5 \text{ kg} \times 3 \times 10^8)^3 \times 1.111 \ldots ^{3/2}]^2 \times (1/1 \times 10^{19})^2 = 10 \times 10^{100}$ kg-m/s

The above implies that momenta in three orientations sum into one orientation; with repetition of low probability of events the input increases very significantly for two pulses. Because of the reference used, SI value is highly inflated by a factor of $1 \times 10^{100}/1 \times 10^{30} = 1 \times 10^{30} \times 1 \times 10^{30} \times 1 \times 10^{10}$, with $(1 \times 10^{-10})^{10} = 10^{-100}$

The other expressions incorporating earth current are:

$$(\mu_{os}\varepsilon_{os})^{-1} \equiv 1/12 \times 2 \times 10^{29} \text{ A}/(1.1111 \ldots \times .5 \times 10^{11} \text{m}) = 1 \times 10^{17} \text{ J-s.}$$

The most important identity **2x10^{30} kg x1/10≡2x10^{29} A** is discussed in Chapter 23   (13.11)

The equivalence between mass and current shown here is the reason for the adoption of abamper = 10 A as the reference in c.g.s. system of units:

$$(3x10^8 \, m/s)^3 \, x1.111 \ldots x1 \, x10^4 \, x \, 2/3 = 2 \, x \, 10^{29} \, A \qquad (13.12)$$

Some Equations in this section were formed years ago, when integer values were not comprehended.

14. Size of hydrogen atom; 1/30, 1/18 and 100 power index probabilities

The density rate of sun pulses is reduced in proportion to the square of the space distance to earth L_{earth}=1.5 $x10^{11}$m. The dimension of the hydrogen atom allows factor 2 for reflection and 1.1111 . . . for relativity correction with particles moving at local frequency 1/5 v_{Lms}=1$x10^{12}$ with quantum correction. The derived hydrogen atom square or spherical size is:

$$2x10^{12}/(1.5\ x10^{11})^2\ x1.1111\ldots x1.04^{1/3} \equiv 1.0006\ x\ 10^{-10}\ \text{m} \qquad (14.1)$$

The square root of three, times the product of number of diameters or cubic dimensions of Bohr atom, $1.5x10^{11}$ m/1 $x\ 10^{-10}$ m, along the mean sun to earth distance and the dissociation energy of the Bohr atom expresses accurately the gravitation force of the sun to earth:

$$F_{Gs} = \sqrt{3}\ x13.6401\ \text{eV}\ x1.5x10^{11}\text{m}/1\ x\ 10^{-10}\ \text{m} \equiv 3.5438\ x\ 10^{22}\ \text{N} \qquad (14.2)$$

The gradient of the mean gravitation force of the sun, that is the force over the distance between the sun and the earth is that of 30 J per m, relativity and quantum corrected:

$$3.5438\ x10^{22}\ \text{N}/1.5\ x10^{11}\text{m} \equiv 23.625/(1x10^{-10})\ \text{N/m}$$
$$\equiv 30\ \text{eV}\ x\ 1/1.2\ x\ 1.04^{1.45}/(1x10^{-10}) = 23.618/(1x10^{-10}) \qquad (14.3)$$
And $\qquad\qquad 30\ \text{J}\ x3x10^8\ x1.1111\ldots \equiv 1\ x\ 10^{10}\ \text{W-m}.$

Equitorial earth circumference of 40000 km is very nearly equatorial radius 6378.2 km time $2\pi/1.04^{1/20} = 39996.9$ km ≈ 40000 km, the difference 3.1 km attributed to $3k_s T_s$ H. u.:
$$40000\ \text{km}/(2\pi\ x\ 1.1111\ldots)\ x1.04^{1.2} = 6005.7\ \text{km} \qquad (14.4)$$

The distance of 5.7 km is the uncertainty $k_s T_{s,}$ making the earth radius 6000 km.

The energy or momentum in atomic kinetic terms applies also to proton, because although its frequency is 1000 time smaller its mass is 1000 times larger.

The pulses of the force of the mass of sun in one direction discharging 10 N per kg decreased by the square of the separation and expressed per square of the sun temperature time 1.1111 . . .^{2} derive exactly 1 N, indicating that the radiation gravitation pressure varies exactly with square of the source temperature:

$$1\ x10^{30}\ x10\ [\text{N/kg}]/[(1.5\ x10^{11})^2\ x\ (6000^2\ \text{K})\ x1.1111\ldots^2] \equiv 1.0000000000\ \text{N} \qquad (14.5)$$

Although [N/m^2] increases with the square of temperature of the source, the rate of oscillations in one pathway increases with the fourth power of temperature:

The geometry of radiation is different for earth becuase
$$6000^4/(1.1111\ldots^2\ x\ 1.04^{6/5}) \equiv 1.0015\ x10^{15}\ \text{Hz} \qquad (14.6)$$

129

frequencies use three planes and with the relativity correction for $1/3 x c_s$ the fourth power of temperature derives again exactly the local frequency of proton oscillations:

$$(3 \times 300 \times 1.1111 \ldots)^4 \equiv 1.00000000000 \times 10^{12} \, Hz \qquad (14.7)$$

Above two relations indicate that the kelvin temperature to power four with corrections derives the particle rate delivery.

Magnetic permiability of space (vacuum) is identified as the energy of the mass m_{es} of the electron in the radiation pulse oscillating at the local infrared frequency of proton $1/5 \, v_{Lms}$:

$$\mu_{os} = m_{es} \times (1/5 \, v_{Lms})^2 = 1 \times 10^{-30} \, kg \times (1 \times 10^{12})^2 \, Hz^2 = 1 \times 10^{-6} \, NA^{-2} \text{ or J} \qquad (14.8)$$

Frequency of 1×10^{12} Hz times the square of the magnetic permiability of vacuum μ_{os}:

$$\mu_{os}^2 \times 1 \times 10^{12} \, Hz \equiv 1 \, W \text{ or J per pulse} \qquad (14.9)$$

Above Equation delivers a correct but incoprehensive result. Integer relativity model atributtes the square to be wave function inreasing velocity by factor of two and amplitude by $\sqrt{2}$, relativity an quantum corrections and the factor 2/3 due to non absorption of electrons:

$$(\mu_{os}) X (\mu_{os}) \times 1 \times 10^{12} \, Hz \equiv \sqrt{2} \, \mu_{os} \times 1.1111 \ldots^{1/2} \times 2/3 \times 1.04^{1/6} = 1.00032 \text{ J or eV} \qquad (14.10)$$

The variance $3.2/1.1111/2 \approx 3 \, k_s. \, \mu_{os}^2$ in Equation 14.9 is a probability product. It expresses vector addition. Both processs are present. Identity is not an equation. Space permittivity ε_{os} is identified as energy of the radiation pulse of proton moving at velocity $1/3 \, c_s$: $\varepsilon_{os} = \frac{1}{2} \, m_{ps} (1/3 \, c_s)^2 = \frac{1}{2} \times 2 \times 10^{-27} \times (1 \times 10^8)^2 = 1 \times 10^{-11}$ F/m or F/m³. Corrected 2 D addition of moments derives: $(m_{ps} \times 1/3 \, c_s) X (m_{ps} \times 1/3 \, c_s) \times$ cor.$= \sqrt{2} \times 2 \times 10^{-27} \times (1 \times 10^8 \times 1.1547^{1/2} / (+0.0111 \ldots) = 3 \times 10^{-19} = 3 \, e_s$ 　　(14.11)

Also $\varepsilon_{os} \equiv 10 \, \mu_{os}^2 = (\sqrt{10} \, \mu_{os}) X (\sqrt{10} \, \mu_{os})$
$= [3 \, J/(10^{12})^{1/2} \times 1.111 \ldots^{1/2}] \times [3 \, J/(10^{12})^{1/2} \times 1.111 \ldots^{1/2}] = 10 \, J/10^{12} = 1 \times 10^{-11}$ F/m 　　(14.12)

Space permittivity ε_{os} is expressed as 10 J (eV) per wave of infrared frequency. Above equations imply that during space oscillation in six orientations the probability of generating pulse in one of six orientations is $1/10^3$ as shown in Fig. 5. The observer sees localized sources arriving from the whole volume of space around him creating the world as we see it. The elementary waves create various atomic and mass structures at different temperatures creating a rainbow of colors expressing frequencies.

Actually with ten orientations and three planes there are 30 possible orientations for radiation from each object or parameter creating variety of parameters. Magnetic permeability with 1/30 orientation probability generates per direction

$$\sqrt{2} \, (\mu_{os})^{1/30} \times 1.111 \ldots \times 1.04^{9/40} \equiv 1.00022 \text{ J} \qquad (14.13)$$

Proton mass generates Planck energy

$$(2 \times 10^{-27} \text{ kg})^{1/30} \times 1/10 \times 1/111 \ldots \equiv 0.01003 \text{ J} \tag{14.14}$$

Electron mass generates also Planck energy $(1 \times 10^{-30})^{1/30} \times 1/10 \equiv 0.0100000 \text{ J}$ (14.15)

The earth with 300 K quanta in each 30 directions generates the force:

Earth attraction $= 300^{300/30} \times 1/10 \times 1/10 \times 1/\sqrt{3} \times 1.04 = 3.5456 \times 10^{22} \text{ N}$ (14.16)

The mass of sun generates $(2 \times 10^{30} \text{ kg})^{1/30} \times 1/10 \equiv 0.999998 = 1 \text{ W or } 10 \text{ W}$ (14.17)

If reflection is included there are 60 beams with probability 1/60, sun mass generating:

$$(1.00 \times 10^{30} \text{ kg})^{1/60} \times 1/1.111 \ldots .^{1/2} \equiv 2.999 \ 999 \ 86 \text{ eV including electron energy}$$
$$\tag{14.18}$$

Above four estimates generate correct values. The sun mass deliver 1 W pulses at 300 K on earth and the proton and electron deliver quantum traveling for 14 billion years, so that the idea postulated in the model is no more fanciful that the black hole formation.

Eighteen is the number of atomic units in the sun pulse derived in Conclussion 64. Validity of power index probability $1/18 = 0.05555 \ldots$, is obtained in the following 0.02% accurate classical charge derivation from temperature:

$$e \equiv (6000 \text{ K})^{1/18}/(1.04^{3/10}) \equiv 1.60246 \text{ J or eV} \tag{14.19}$$

Integer electron charge expresses integer cubic SI proton mass moment at local velocity $1/3 \ c_s$:
$$e_s \equiv 1/2 m_{ps} \times 1/3 c_s = 2 \times 10^{-27} \text{ kg} \times 1 \times 10^8 \text{ m/s} \equiv 2 \times 10^{-19} \text{ C} \tag{14.20}$$

Larmor formula expressed in integer cubic SI units is dealt in Equations 22.58-9. These have shown that $(\sqrt{3})^3$ with relativity correction $(\sqrt{3})^3 \times 1.154700538 = 5.999 \ 999 \ 998 = 6 \text{ W}$ and so Larmor formula power expresses power from charge e_s in six orientations. With 1 W per direction the value $3G_s$ is created by charge $e^{1/2}$ because two pulses are needed to generate light velocity. In the classical SI units $3G_s$ is:

$$(1.6021773 \times 10^{-19})^{1/2} \times 1/1.15470538^2 \times 1/1.04^{1/60} = 3.000079 \times 10^{-10} \equiv 3G_s. \tag{14.21}$$

In integer cubic model the result is exact on the calculator

$$(1.0000 \times 10^{-19})^{1/2} \times 1/1.111 \ldots .^{1/2} = 3.0 \times 10^{-10} \equiv 3G_s \tag{14.22}$$

Inserting the velocity c_s in the Larmor formula incorporating integer model SI units derives result that is ten times too high. The accurate derivation of the above six equations derives other application. Data with c_s^3 caused historical sensation in physics, which can be now explained. Two pulses at 90^0 are needed to propagate one pulse at the velocity of light in one orientation. With radiations occurring in six orientations this is taken to represent total power. The third c_s term converts with quantum corrector atomic terms to integer SI units. Space is absent in the probability integer model, which makes velocity of light just the pulse rate this being the resultant pulse rate of three or more orientations:

$$(c_s \times 6)X(c_s \times 6) \times c_s \times 1.04^{7/10} \times 1.04^{1/40} = 1.0000355 \times 10^{27} \text{ pulse rate} \qquad (14.23)$$

With pulse energy 1×10^{-27} J, (Equations 2.81-2), the input is 1 W.

The Julian year duration with eight decimal places is 0.3155780×10^8 s. This may express the pulse rate distributed into 3D by pulses brought in 1D, the rate represented by value $1/3 \; c_s$:

$$3 \times 0.3155780 \times 10^8 \times 1.111 \ldots \times 1.04^{1/40} = 0.9998978 \times 10^8 \equiv 1 \times 10^8 \equiv 1/3 \; c_s \text{ with } 1 \times 10^{-7}$$
$$\text{accuracy} \qquad (14.24)$$

The source of high sun powers in SI units arise due to high energy input required by wave fuction format of two vectors, which transform them into $\sqrt{2}$ or $\sqrt{3}$ lager vector moving with velocity of light. Conversion of each vector initially at atomic units requires a probability (chance) multipier 10^{30} and for two a multiplier 10^{60}. The atomic to SI conversion needs a multipler derived as the ratio 1 J/1 $\times$ 10^{-15} J, from Equations 2.36-7 applying both for electrons and protons. Both value generate the total converter value 10^{75}. This value should be applied to minimum energy value of 10^{-45} J generating 1×10^{30} W. Thus with quantum corrector of $1.04^{29/40}$ the total sun power at the source in 60 orientatios is:

$$\text{Total sun power} = \sigma_s T^4_{6000} \times A_{cirsun} \times 6 \times 1/10 \times 10^{15} \approx 60 \times 10^{100} \text{W} \qquad (14.25)$$

Such huge power should be treated with a lot of skepticism. Above $60 \times 10^{100}/10^{75} \times 10^{-28} = 0.06$ J expressing energy of Planck radiations in six directions.

Because of its 1 m reference SI system hugely oversizes atomic dimensions and powers. The sun interacting power with earth 4×10^{29} W is given in the following expression:

$$\sigma_s T^4_{6000} \times A_{cirsun} \times 1/2 \times 6\,000 \times 1.11111111^3 = 3.9999999999 \times 10^{29} \text{ W} \qquad (14.26)$$

And it might be related to the total SI sun power in the following relation: $\qquad (14.27)$

6(orientations)$\times$(4$\times10^{29}$ W)$\times$1.5(with electron contribution)$\times$1$\times10^{12}$ Hz$\times$1.04$^{1.1} \equiv$ 3.75 $\times10^{42}$ W. The value of 60 is an invitation to validate the sun temperature value 10000 K.

Pulse generated from 1kg of sun at 10000 K is $10000^{1/60}/1.154700/1.04^{1/4} \equiv 0.99986$ W $=1$ eV-1.5k_s; 10 000$^{1/32}$/1.154700^2 $\times1.04^{1/300}=1.0000093$ W or J/m or eV per pulse with variance $1/10 \; k_s$ $\qquad (14.28)$

Circular sun area of 1.5 $\times 10^{18}$ m^2 derives sun volume (vol) at 1.3819788×10^{27} m^3, 1/6 of which interact with earth, and is sending $(1/6 \times \text{sun vol})^{1/30} \times 1.33333 \times 1.04^{9/40} = 9.9964$ J/m or eV/pulse with variance $6 \; k_s$. $\qquad (14.29)$

The mass of sun with probability 1/30 generates $(2 \times 10^{30} \text{ kg})^{1/30} / 1.04^{3/5} = 9.9957$ J or eV with variance $6k_s$ (14.30)

Circular area of earth 1.27800510^{14} m^2 derives earth volume at 1.08682710^{21} m^3 reacting with its whole volume with sun with factor two because two earth pulses 300 K generate first 600 K radiation $\equiv 2(\text{volume})^{1/30}/1.04^{1/8} = 10.00084$ J or eV with variance of $8 \ k_s$. The whole earth mass interacts with sun delivering with probability 1/30 radiation of

$(6 \times 10^{24})^{1/30} \times 1.15470^2 \times 1.11111 \times 1.04^{1/5} \equiv 10.00094$ J/m or eV per pulse with variance $9 \ k_s$ variance (14.31)

There are 32 probabilities as long as source-receiver optional orientation is not active:

One kg of sun at $(6000 \text{ K})^{1/32} \times 1/1.154700538^2 \times 1.04^{5/12}$ generates 1.00052 J/m or eV/event.

Using now the effective mass of sun $(1 \times 10^{30} \text{ kg})^{1/32} \times 1.1547 = 9.99929$ J or eV displaying from 10 J variance equal $1/3 \ k_s$.

Radiation from earth mass $2 \times (6 \times 10^{24} \text{ kg})^{1/32} \times 1/(1.1547 \times 1.04^{36/50}) = 10.00228$ J with variance of $23 \ k_s$ expected because of presence of 34 orientations $23/34 \ k_s \approx 2/3 \ k_s$.

The radiation from proton $6 \times (1 \times 10^{-27})^{1/32} \times 1/1.1547 \times 104^{1/5} = 1.00062$ J (14.32)

Probability 1/32 is related to two probabilities 1/64 of the sun generating the pulse:

$(1 \times 10^{30} \text{ kg})^{1/64} \text{X} \ (1 \times 10^{30} \text{ kg})^{1/64} \times 1/1.1547 \equiv (10.000 - 0.00065)$ J with variance $6 \ k_s$ (14.33)

The total probability is 1/4095 with each pulse having probability 1/64. With the pulses discharging together the input pulse energy increases probably with corrections to: $4095 \times 1.2 \times 1.1547 \times 1.1111^{1/2} = 5000.16$ J.

Structure of the basic pulse source satisfying minimum energy conditions is a cube of 3x3x3 = 27 quantum steps of Bohr atoms as derived in Chapter 11. It discharges from each plane nine quantum radiations, which after relativity correction 1.1111 generate $3^2 \times 1.11111 = 10$ J. These are directed in three orthogonal orientations away from the cube centre generating together $\sqrt{3} \times 1.154799 = 2$ twice larger velocity and three times larger pulse rate. The total energy $3^3 \times 1.1111 = 30$ J. The three planes generate each nine pulses and for instantaneous radiation establishes 1/27 probability and 1×10^{-27} kg mass of proton in one direction. The three planes, generating the pulses, have 19 quantum orientations (six directions in three planes + one total) establishing probability of 1/19, which establishes the electrical charges for the electron and the proton.

The probability 1/19 applied to the proton mass with 10 input units and quantum correction derives the Planck Radiation term 0.01 J or eV with var $= k_s/30$:

$$(2 \times 10^{-27} \text{ kg})^{1/19} \times 1.04^{3/4} \times 10 \equiv 0.0100038 \text{ J or eV with var.} = k_s/30 \qquad (14.34)$$

Mass of sun generates a force of 1 N/kg with temperature increasing that force to power four and two moments in wave format being needed to produce the velocity of light. This assertion with relativity and quantum correction derives the mass of sun and the mass of earth as the uncertainty of this relation:

Mass of sun $\equiv [\{(600 \text{ K})^4 \times 1 \text{ N/kg}\} X \{(600 \text{ K})^4 \times 1 \text{ N/kg}\}] \times 1.1547^{3/2}/1.04^{1.05}$
$\qquad = 1.99999428 \times 10^{30} \text{ kg} = (2 - 0.0000058) \times 10^{30} \text{ kg} \quad (14.35)$

The variance from the mass of sun 2×10^{30} kg $\sim 6 \times 10^{24}$ kg representing for an unknown reason the mass of earth.

Unity earth mass creates first 600 K pulse from 300 K which generates 1/2 pulse of attraction

Unity earth mass $\times (600^8 \times 1.111 \ldots^{3/2}) = 1.7705 \times 10^{22} \text{ N} \approx \frac{1}{2} \times 3.5438 \times 10^{22} \text{ N} \qquad (14.36)$

While 1 kg at $600^4 \times \varepsilon_{os}/(1.1547^{3/2} \times 1.04) = 1.0043 \approx 1$ eV, J or W $\qquad (14.37)$

Comparing above with $\sigma_s T^4 = 405$ W we find that $\boldsymbol{\sigma_s/\varepsilon_{os} \times 1.2 \equiv 6000 \text{ K}} \qquad (14.38)$

Attributting to earth probability corrected by no-reacting factor and quantum: 1kg at temperature 300 K with probability delivers:

$300^{1/30}/(1.2 \times 1.04^{1/2}) \equiv (1 - 0.00004)$ J or eV expressing variance $0.4 \times 10^{-23} = 1/3 \; k_s \times 1.2$.

15. Components of velocities of light, relativity corrections and time constant of the gravitation pulse

Attraction and repulsion has to be explained in better ways than by positive and negative charges. Attraction is now represented by the loss of potential energy; repulsion is represented by the gain of potential energy. The model postulates that initially the particles were electrons. Electron is really not a particle; it is a set of parameters characterizing wave motion: momentum (impulse), energy, frequency, power and energy density distribution including maxima and minima characterizing the wave of physical dimensions. Above parameters with 2001 units of proper probability and distribution terms generate emission of proton mass expressed by 2×10^{-27} kg value.

The basic radiation energy $U_{ps} = e_s^2/\varepsilon_{os} = 1 \times 10^{-27}$ W, is expressed by mass of proton m_{ps}, and augmented by five probability pathways of space quantization over light velocity. That total energy per meter, derives the graviton energy $g_J = 1/3 \times 5 e_s^2 \times 1/(\varepsilon_{os} \times 1/3 \ c_s) \equiv 1.66 \ldots \times 10^{-35}$ J/m. **For an event to occur with unity mass displaying the probability of unity the measured velocity of light c_s is derived from the probability P_v = [Invert of space quantization (10)].x(unit length 10^{-10} m of space cube)/(1 m of SI) = $10/10^{-10}$ = 10^{-9}:**

$$3 \ P_v \ x \ c_s \ x \ 1.1111 = 1.000 \text{ deriving } c_s = 3.000 \ldots \times 10^8 \text{ m /s} \tag{15.1}$$

This is the fundamental requirement for unit event display and the reason for light velocity having value $3.000 \ldots \times 10^8$ m /s.

To include the possibility of protons moving with velocity $1/2 \ c_s$, the event of unity may be attributed to represent, one event of splitting density mass in both directions, and so:
$$3 \ P_v \ x 1/2 \ c_s \ x \ 1.1111 + 3 \ P_v \ x 1/2 \ c_s \ x \ 1.1111 = 1.000$$

Similarly
$$1/2 \ c_s = 1/2 \times 3.000 \ldots \times 10^8 \text{ m} \tag{15.2}$$

The value of the velocity of light has other specific derivations given by
$$2\pi v_{Hms} \ x 2a_{os} \ x 10 x 10/1.111^{1/2} = 2\pi \ x5x10^{15} \ x1 \ x10^{-10} x100/1.054087 = 2.98038 \ x10^8 \text{ m/s}$$

And
$$c_s = 3F_{atoms}/m_{es}^{1/2} = 3 \times 10^{-7}/10^{-15} = \mathbf{3 \times 10^8} \text{ m/s} \tag{15.3}$$

The velocity of light is related in the following manner with the electron frequency:

$$c_s = 3x10^8 = (5 \ x10^{15})^{2/3} \ x1/10 \ x \ 1/10 \times 1.039742^{2/3} \approx 3.001 \ x10^8 \text{ m/s} \tag{15.4}$$

Identities related with sun power, temperature and elementary energy are:

$$(10 \ x \ 1/3 \ x \ c_s)^{1/3} = 1000 \text{ W/m}^2 \text{ and } 1000 \times 1/10^3 = 1.0000 \ldots \text{ eV} \tag{15.5}$$

And
$$3 \ x \ 3 \ x \ (c_s)^{1/3} \equiv 6021 \text{ K} \tag{15.6}$$

And
$$(c_s \times 1/10^2 \times 1/10^2 \ x1/10^2)^{1/3}/(1.111 \times 6) \equiv 1.004 \text{ eV} \tag{15.7}$$

135

The value 6021 comprises temperature 6000 K, plus energy 7 kT for each of the three planes with $1/2 \times kT$ uncertainty along each axis and four energies of $3.2kT = 3\,kT + 1/5\,kT$ in each direction.

The model proposed presents values of physical parameters in three forms as one, two and three dimensional presentation: 1D, 2D and 3D. They express the 'dice' probability of the result as the product of two, three, four and even six probabilities of individual events produced by product of two momenta with each momentum being formed by product of two or three space probabilities. Applying 1D presentation to Hartree energy of cubic space (30 eV) obtains for 3 planes

$$30^{1/3} \approx 3 \text{ eV} + 0.1 \text{ eV, and } 27.21381^{1/3} = 3.0079 \text{ eV} \tag{15.8}$$

Applying 1D presentation to velocity subjected to two stages of space quantization derives the sun radiation power:

$$(c_s \times 1/10 \times 1/10)^{1/3} \equiv (3 \times 10^8 \times 1/10 \times 1/10)^{1/3} = 1441 \text{ W/m}^2 \tag{15.9}$$

$$(c_s)^{1/3} = (3 \times 10^8)^{1/3} \equiv 669.39 \text{ eV} \equiv 669.39 \text{ W/m}^2 \tag{15.10}$$

The above energy of the unity mass moving at 1D velocity of light time relativity $1.1111^{1/2}$ and divided by quantum correction $1.04^{1/9}$ expressing 702.5 W is identified with the thermal component of the sun pulse and the balancing thermal component of the earth pulse of Equation 10.4.

In the proton mass of 2000×10^{-30} kg there are about 666.66 electrons rotating in each plane. The difference $(669.39 - 666.66) \times 1.111 \ldots \equiv 2.73$ eV $\times 1.111 = 3$ eV, while 2.73 eV is the space radiation energy. This equilibrium condition may need an interval of time to be reached.

On the other hand the energy of 666.66 electrons oscillations of mass 10^{-30} kg at 1×10^{-15} Hz of 3 planes in one of two directions, expressed by ½, corrected for volume derives:

$$\text{Electron energy per K} = 1/2 \times 3 \times 666.66 \times 1 \times 10^{-30} \times (1 \times 10^{-15})^2 \times 1 \times 10^{-33} \times 1 \times 10^{-30} = 1 \text{ J} \tag{15.11}$$

The classical SI Hartree energy $[1/4\pi\varepsilon_{os} \times e/\varepsilon_o^2] = 27.2138\text{eV}$, multiplied by $1.111 \ldots$ and divided by $1.04^{1/5} \times [1+1/(80 \times 300)]$ derives $30.00000+0.000046$ J or eV converting Hartree energy to space value of 30 eV with uncertainty $\sim 1/2\,k_s$.

Velocity of light can be derived by expressing cubic mass of earth in one dimension over $\sqrt{3}$ with quantum corrector

$$c_s \equiv \sqrt{3} \times (6 \times 10^{24} \text{kg})^{1/3} \times 1/1.111 \ldots^{1/2} \times 1.04^{1/7} = 3.00260 \times 10^8 \text{ m/s} \tag{15.12}$$

or the local velocity $1/3\ c_s \equiv (6 \times 10^{24} \text{kg})^{1/3} /\sqrt{3} \times 1.039742^{5/4} = 1.0007 \times 10^8 \text{ m/s} \tag{15.13}$

Using classical SI mass of earth in 1D display with relativity and quantum correctors derives velocity of light with acceptable accuracy

$$c \equiv \sqrt{3} \times (5.9742 \times 10^{24}\,\text{kg})^{1/3}/1/1.111\ldots^{1/2} \times 1.04^{1/7} = 2.9963 \times 10^{8}\,\text{m/s} \qquad (15.14)$$

The product of two unity vectors of probability $(3 \times 10^8)^2$ corrected by relativity factor $1.111\ldots$ forms energy of space parameter $E_{sspaceSI}$:

$$E_{sspaceSI} = 1\,\text{kg} \times c_s^2 \times \text{rel. cor.} = (3 \times 10^8)^2 \times 1.111\ldots = 1 \times 10^{17}\,\text{J-s} \qquad (15.15)$$

This frequency is related to values in Equations 7.10 and 13.7.

Integer model Planck constant h_s in 2D display, over relativity correction, derives exactly pulse energy per orientation expressing the invert of the space pulse :

$$1/E_{sspaceSI} \equiv 1/3\,h_s^{1/2}/1.111\ldots^{1/2} = 1.000\,000\ldots \times 10^{-17}\,(\text{J-s})^{-1} \qquad (15.16)$$

In 1D display Planck constant h_s displays permittivity of space ε_{os}:

$$\text{Permittivity of space } \varepsilon_{os} \equiv h_s^{1/3} = (1 \times 10^{-33})^{1/3} = 1 \times 10^{-11}\,\text{Fm}^{-1} \qquad (15.17)$$

The corrected value of velocity of light displayed in two dimensions expresses the energy of the electron mass converted to energy:

$$c_s^{2/3} \times 1.111\ldots \times 1.04^{5/4} \equiv 1/2 \times m_{es}c_s^2 \times 1.111\ldots/1 \times 10^{-19} \qquad (15.18)$$
$$0.49989 \times 10^6 \approx 0.5 \times 10^6$$

The corrected light velocity in one dimensional display expresses the value of space impedance:

$$c_s^{1/3}/1.039742^{1/10} \equiv 666.82\,\text{ohm} \qquad (15.19)$$

One dimensional display of velocity of light in three planes corrected by squared relativity factor and by quantum derives the temperature of the sun in one rotational orientation of $\tfrac{1}{4} \times 6000\text{K} = 1500\,\text{K}$ plus energy of one degree of freedom $\tfrac{1}{2}\,k_sT_{600}$:

$$3\,c_s^{1/3}/(1.1547^2 \times 1.04^{1/10}) \equiv 1506.22/1.04^{1/10} \equiv 1500 + 0.30\,\text{J} = 1500 + \tfrac{1}{2}\,k_sT_{600} \qquad (15.20)$$

Velocity of light to power $\tfrac{1}{4}$ generates with quantum correction the invert of the fine structure α_s:

$$c_s^{1/4} \equiv 1.04^{1.04}/\alpha_s \qquad (15.21)$$

Three pulses of equivalent mass $\sim 1 \times 10^{-30}$ kg with squared velocity $1/3\,c_{se}^2$ m/s derive energy $1 \times 10^{-30}\,(1/3\,c_s)^2 = 1 \times 10^{-14}$ J with energy from three orthogonal presentations being subjected to three probability events generating output $(1 \times 10^{-14})^3 = 1 \times 10^{-42}$ J. This energy is subject to effective proton frequency 1×10^{12} Hz generating energy per cycle of 1×10^{-54} J. With number of protons per kg being 1×10^{54} the power per cycle is 1 W, while power 1×10^{-42} Jx1×10^{12} Hz $= 1 \times 10^{-30}$ W. This calculation in SI units is actually done at atomic level because some classical SI parameters refer to atomic data so that at SI level power is 1×10^{-30} W x$1 \times 10^{30} = 1$ W. Two such pulses are subjected to conversion from 1 W atomic to 10 W classical S1 1 Wat x1 x$10^{20}/1 \times 10^{-19} = 10$ WSI or 10 J.

The set of the proposed integer relativity constants of space were derived solely from the properties of the electron fitting the elementary space with other observers seeing possibly different views of particles. Large part of reality can be quite well represented by the mean values of electron and of proton frequency and by various probabilities.

Time constant of gravitation radiation pulse.

There is no time function in Equation 6.1, so that in classical SI units sun gravitation is a static force. Relativity corrected cubic integer space units converted to SI expressed initially the gravity force also as a static force, its time function being considered presently.

The velocity of light was derived in various expressions. Surprising 10 year period or inverse of 1/10 probability derives also c_s:

$$c_s \equiv 10 \text{ years in seconds}/1.111\ldots^{1/2} \equiv 10x0.31557 \text{ x}10^8/1.111\ldots^{1/2} \equiv 2.9938 \text{ x}10^8 \text{ m/s} \quad (15.22)$$

The other expression defining a possible source of c_s is derived by the examination of the incoherent radiated energy expression using the linear probability 10^{-10} and Z_{ss} of 333.33 ohms, $I_{os} = 5 \text{ x}10^{-11}A \equiv 5\varepsilon_{os}$ [46, Equation 9.13]:

$$5 \text{ x } 1.111\ldots^2 \text{x } 10^{-20} I_{os} \text{ x } Z_{ss} \equiv 1.0286 \text{ x}10^{-27} \text{ J} \quad (15.23)$$

The above expression will be compared with its equivalent, where $I_{os} = 1/2 \text{ x } 10^{-27} \text{ x } c_s^2 x$ 1.1111:

$$5 \text{ x } 1.111\ldots^{1/2} \text{x}10^{-20}\text{x} \int_o^\infty I_{os} e^{-t/\tau} dt \equiv 1.0286 x10^{-27} \text{J} \quad (15.24)$$

Comparison of these two identities with I_{os} = unity at atomic level derives $\quad (15.25)$

$$I_{os} xZ_{ss}=5x10^{-11} \text{ x } 333.3=1.666 \text{ x } 10^{-8}=1/2 \text{ x } 10^{-27}c_s^2 \text{ x } 1.111 \ldots \text{ x } 333.33=\int_o^\infty I_{os} e^{-t/\tau} dt = I_{os}\tau$$

finding surprisingly that **the velocity of light is 1/2 of the inverse of the time constant τ_s =1.6666 x 10^{-8} s of the radiation packet**. The pulse wave length is 5 m, but multiplying by space quantization of 1/5 and linear probability derives 1 x 10^{-10} m as the diameter of Bohr atom. In cubic SI units the pulse length is one meter becoming 1 x10^{-10} m in atomic cubic units. The time constant τ [s] value compares well with Eisberg's [33, Eq. 13.118] derivation of the duration of excitation states at $\tau_s \approx 10^{-9}$ s. The derived now velocity of light:

$$c_s = 1/(2\tau_s) = 3 \text{ x}10^8 \text{ m/s} \quad (15.26)$$

Relativity correction $K_{rel} = 1/[1-(1/10^4)^2]^{1/2}=1.0005$ time $10\sqrt{2}x1.02$ derives 1.00721, which compares well with 30000/29786 =1.00721, indicating that allowing for energy in 10 pathways and quantum corrector **derives a round value of 30000 m/s** for earth linear velocity. It is suggested also that after similar corrections are made to 2 x10^{30}/1.989 x10^{30} = 1.0055 and to 6 x10^{24}/5.9742x10^{24} = 1.0045 the SI mass of sun interacting with earth should be treated as the integer **2 x10^{30} kg** and the mass of earth as **6 x 10^{24} kg.**

Since radiations from the stars are derived from iron mass to energy conversion the above time constant of the elementary radiation pulse and the related velocity of light is likely to apply to the whole universe.

It is suggested that the exponential pulse $e^{-t/\tau s}$ from Bohr atom is followed by reflected pulse, $e^{-t/\tau s}$. It radiates for 10×10^{-19} s, by which time the converter, 1×10^{19}, generates 10 eV. The elementary wave vector force is formed on the sun or planet surface **and it is sustained until the reflection covering distance $3a_o$ cancels it.** Forward pulse and its reflection oscillate at 5×10^{12} Hz forming the radiation wave carrying heat energy, but no directed force. Reflection may be from mass at the back and from an earlier pulse. Multiplying 10 eV by 1.154572 and adding kT terms forms ionization energy of the hydrogen atom. Maxwell waves are believed to be produced by multitudes of elementary pulses or waves.

Since all 1D signals are generated by movements of a dipole along its axes away from the receiver they are sourced from a particle receding at a large fraction of velocity c_s away from the receiver. Consequently **it is impossible to assess firmly any part of infrared shift, that can be contributed to the expansion of the Universe.** Paul Marmet [41] expressed other objections of the very common presumption that the Universe is spreading with velocity of light.

The square of the velocity of light converts the signal from its static state described by ε_{os} to the dynamic state of μ_{os} with $\varepsilon_{os} c_s^2$ **x 1.111 . . .=$1/\mu_{os}$.**

Individual pulse values would be distributed at Boss-Einstein statistics or Boltzmann statistics within limits of Heisenberg uncertainty. Fractions of Bohr atom ionization energy of 1 eV form 10 eV, making 13.63 eV after relativity correction with three or four units making a total of 50 eV delivered by the sun. This value represents energy of Bohr atom:

$$F_{as} \text{ x } a_{os}/e_s = 1 \times 10^{-7} \times 1/2 \times 10^{-10}/1 \times 10^{-19} = 50 \text{ eV} \qquad (15.27)$$

The transmitted variations within the radiation pulse occur only in the direction of progression and include probably different components of the velocity of propagation of infrared frequency. There is negligible high electron frequency vector in the earth gravity signal.

Relativity increases the mass of radiation particles and it decreases in the same way their velocity so keeping the momentum constant [Beiser, 34]. Decreasing velocity causes time dilation modifying the length observed by different observers.

The vector of gravitation radiation is believed to carry mass and hence '*mass(kg)* xc^2' carries also mass. It **is accepted that oscillations in photons cancel opposing momenta. Hence it is suggested that in gravitation radiation opposing momenta can be also nulled.** The effect on total energy is discussed in the Appendix.

The proposed local velocity $1/3\ c_s$ in three orthogonal orientations derives the light velocity c_s as their vector sum, relativity and quantum corrected:

Integer model velocity of light, accuracy limited by the calculator

$$=\sqrt{3} \text{ x} 1 \times 10^8 \times 1.154700538^3 \times 1.1111 \ldots \times (1+1/80) = 2.999\ 999\ 97 \times 10^8 = 3 \times 10^8 \text{ m/s} \quad (15.28)$$

Another derivation of velocity of light

Three local light velocities of 1×10^8 m/s in orthogonal orientations subject to relativistic correction add vectorially generating

$$\sqrt{3} \times 1/3 \times c_s \times 1.154700 = 1.9999991 \times 10^8 \text{ m/s} \tag{15.29}$$

Two such pulses displaced in space by $90°$ are in wave format generating velocity of light:

$$\sqrt{2} \times 1.9999991 \times 10^8 \text{ m/s} \times 1.1111 \ldots^{1/2} \times 1.04^{1/6} = 3.00097 \times 10^8 \text{ m/s} \tag{15.30}$$

The variance is the accuracy limit of $\frac{1}{2}$ of 3×10^{-5} s/m derived in Conclusion 64.

16. Gravitation force within hydrogen atom and relativity dilating physical parameters of space

Two energy states of the hydrogen atom will be considered: full excitation and 'zero' energy level. At 'zero' energy level the three oscillators i.e. the electron, the proton and the pair (proton + electron) have each energy 0.05 eV. At full excitation the relativity corrected radiated energy of the atom in three directions in terms of space cubic units converted to SI is $13.62 \times 3 \times 1.2 \times 1.04^{1/2} = 50.003$ eV. This $\sqrt{3} \times 30/1.04 = 49.963$ eV is attributed to total conversion of Bohr atom to radiation. With other three components the total energy is 50.15 eV. Helium has ionization energy ~25 eV so it molecule's ionization energy is also ~50 eV. The product of one half of minimum energy and the square of local velocity of light $1/2 \times 10^{-14} \times (10^8)^2 \equiv 50$ eV$\equiv 50$ J is identified also for six orientations to express the earth temperature 300 K. The electron mass interactions with the proton transfer momentum from electron to proton in steps of 0.001 eV expressed by

$$G_s \times 10^7 = 1 \times 10^{-10} \times (1/10 \times 1/10 \times 1/10)(1/10 \times 1/10 \times 1/10) \times 1/10 \equiv 0.001 \text{ eV} \qquad (16.1)$$

The converter 10^7 becomes $1 \times 10^{-7} = 1 \times 10^{-10} \times 10 \times 10 \times 10$ on converting in the opposite direction.

The internal atom radiation pulse of duration ~10^{-19} s forms a vector force, which pulls proton away from the center of the atom by the formation of the gradient dW/dx. The pulse generates 1 eV per kg of mass in one orientation leading to equipartition of 50 eV energy between the three oscillators discussed earlier. Individual energies corrected by relativity become:

$$1/3 \times 50/1.111 \ldots = 15 \text{ eV} \qquad (16.2)$$

At this stage $dW/dx = 0$, the atom becomes unstable generating radiation exciting other atoms, which are at zero energy level. At level of 0.05 eV the energy can be expressed by

$$v_{pr} h_s/e_s = 5 \times 10^{12} \times 1 \times 10^{-33}/1 \times 10^{-19} = 0.05 \text{ eV} = 3 \times 1/5 \times k_s T_{300}/e_s \times 1.111 \ldots \qquad (16.3)$$

The graviton energy measured by the author some 35 years ago as quantum $1/3 v_{pr} h_s/e_s = 0.01666$ eV is described in [47] and in the Appendix.

One could argue that the proton is being further heated by electron bombardment by a further 300 K because the final energy ratio 15 eV/0.05 eV = 300.

The gravitation force of attraction between the mass of electron and mass of proton at atomic level is evaluated at 0.006 N. It is derived from the base force of 10 N, power of 10 W and energy of 10 J radiated per kg mass of earth per kelvin in each of three planes converting to 1 N per orientation: $10/6 \times 2/3 \times 1/1.111 = 1$ N. The density decreases with the square of the distance from the source and applies a force on the mass at the other location. The force between the mass of electron and mass of proton appears to be the same as the gravity force of the sun per kilogram of the mass of earth.

Atomic F_{Gsa} = √3 x10 [N/kg] x $m_{es} m_{ps}/(a_{os})^2$ x3 x300 [K] x1x10³⁰/1.04 = 0.0059956 N
≡ 0.006 N = 3.6000x10²² N/6x10²⁴ kg of earth $\qquad$ (16.4)

Since in kinetic terms of Equation 2.52a Force ≡ Energy, force of 0.006 N in three orthogonal orientation with quantum corrector derives 1ˢᵗ term of Planck Radiation Energy:

1ˢᵗ term of Planck Radiation Energy = √3 x 0.006 N/1.04 = 0.0099926 J ≈ 0.01 J $\quad$ (16.4a)

$\qquad$ Above definition applies to the force of sun attraction on earth of 1 N/kg spread over surface of the sphere along which earth is rotating, with π replaced by 3. The force, in model integer SI units, delivers a surface pressure generated by the sun mass: 2 x10³⁰ kg/[4 x 3 x $(1.5x10^{11} m)^2$] = 7.407x10⁹ kg/m². That value, times relativity squared and extending over six circular areas of earth, expresses the force acting on earth from all sides, as shown in Fig. 7, 7.407x10⁹ kg/m² x 1.1547² x 6 generating times 1 N of G resultant 5.92599 x10¹⁰. To obtain 0.0059 N it has to be multiplied by linear probability 10⁻¹⁰ and space quantization 1/10³. The converter 1/10⁷ = 10⁻¹⁰ x 1/10³ is used in Table 14. The force 0.00592599 N/kg multiplied by earth 6x10²⁴ kg over quantum correction 1.04^{1/12} generates as shown earlier **the reference force of gravitation 3.5438 x10²² N.**

$\qquad$ Temperature corrected energy per plane is the Boltzmann constant

Boltzmann constant k_s ≡0.006/3 x 300/6000 = 0.0001 N ≡ 0.0001 Jx1x10⁻¹⁹=1x10⁻²³ J $\quad$ (16.5)

$\qquad$ Force 0.006 [N/(kg-K)] x 6 directions x 300 [K]/1.04² derives 10 N/kg.
$\qquad$ Force 0.006 N/(kg-K) time (300 K x1.111 . . .) delivers on earth 2 W ≡ 2 eV, equal also the gravitation force spread over sun circular area, requiring two step quantization and relativity with quantum corrections:

Force density=3.5438 x10²² N/1.52179x10¹⁸ m²x1/10²x1/10²x1/(1.1547x1.04^{1/5})
$\qquad$ =2.00094 N/m²≡2 W/m² $\qquad$ (16.6)

The atomic force F_{atoms} =10⁻⁷ N at 300 K for 2 oscillators increases with the cube of temperature; square value indicated in Equation 2.52a is the formative step increasing local velocity of light and forming a new magnitude requiring now temperature multiplier:

$$F_{atoms\,300} = F_{atoms} \text{ x300}^3 [K^3]/(1.1547^2 x1.04^{1/3}) = 1.999 \text{ N/kg} \approx 2 \text{ N/kg} \qquad (16.7)$$

The component of Equation 16.4 equals sum of atomic forces in three orientations

$$3 F_{atoms} = m_{es} m_{ps}/(a_{os})^2 x1x10^{30}/1.1547^2 = 3x10^{-7} \text{ N/m} \qquad (16.8)$$

It is suggested that since 3x10 eV is the total energy of the Bohr atom, 30 eV at 300 K can be expressed as delivering:

$$F_{Gsa\,300} \equiv \sqrt{3} \text{ x30 x300}^2 \text{ x } 3F_{atoms} \text{ x1.2 x } 1/1.1111^{1/2} x1.04^{1/3} = 2.00 \text{ N/kg} \qquad (16.9)$$

Thus generally the gravitation force between the proton and the electron is proportional to the square of temperature. At 6000 K the temperature is twenty times higher increasing the force and energy by 400, but since the signal comes from H_2 in the sun the increment increased by another factor of 2 making a total energy and power 1600 W/m^2. The 1600 W/m^2 value divided by relativity correction 1.1547 and quantum correction $1.04^{1/4}$ derives the sun constant 1372.1 W/m^2, the variance being ½ *kT*.

The finding that parameter identified earlier as a force in Equation 16.7 is energy 30 eV expressing energy of the Bohr atom inclusive of the two infrared oscillators and of the electrons requires further consideration; this is in view of the force produced being only 2 N/kg. It is neccessary to go back to the definition of temperature in Chapter 8. The force of 1 [N/kg-K] is actually derived from the room temperature, energy of ionization of the Bohr atom, representing atomic forces in three planes augmented by the addition of three orthogonal vectors by √3 and corrected by relativity and by quantum. It is important to identify, which atomic energies are quoted in electron-volts and which frequencies refer per kelvin and which refer to 300 K. Equation 16.7 shows that the force generated is 2 [N/kg-K] per atom with v_{es}=5x10^{15} Hz deriving energy equals ½ m_{es} v_{es}^2x1.2 = 25 eV x 1.2 = 30 eV or J. Energy 30 eV becomes a gravity vector after one step quantization and factor 2/3 eliminating electron oscillation component:

$$30 \text{ x } 1/10 \text{ x } 2/3 = 2 \text{ [N/kg-K]} \tag{16.10}$$

The subsequent calculation confirms that **frequency** v_{es} refers to 300 K and converts to the new frequency **at 1 K** in five quantization pathways taking the values 5x10^{10} Hz and converting to **linear frequency 1x10^{10} Hz**

$$v_{es}(10^{10})=1/5 \text{ x } 5\text{x}10^{15} \text{ [Hz]}/(1.111 \ldots \text{x}300^2 \text{ [K}^2\text{]}) = 1\text{x}10^{10} \text{ Hz} \tag{16.11}$$

and defines the gravitation constant G_s =1x10^{-10} as energy (N-m) per unit mass (kg) per cycle of electron frequency (5 x10^{15} Hz) at 1 K per gradient of distance (m) per unit mass (kg). The frequency 5 x10^{15} Hz can be applied to protons as a congregant of electrons.

The forces of the integer gravitation model were successfully applied between the proton and the electron of the Bohr atom in Chapter 23 of author's book 2 [46]. Force, energy, power and energy gradient expressed in terms of the square of Bohr atomic radius relativity corrections and linear probability is:

$$F_{atSI} =(1/2 \text{ x}10^{-10})^2\text{x}10^{20}/(1.111 \ldots^2 \text{ x } 1.0328^{1/2}) \text{ x}10 \approx 2 \text{ N} \equiv 2 \text{ J} \equiv 2 \text{ eV} \equiv 2 \text{ J/m} \equiv 2 \text{ W} \tag{16.12}$$

These elementary particles from sun are in dynamic equilibrium with the gravity of earth of 10 N/kg because, besides other forces, they are subjected to a radial outwards acceleration by the force of gravity from the mass of earth located at the centre:

Outwards acceleration: F_{acc}=10 m_{earth} (2/3 rad_{earth}) x 10^{-27} J/kg/(1/4 time of revolution) (16.13)

$$= 10 \times 6 \times 10^{24} \text{ kg } (2/3 \times 6 \times 10^6 \text{ m}) \times 10^{-27}/(1/4 \times 24 \times 3600) = 10 \text{ N/kg}$$

Please note that $6 \times 10^{24} \times 10^{-27} = 0.006$ N/kg and that 10 is the converter atomic to SI units $10^{20} \times 1 \times 10^{-19}$.

The forces of repulsion cancel the forces of attraction inducing long term stability to the Universe. Repulsion takes place in different ways: a planet rotating around a sun or a sun around a black hole or center of galaxy; rotation of any atomic particle around another particle sole or in groups; revolution of a group of particles around themselves forming a planet or a sun. The balance can survive billions of years or be over in seconds as stability of protons is ended, or rotation terminated. In the final stage of research it was concluded that any rotation of particles having a central equivalent mass induces instability to some protons, inducing radiation by mass to energy conversion, which balances the gravity forces of the central mass.

The force of repulsion F_{Rs300} between the electron rotating at speed 1×10^7 m/s and the proton is expressed by the Equations 13.4-5 derived for the sun to earth couple. It is doubled because the spherical geometry is replaced by a planar geometry.
The current $1 \times 10^7 \times 10^{-19} = 1 \times 10^{-12}$ A and converter 1×10^{20} changes atomic units to SI units

$$F_{Rs} = 1/3 \times \mu_o I/2\pi \times 1 \times 10^{20} \times 1/1.04^{7/10} = 1/3 \times 1/9 \times 1 \times 10^{-8} \times 1 \times 10^{-12}/2\pi \times 1 \times 10^{20} \times 1.04^{9/20} = 0.0059996 \text{ N}$$
(16.14)

Above identity of Equations 16.4 and of 16.14, between the force of attraction per kilogram of the sun to earth force with the force of electron attracting proton per unit atom applies both to the force of attraction and to the force of repulsion. This is an extraordinary finding and justifies expressing units of energy in joules as electron-volts per atomic event between atoms.

The gravitation constant is the space gradient of the infrared frequency energy contained within the Bohr atom radius numerically equal:

$$G_s \equiv h_s v_{Lms}/a_{os} = 1 \times 10^{-33} \times 5 \times 10^{12}/0.5 \times 10^{-10} = 1 \times 10^{-10} \text{ N-m}^2/\text{kg}^2 \qquad (16.15)$$

The value obtained is the product of 1 eV times the linear probability 1×10^{-10}.

In five space quantization directions and in three planes the product of proton mass and 10 W/kg supplies to the electron energy of 30 eV.
Energy from atomic mass:

$$E_n = 5 \times 3 \times m_{prs} \times 10 \text{ eV/kg } /(0.5 \times 10^{-10}) \times 10^{-30}/1 \times 10^{-19} = 30 \text{ eV} \qquad (16.16)$$

is divided equally at 10 eV between oscillations of electron, of proton and of (proton + electron) pair.

The energy in three planes per K is

$$dE_{ns3planes} = h_s v_{Lms} \times \ln 2 /(1.111 \ldots \times 1.04 \times e_s) = 0.029992 \text{ eV}$$

$$= 0.03 \text{ eV} \pm 8 \text{ parts per } 3 \times 10^4 \text{ parts} \qquad (16.17)$$

Thus the force 0.006 N is generated by event probability of 5 quantization pathways each displaying energy 0.03 eV.

At 300 K time 1.111 . . . above energy derives 10 eV: 0.03 eV x 300 K x 1.111 . . . =10 eV

Atomic force for Bohr atom is $F_{atomic} = \sqrt{(h_s/e_s)} = 1 \times 10^{-7} \text{ N}$ $\qquad (16.18)$

To derive energy requires multiplication by a_{os} x (2/3) /1.111 . . .

Hence Bohr atom energy i.e. Hartree energy=F_{atomic} x a_{os} x (2/3)/1.111 . . . $e_s =$ 30.00000 eV

The relation is believed to integrate energies in five local directions F_{atomic} x $a_o/e_s =$ 5 x 10 eV, each expressing after relativity and quantum corrections the dissociation energy 13.62 eV. Energy 50 eV is the sun pulse energy and

$$\text{Energy } E_n = F_{atomic} \text{ x } a_o/(1.15474^4 \text{x } 1.04^{1/5}e_s) = 30.00 \text{ eV} - 1/2 \text{ x } k_s T_{300}/e_s \qquad (16.19)$$

Force times velocity is power. Hence the force of 1 N/kg acting on the electron mass, m_{es} with electron circular motion at velocity $v_e = 1 \times 10^7$ m/s, derives power equal numerically to the Boltzmann constant k_s:

$$m_{es} \text{ x 1N/kg x } v_e = 10^{-30} \text{ x } 10^7 = 1 \times 10^{-23} \text{ W} \equiv k_s \qquad (16.20)$$

The proton attracts the electron with the same force directed in the opposite direction making a total of 2 N. The force of sun gravity per kelvin per atom corrected for non-interacting radiation and for quantum is

$$3.5438 \times 10^{22} \text{ N}/300 \text{ x} 1 \times 10^{-19} \text{x} 1/1.2 \times 1.04^{2/5} \equiv 9.99953 \text{ N} \approx 10 \text{ with variance} \approx 5 \text{ } k_s \qquad (16.21)$$

Thus in consequence of Equations 16.4 and 16.14 equating atomic and gravitation forces, the 10 eV energy of Bohr atom defines the gravitation force between the sun and earth. This most interesting finding is expressed now in a simple equation!

Power index expressing probability of 1/(3 planes) in 1/(4 orientations) derives accurately 1 N:
$$(3.5438 \times 10^{22} \text{ N})^{1/12} \equiv 1.000072 \text{ N} \qquad (16.22)$$

In two directions 2 x10 eV = 20 eV per Bohr atom, the same value as derived in SI classical units.

Atomic forces in three orthogonal directions over electron mass in 2D (rotating) define the velocity of light:

$$c_s = 3F_{atomics}/m_{es}^{1/2} = 3 \times 10^{-7}/(1 \times 10^{-30})^{1/2} \equiv 3 \times 10^8 \, m/s \qquad (16.23)$$

Mass of 1 kg generates force of 1 N decreasing with square of the distance and on application the force acting on the attracted object becomes proportional to its mass. That argument can be applied starting with either mass. Thus each mass attracts the other mass with the same force requiring hence no reflection. In the vicinity of earth vectorized thermal flux of earth nulls vectorized flux of sun. Nulling could be interpreted in different ways. All along the long sun beam from the sun to earth the sum momenta are equal and opposite to momenta from earth released as the consequence the centrifugal force inducing acceleration of the earth. The acceleration of earth induces release of energy from instability of protons which is balanced by one quantum location from the sun. There is no firm indication whether sun is also accelerated in a similar process or of the fraction of atomic conversions within the sun that become the source of mommenta in the sun beam.

Application of the above argument to the integral values of masses of sun and of earth and integer distance between them generates the following Special Relativity gravitation force after correcting for relativity:

$$F_{Gs} = G_s \times m_{sun} \times m_{earth}/(Learth)^2 \times 1/K_{rel}^2$$
$$= 10^{-10} \times 2 \times 10^{30} \times 6 \times 10^{24}/(1.5 \times 10^{11})^2 \times 1/1.1547^2 = 4.0000035 \times 10^{22} \, N \qquad (16.24)$$

corrected for quantum factor and relativity of $1/3 \, c_s$ derives the exact SI reference value:

$$4.0000035 \times 10^{22} \, N/(1.1111\ldots)^{3/2} \times 1.04^{1/10} = 3.5438 \times 10^{22} \, N \qquad (16.25)$$

As the consequence of the Special Relativity Theory different observers see large variations in measured time, and in distance due to variations in the observer speed relative to the reference. Presently these variations are observed in the force and in the distance.

The mutual force of attraction contests the need for the force of reflection produced by the centrifugal force. Several different forces take part in balancing various part of the large structure of the sun and planets. The earth radiating in all orientations seems to be unable to balance the sun gravitation concentrated within its beam attracting the earth unidirectionally. The earth needs rotation and outward directed acceleration redirecting vectors opposing the sun. The balanced gravitation between the sun and planet historically does not consume energy, but bodies must have energy sources to sustain balance. Rotation in four planets unwinds electron spin of electron momenta forming the protons, causing protons to release some energy balancing sun gravitation.

Attraction implies release of energy from the mass and it requires source of energy. That may be by regeneration of absorbed energy or a limited life of mass, which has to be also considered. Corrections involve relativity, $1/3c_s^2 \times K_{rel}$, volume probability, and quantum factor and space quantization incorporating three orthogonal vectors. To generate an output vector pulse rotational symmetry for both particles requires due to space quantization probability 10^4 time higher input energy than the input energy

$$\text{Input energy} = 0.001 \, eV \times (10 \times 10) \times (10 \times 10) = 10 \, eV \qquad (16.26)$$

$$\text{Otherwise Energy} = 1/12 \times 10 \, eV \times 1.111\ldots^2/1.04^{7/10} = 1.00009454 \, eV \qquad (16.27)$$

The force of attraction requires energy (momentum in atomic terms) derived from masses of electron and of proton generating forces of 1 N/kg from mass to energy conversion on acceleration.

During radiation initiated by an external turbulence

$$\text{charge } e_s = 1/2 \; m_{pds} \; \text{x} 1/3 c_s = 1 \text{x} 10^{-27} \; \text{x} 1 \text{x} 10^{8} = 1 \text{x} 10^{-19} \; \text{C} \tag{16.28}$$

The above relation implies that momentum of Bohr atom m_{pds} x1/3 x c_s is split in two opposite flow directions (see Equation 12.2), such event being plausible because mass in atomic units is expressed by momentum.

With 1/3 c_s being the converter to SI units velocity 1/3 x c_s derives the elementary energy $1 \text{x} 10^{-14}$ J in

$$\text{Energy per each of 10 orientations } 1/3 \text{ x } c_s \text{ x } 1/3 \text{ x } c_s \text{ x} 10^{-30} \equiv 1 \text{x} 10^{-14} \text{ J} \equiv 1 \text{x} 10^{-14} \text{ eV}$$
$$\text{requiring input of } e_s \; c_s^2 \; ....x1.111 ... = 1 \text{ x} 10^{-19} \text{ x } (3 \text{x} 10^8)^2 \; x \; 1.111 ... \equiv 1 \text{x} 10^{-13} \text{ eV} \tag{16.29}$$

Now energy of five quantization paths each carrying $1 \text{x} 10^{-14}$ J over the distance $0.5 \text{x} 10^{-10}$ m adds to

$$5 \text{x} 1 \text{x} 10^{-14} \text{J} / 0.5 \text{x} 10^{-10} \text{m} \equiv 0.001 \text{ N, J, eV} \tag{16.30}$$

Atomic energy of 0.001 eV corresponds to the SI input energy per quanta measured some 30-40 years ago by the author in a divalent impurity of NaCl crystal representing Bohr atom:

$$\sqrt{3} \; F_{Gs} a \text{ x} 10 = \frac{1}{2} k_s T_{293} / e_s \; \text{x} 1.04^4 = 1 \text{ x} 10^{-23} \text{x} 293 / 10^{-19} \text{ x } \ln2 / 1.04^4 = 0.0174 \text{ eV} \tag{16.31}$$

The mass of $1 \text{x} 10^{-27}$ kg of one way radiation particles, with linear quantization and $1/3 c_s$:

$$\frac{1}{2} x \; 2 \; x 10^{-27} v_{space}^{\;2} = 10^{-27} x 1/10 \; x (10^{8})^2 \equiv 1 \text{x} 10^{-12} \text{ J} \tag{16.32}$$

The energy of the radiation pulse $1 \; x 10^{-27}$ J in 3D display is:

Pulse energy in 3D $= (1 x 10^{-27})^{1/3} = 1 \; x 10^{-9}$ J, with $3 \; x 10^{-9}$ J for 3 planes $\tag{16.33}$

Energy in SI units using converter c_s and relativity correction:

$$3 \; x 10^{-9} \text{ J } x \; c_s \; x 1.111 ... \equiv 1 \text{ J} = 1 \text{ eV} \tag{16.34}$$

Use of rounded up values of the mass of sun, earth and their separation, distance and velocity of earth derives interesting results, because gravity force becomes also an integer round value. Corrections involve linear probability, relativity for 1/3 c_s and quantum:

$F_{Gs}' = 2/3 \; x \; G_s \; x \; m_{sun} \; x \; m_{earth}/(\text{Learth})^2 \; x \; \text{corrections}$
$= 2/3 \; x1 \; x10^{-10}x2x10^{30}x6 \; x10^{24}/(1.5 \; x10^{11})^2/1.111 \ldots x \; 1.04^{3/2}/1.04^{1/7}$
$= 3.00003 \; x \; 10^{22} \; \text{N} = 1.1547^2 \; x \; (1.5 \; x \; 10^{11})^2 \, \text{N}$ (16.35)
$= (1.73205 \; x \; 10^{11})^2 = (\sqrt{3} \; x \; 1.0000 \; x \; 10^{11})^2 \; \text{N}$

Force of 3 x 10^{22} N over mass 6 $x10^{26}$ kg generates acceleration 0.005 m/s, which time $1.04^{1/6}/1.15472$ **derives the measured SI value 3.7755 mm/s.** Variance 0.00003 x 10^{22} N $\equiv$ 3 x 10^{17} J pulse.

Using rounded integer values uncovered above new facts. The high accuracy of the result, one part per 10^5 justifies the calculation with **1.04$^{1/7}$** quantum corrector. The relativity corrected force based on integer value of parameters is equal to K_{rel} time square of the distance sun to earth separation. Actually it is suggested that the relativity corrected distance to earth is **1.5$x10^{11}$x1.111 . . . x1.04 = 1.7333$x10^{11}$≈1.73205$x10^{11}$** m, the value 1.111 . . .1 being the corrector for velocity 1/3 c_s. Identity **3.00003$x10^{22}$ N = (1.73205$x10^{11}$)^2** is believed to be derived from Force F_{Gs}/Distance L_{earth} = Distance L_{earth} expressing a differential equation explained subsequently:

$$L_{earth} - \partial F/\partial L_{earth} = 0 \qquad (16.36)$$

The integral of linear energy density in the sun signal of 2 J/m over the sun to earth distance derives the total force of gravitation:

$$F_{Gs} = \int L_{earth} \, (2 \, J/m) \; \text{x Learth x} \; \partial L_{earth} = L_{earth}^2 \qquad (16.37)$$

At atomic level 2 J/m $x10^{-10}$ x 3 $x10^8$= 0.06 W-m $\equiv$ 0.06 eV = 60 x0.001 eV or J

Please note $\varepsilon_{so} \equiv 6 \, k_s \, L_{earth}$ x 1.111 . . .= 6 $x10^{-23}$ J x 1.5$x10^{11}$ m x1.1111 . . .= 1$x10^{-11}$ J-m

The above equation explains the next two identities comprising also the quantum corrector. The gravitation force is derived using Equation 16.38 in classical SI units time 2, because gravity in sun beam is in one plane only, by adding relativity and quantum corrections:

$$F_G = (1.4961x10^{11})^2 \; x \; \sqrt{2} \; x \; 1.111 \ldots x1.04^{1/5}=3.5446 \; x10^{22} \; \text{N} \qquad (16.38)$$

In rounded up space cubic units converted to SI

$$F_{Gs} = (1.5x10^{11})^2 \; x \; \sqrt{2} \; x \; 1.111 \ldots x1.04^{1/16}=3.5444 \; x10^{22} \; \text{N} \qquad (16.39)$$

and force density of 2 N/m^2 is obtained at the surface of sun from the radiation from earth because electron frequencies are not radiated from relatively cold earth:

$$5.9742x10^{24}\text{kg}/[(1.4961x10^{11})^2]x1/10x1/10x1.111^{1/2}x1.04^2=1.999 \; \text{N/m}^2 \qquad (16.40)$$

Integrating again $\int_s \partial L_{eaths} = 1/6$ x $L_{eaths}^3 = 1/6$ x $(1.5 \times 10^{11} m)^3 = 5.6250 \times 10^{32}$ m^3 derives cubic meters in the circulatory 1/6 of the volume of earth around the sun directed at earth. The fraction 1×10^{-10} of the beam over relativity 1.2 x1.1547^2 x$1.04^{1/5}$ in this orthogonal direction obtains the gravitation force $F_G = 3.5433 \times 10^{22}$ N.

The gravitation square relations on separation expressed by Equations 16.38-41 apply only to earth, with the reference Equation 6.1 indicating that the sun gravitation force for all the planets is inversely proportional to squared *L*.

The Special Relativity force 3.0000×10^{22} N times relativity correction 1.1547 for ½ x c_s and $1.04^{23/40}$ quantum corrector derives 3.5432×10^{22} N, a value almost within the reference limits.

The sun Special Reativity gravitation force of 3×10^{22} N is derived with the limits of calculator performance in two dimensional display of relatively corrected sun mass in one direction:

$F_{Gs} = (1 \times 10^{30})^{4/3}/1.1111\ldots^{1/2} = (1 \times 10^{30})^{2/3}$ x $(1 \times 10^{30})^{2/3}/1.1111\ldots^{1/2}$
    = 3.000000001 $x10^{22}$ N, generating $F_{Gs}/L_{earth} = 2 \times 10^{11}$ N/m, which times 1 x 10^{-10}
    derives 20 eV/m   (16.41)

Displaying F_{Gs} as product of two equal components displaced in space by 90° expresses the wave function format. The product creates an area of magnetic flux generating a √2 larger component moving from the origin with doubled velocity of light creating a progressing pulse.

The component $(1 \times 10^{30})^{2/3}$ converted by multiplying by e_s to electron volts expresses:

The energy at the atomic level=$(1 \times 10^{30})^{2/3}$x1.0 $x10^{-19}$=10×10^{19}x1.0 $x10^{-19}$=10 eV   (16.42)

Using linear size converter in atomic units the pulse energy = 10×10^{19} J x 1×10^{-10}
=1 $x10^{10}$ J (invert of the Gravitation constant), which times 1×10^{-10} derives 1 J   (16.43)

Thus the product of two components $(1 \times 10^{30})^{2/3}$ generate in model integer cubic SI unity energy gradient 1 $x10^{10}$ J/m taking value 1 J/m at the atomic level. The energy gradient remains constant along the whole sun-earth integer space distance of 1.5 x 10^{11} m along the sun-earth axis. It creates an apparent conflict with the earlier view that gravitation force decreases with the square of the distance $L_{earth.}$ The energy gradient values quoted above are partial differentials, which are not sensitive to presence of components at right angle to these components. These explanations are consistent with data in Equation 7.4.

The probability of sequential discharges in four orthogonal directions followed by sequential three planes is 1/4 x1/3 = 1/12 and the power of 1 $x10^{30}$ W generated from the sun mass of 1 $x10^{30}$ kg is:

$$(1 \, x10^{30})^{1/12}/1.1111\ldots^{1/2} = 299.9999946 \, \text{K} = 300 \, \text{K} \qquad (16.44)$$

The two plane power of sun mass directed in one direction time surface probability 10^{-20} derives generated power 1 W/m^2

$$(1 \, x10^{30})^{2/3} \, x \, 10^{-20} \equiv 1 \, \text{W/m}^2 \qquad (16.45)$$

On earth side obtaining 1 W/m^2 requires taking the sun pulse to be 1000 W:
$$1 \, \text{W/m}^2 = 1/5 \, x \, (6_s \, x1/3 \, c_s)$$
And:
$$1.0 \, \text{W/m}^2 \, x \, 300^4 \, x \, 1.1111\ldots^2 \, x \, 10^{-10} \equiv (1 \, x10^{30})^{1/3} \, x \, 10^{-10} \qquad (16.46)$$
$$0.999999999 \, \text{W/m}^2 = 1 \, \text{W/m}^2$$

The result is within the accuracy of the calculator. To obtain 1 W/m^2 from the earth mass, space quantization has to have value $(1/10 x 1/10) X (1/10 x 1/10)$ with quantum and relativity corrections:

$$\text{Energy} = (6 \, x10^{24})^{1/6}/[\, 1.1547^2 x(1/10 \, x1/10) \, X \, (1/10 \, x1/10) \, x1.04^{1/4}] \equiv 1.001 \, \text{eV} \qquad (16.47)$$

The equality of the energy derived from different expressions are taken to represent equipartition of energy in all degrees of freedom. Equiparition of energy applies to SI system, the rules of kinetics seem to apply also for inside the atom.

The pulse energy in six space directions is 6 J and this value and the Equations below suggest the distance between the sun and earth expresses quarter wave length of gravity radiation $1/4 \, \lambda = 1.5 \times 10^{11}$ m which is spread between sun and earth. Both sun and earth are the centers around which other body can circulate in a circle having a diameter 3×10^{11} m $= 1/2 \, \lambda$.

Sun-to-earth distance $= 1/4 x \lambda = 1.4961 x 10^{11} \, x1.04^{1/15} = 1.5000167 \, x10^{11}$ m $\approx L_{earth}$. $\qquad (16.48)$

Equation 1.15 states that electrom charge e_s is local proton momentum and the product λ $x1/3 \, x \, c_s = 1 \times 5 \times 10^{11} \times 10^8 = 1 \times 5 \times 10^{11}$ m^2/s is reduced without electrons to $1 \, x10^{11} \rightarrow x$ $10^8 = 1 \, x10^{19} = 1/1 \, x10^{-19} = 1/e_s = 1/(\text{local proton momentum})$ confirming its infrared source.

The related frequency f of protons is:

$$f = (\text{velocity } 1/3 \, c_s)/\text{wave length } \lambda = 1/3 x3 x10^8/6 x10^{11} = 1.666 \, x \, 10^{-2} \, \text{Hz} \qquad (16.49)$$

The energy identities in six space directions supporting display of such wavelength are:

$$\lambda \, x1 x10^{12} \, \text{Hz} \, x \, h_s = 6 \times 10^{11} \, x1 x10^{12} \, x1 x10^{-33} = 6 \, \text{J} \qquad (16.50)$$
$$\lambda \, x \, e_s \, x \, 1/3 \, x \, c_s = 6 \times 10^{11} \, x \, 1 x10^{-19} \, x1 x10^8 = 6 \, \text{J} \qquad (16.51)$$
$$\varepsilon_{so} \, x \, \lambda = 1 \, x10^{-11} \, x \, 6 \times 10^{11} = 6 \, \text{J} \qquad (16.52)$$

An integer force of **4.00 . . . $x10^{22}$ N** is derived by attributing integer values of 6 $x10^{24}$ kg to the mass of earth, $1.5x10^{11}$ m to its distance to the sun with a force of 1 N generated per kilogram of mass and a linear probability 10^{10} with space quantization of 1/10. It converts to $3.6000x10^{22}$ N, the cubic integer gravitation force on dividing by 1.111 . . .:

$$1/10x6\ x10^{24}/1.5x10^{11}\ x10^{10}= 4.000\dots x\ 10^{22}\ N \rightarrow x1/1.111\dots= 3.6000\dots x10^{22}\ N \tag{16.53}$$

The value **4.00 . . . $x10^{22}$ N** suggests that SI carry relativity corrections, which makes also lengths contracted and times dilated.

The above identifies $m_{earth}/F_{Gsun} = 6\ x10^{24}/4\ x10^{22} \equiv 1500$ kg/N $\equiv$ ¼ x 6000 K $\equiv$ 1500 W/m²

Energy 1500 W/m² times relativity and quantum correctors derives the sun constant:

$$1500\ x\ 1/1.11\dots x\ 1.04^{2/5}=1371\ W/m^2 \tag{16.54}$$

The power 1500 W/m² times one step quantization 1/10 and relativity correction derives 150 W/m² $x1.1111$ equal inverted 0.006 N/ kg.

The gradient of energy of the sun beam converted to atomic level requires another division by squared relativity generating:

$$F_{Gs}/[(L_{earth}^2\ x\ 1.1547^2)\ x\ 1\ x10^{-10} \equiv 20\ eV = 2x10\ eV \tag{16.55}$$

This represents oscillator energy of the proton and of the (proton+electron) pair, which following one step quantizatiion becomes $2x1$ eV.

The Special Relativity gravity force of $3.0000\ x\ 10^{22}$ N expresses very closely the reference by being multiplied by relativity and quantum corrector:

$$3.\ 00000\ x\ 10^{22}\ x\ 1.1547\ x\ 1.04^{3/5} = 3.5466\ x10^{22}\ N \tag{16.56}$$

The value of force of 0.006 N/kg applies solely to earth, because it expresses both the temperature and the energy density of the sun energy at the earth to sun separation. The distance of the planet to sun is a function of its temperature and its ability to generate a balancing force opposing gravitation.

External perturbation may be needed for activating the radiation process. Perturbation brings probably the fourth 10 eV energy components into the 50 eV relativity corrected radiation pulse. Excess energy is not considered here. Perturbation breaks the coupling between the electron and the proton moving at 1/3 c_s. They start to move in the opposite directions with equal momenta. With zero total momentum the rule of conservation momentum remains satisfied for the one dimensional signal. The backwad wave, reflected by the atomic potential barrier, changes direction and subtracts from the forward wave producing a pulse. Thus **the model does not negate the well accepted**

concept that a mass is a source of an infinitely long wave. Two infinite anti-phase waves, one started with delay and reflected, produce a pulse. Hence **the model clears up the contradiction, the big conflict, between the old wave theory of light and the quantum theory.** Pulses are postulated to be the main source of interaction. Several pulses may form chain of wave like pulses.

Within the Bohr atom the force of attraction of 1 N/kg energy is discharged in one direction. In comparison the energy radiated is that of 1 N per each of 6 directions producing a gravity vector reduced by probability by 10^3 deriving:

$$F_G/mass_{earth} = 3.5438 \times 10^{22} \text{ N}/5.9742 \times 10^{24} \text{ kg} \times 1.04^{1/20} = 0.00599 \approx 0.006 \text{ N/kg} \qquad (16.57)$$

In gravity space the main variable is the momentum and the force per kilogram is the space gradient of momentum, which is the acceleration. With corrections these simple definitions balance gravity forces between the two bodies.

The definition of energy of the hydrogen atom becomes complicated with three oscillators contributing to energy. Text book value of 21.76×10^{-19} J for electron hydrogen energy from n =1 to n =∞ corresponds to the model value 2×10 eV$\times 1.1547^{1/2} \times 1.04^{1/3} =$ 21.77 eV. Author's experiments require 10 eV to expand the electron in one direction to its dissociation length and energy of 8.1 eV has confirmations from other sciences [53], if relativity is neglected. Hartree energy of 30 eV of the model has been earlier equated to the classical value of 27.21 eV both value defining the total energy of the three oscillators in the Bohr atom including correctors for relativity and quantum. The multiple of 10 eV basic pulse energies could be also attributed to conversion of two or four atoms of hydrogen into radiation. High energies with long and infinitely long waves obeying Maxwell Equations representing large groups of pulses might be produced if numerous pulses are exited by turbulence or by energy source.

According to Wikipedia Encyclopedia (see Natural Units) the product u_o c, listed among expressions used in natural physical systems, derives

$$\text{Characteristic impedance } Z_o = \mu_o c \qquad (16.58)$$

Using now integer cubic SI units c_s and u_s derives

Z_{os} =1.0 x10^{-6} x 3 x10^8 = 300 ohms, times 1.111 . . . x1.1547/1.04 = **370.1** ohms $\qquad$ (16.59)

Alternatively **Surge impedance of space** $\sqrt{(\mu_{os}/\varepsilon_{os})}$ x1.1547x1.04$^{1/2}$=316.227766 x1.1547x1.04$^{1/2}$ = **372.38** ohms obtained using relativity and quantum corrections.

Coulomb constant $1/(4\pi\varepsilon_o)$ in integer space units is $1/(12\varepsilon_{os})$ time relativity correction:

The integer cubic Coulomb constant = 8.333333333 x 10^9 x 1.2 =1x10^{10} is the exact inverse of the gravitation constant $\qquad$ (16.60)

The classical value measured experimentally by the author [2, 3] of the minimum energy step of 0.0173 eV in Bohr atom model created by impurity in NaCl crystal is according to the model proposed now 1/5 permittivity of space ε_{os}, expressing its local value in one of 5 pathways times uncorrected Coulomb constant, $1/(12\varepsilon_{os})$, times 1.04 quantum correction:

$$1/5 \text{ x } \varepsilon_{os} \text{ x (Coulomb constant) x1.04} = 1/5 \text{ x } 1/12\varepsilon_{os} \text{ x1.04} = 1/60 \text{ x1.04} = 0.01666 \ldots \text{x1.04}$$
$$= 1/5 \text{ x1x10}^{-11} \text{ x } 8.333\text{x10}^9 \text{ x1.04} = 0.0173 \text{ eV} \quad (16.61)$$

Above Equation displays all its components in detail to explains the presence of the factor $1/60=0.01666\ldots$x1.04 in other equations. The above energy of 0.0173 eV indicates that the energy values measured in classical SI experiments are the same as calculated using integer values of the model This equality does not extend to forces and distances.

Energy $60k_sT_{s6000}/e_s = 36$ eV is eight $6k_sT_{s6000}/e_s$ from 8 corners of the cube plus $1.5k_sT_{s6000}/e_s =$ uncertainties in these eight corners that is $12k_sT_{s6000}/e_s = 12$ x 0.6=7.2 eV making a total of 43.2 eV, divided by 1.04^2 derives $39.94 \approx 40$ eV, which fits well H2 molecule.

Total energy in the sun pulse from cubic structure including uncertainties and correction:

$$\text{Energy of H2 sun cubic structure} = 8 \text{ x} 6k_sT_{s6000}/e_s + 12k_sT_{s6000}/e_s \text{ x1.04 }^{1/40} \approx 39.9992 \text{ eV} \quad (16.62)$$

The model α_s fine structure constant expressing electron spin has not been discussed so far in the model. It's α_s value in space cubic units converted to SI is very close to SI value 1/136.036. It is identified to be the geometric factor of the Coulomb constant corrected for relativity and quantum:

$$\alpha_s = 1/12^2 \text{ x1.111} \ldots \text{ x } (1+1/80)/1.04^{1/10} = 1/136.069 \quad (16.63)$$

Ten values of $1/\alpha$ times quantum correction are now identified as the solar constant:

$$10 \text{ x } 1/\alpha \text{ x1.04}^{5/24} = 1371.9 \text{ W/m}^2 \quad (16.64)$$

Forward 6000 K pulse divided in five quantized orientations delivers:

$$\text{Sun constant W/m}^2 = 6000/5 \text{ x1.1547/1.04}^{1/4} - k_sT_{s6000}/e_s = 1372.1 - 0.6 = 1371.5 \text{ W/m}^2 \quad (16.65)$$

In the proposed model $1/\alpha$ is identified with the sun pulse of 6000 K over 50 eV with quantum and relativity correctors $1.111 \ldots$ x $1.04^{52/100}$:

$$1/\alpha_s \equiv 6000 \text{ x } 1/50 \text{ x } 1.111 \ldots \text{ x } 1.04^{52/100} \equiv 2 \text{ x } 60 \text{ x correctors}$$
$$= 2 \text{ x } 1/0.016666 \text{ x } 1.111 \ldots \text{ x } 1.04^{52/100} = 136.081 \quad (16.66)$$

Dimensionless constants have the same values in all systems. Such constant is the fine structure constant. The equations quoted on WEB pages expressed now in integer space units, $\alpha = e_s^2 2\pi/[h_s c_s(4\pi\varepsilon_o)]$ and Coulomb constant $e_s^2/(h_s c_s)$, $e_s^2 c_s\mu_{os}/2h_s$, did not generated meaningful values. Further analyses found that 6000 K x 1/50 is twice the invert of the quantum value 1/0.01666 eV. Further consideration discovered unexpectedly that 1/αs corrected for relativity and quantum derives Planck radiation energy:

Inverted fine structure constant $1/\alpha_s$ corrected for quantum and relativity derives 0.01 eV identified now as dashed fine structure constant $1/\alpha_s' \equiv$ Planck radiation energy

Planck radiation energy = $1/\alpha_s$ x1.1547^{2}x1.04$^{52/100} \equiv$ 0.01 eV- 0.0000004 eV   (16.67)
the variance being 4 x 10^{-27} J.

Expressions in classical SI system apply to masses of one kilogram, while values at the atomic level in SI units refer to the value per event or per atom, varying at times by the factors of 10.

It is suggested that several calculations in this section express well proven rules of Special Relativity (SR). There is no privileged reference frame according to SR. At velocities expressed by a fraction of velocity of light the measured length may be contracted and the time may be dilated. Mass-energy equivalence remains sustained. Global Position Satellite System performs exact periodic corrections on the on-board clocks to correct relativistic effects.

There are several accurate values of the calculated gravity forces expressed by the rules of SR. At high velocities the measurements of the same event will be different for observers in relative motion. Consequently space distances between the sun and earth may take different values. The displayed length may be contracted leading to generating an increased force to keep energy and momentum constant. Since the force is the mass over the square of the distance the smallest force, that is 3.000000x10^{22} N, is according SR the most representative, the other values having different relativity and quantum correction attributed to their motion. This effect is caused by relativity gravity force dilation.

Since all particles attract each other **the same gravitation force is responsible for the presence of all solids.** Since **the gravitation force is balanced by rotation the same process must occur in all solids,** although it may be quantized.

Essential values 10 J, 1 J, 0.1 J, 0.01 J, 0.001 J express per oscillator the same vectorized radiation energy; they may produce components with factors 10^{-19}, 10^{-30}, 10^{-10}, 10^{-20}, 1000 K, 300 K. All the figures express the same energy of the same pulse originating from the mass of (proton+electron) or the proton of the hydrogen atom or molecules converted to energy local velocity of light several pulses generating package. The other two essential vectorized energies are $F_{sspaceSI}$ = 10^{17} J and $E_{ssearthSI} = E_{ssunSI}$ = 10^{21} J.

The author has found two interesting presentations on WEB pages recommended for reading relating to Sections 16-18 by R. Bradley [50] and by D. McManon [51].

17. Gravitons, muo10ₛ and other energies; geometric mean of masses, balanced probability, and wave function formation and uncovered conclusive gravitation equations

The graviton energy of 0.016666 . . . eV of Table 1 divided by the correctors of quantum, 1.04^3, and of relativity, K_{rel}^2 and K_{rel} 1/3c, derives radiation energy, 0.01 eV, expressing the first value of Planck radiation law confirming validity of correctors:

$$0.016666\ldots/(1.1547^2 \text{ x } 1.111 \ldots \text{ x } 1.04^3) = 0.0100011 \text{ eV} \qquad (17.1)$$

Experimental data in the Appemdix, Equation 6, finds that $kT_{294}ln\ 2 = 0.0174$ eV is the smallest energy change that can be measured and $0.0174/1.04 \approx 0.01666$. In the integer model it is the sum of two orthogonal pulses 0.01 eV equal $\sqrt{3}$ x0.01 = 0.0173 eV.

Planck radiation energy per kelvin and per plane is generated by proton radiation mass m_{psID} moving with local velocity 1/3 c_s converted to electro volts and subjected to two quantization steps 1/10 x 1/10:

$$1 \text{x} 10^{-27} \text{x} 1 \text{x} 10^8 / 1 \text{x} 10^{-19} \text{x} 1/10 \text{ x} 1/10 = 1 \text{ Jx} 1/10 \text{ x} 1/10 = 0.01 \text{ eV} \qquad (17.2)$$

The energy from the whole volume of the elementary cubic structure comprises three 10 eV units in three planes adding to 30 eV, the Hartree energy in cubic space units and in some evaluations may reach 40 eV as shown in Equation 16.62.

Classical physics [Beiser, 34] assume equivalence between the mass and the generated wave. Generated wave must be supplied by energy; in the sun that involves mass to energy conversion; radiation energy in space may occur from instability of the proton or electron. In the integer model the unity reference signal at 300 K is the rms or average value of a distribution ranging from maximum excitation to "zero" level expressed by SI Planck radiation at 1 K decreasing at the atomic level by factor of 10 to 0.001 J. The quantum physics instead of wave considers packets of energy. For large masses these packets may well combine into a continuous wave. The conversion of mass into energy is considered by many scientists to be the source of power in all exothermic processes of physics and of chemistry.

Present research suggests that mass of 1 kilogram generates force of 10 N decreasing for one orientation to 1 N and density of such pulses decreases with the square of the distance. The force applied acting on the attracted object becomes proportional to its mass. Corrections involve relativity, $K_{rel}^{1/3}c_s^2$, volume probability, quantum factor and space quantization incorporating three orthogonal vectors. Rotational symmetry for both particle components (e.g. B and E) generating power requires due to space quantization probabilities, 10^4 higher input energy to generate a vector pulse:

$$\text{Input energy} = 0.001 \text{ eV x } (10 \text{ x} 10) \text{ x } (10 \text{ x} 10) = 10 \text{ eV} \qquad (17.3)$$

Otherwise $\mu_{os}/(\varepsilon_{so} \times 1/4 \times c_s) \times$factors$=(0.001/(1 \times 10^{-11}) \times 10^{-19}/(1/4 \times 3 \times 10^8) \times 10^{20}/1.1547^2 = 10$ eV.

$$(17.4)$$

During radiation initiated by an external turbulence

$$\text{Charge } e_s = 1/2 \, m_{pds} \times 1/3 c_s = 1 \times 10^{-27} \times 1 \times 10^8 = 1 \times 10^{-19} \text{ C} \qquad (17.5)$$

The above relation implies that momentum of Bohr atom $m_{pds} \times 1/3 c_s$ is split in two opposite flow directions (see Equation 12.1), such event being plausible because mass in atomic units is expressed by momentum.

The sun has only a limited amount of mass suitable for conversion to iron. Exhaustion of that mass leads to sun expansion and explosion observed in the Universe for other stars. Stores of energy in the masses of proton and of electron to generate the force of 1 N per kg must be sustainable for long periods.

Next finding is considered important because it uncovers the original age of the Universe derived by CERN is the relativity corrected longevity of proton based on minimum energy 1x10-[14]** J converted to atomic level and spread in 4 directions.** A simple calculation in which radiation pulse mass converted to energy over the product of radiation signal of 4×10^{-44} Wat power and a year duration in seconds show that energy can be generated from the proton for a period t which in Chapter 21 the integer model identifies with the age of the Universe:

For proton $t = 2 \text{ J} \times 1.111 \ldots /(4 \times 10^{-44} \times 3600 \times 4 \times 365.242) \times 10 \times 10^{-28} \times 1.04^{1.2}$
$$= 1.0077 \times 10^{10} \text{ years} \qquad (17.6)$$

The gravitation force per kilogram per vector in one direction is

$$F_{Gs}/\sqrt{3} m_{sun} = 3.5438 \times 10^{22}/\sqrt{3} \times 2 \times 10^{30} = 1.0213 \times 10^{-8} \text{ N/kg} \qquad (17.7)$$

This force is applied at the average speed $1/3 \, c_s$ generating power

$$F_{Gs}/\text{kg} \times 1/3 \, c_s/\text{cor.} = 1.0213 \times 10^{-8} \times 1 \times 10^8/1.04^{1/2} = 1.0015 \text{ W/kg} \qquad (17.8)$$

The gravitation force of attraction per thermal radiation in one of six orientations of the sun derives the infrared frequency 5×10^{12} Hz indicating clearly the infrared frequency being responsible for the force of attraction. The gravitation force impulse, expressed by Newton-seconds is the power of watts/m²-cycles:

$$3.5438 \times 10^{22}/[6 \, \sigma_s (6000-300)^4]/1.04^{1/5} \equiv 4.9964 \times 10^{12} \text{ Hz} \qquad (17.9)$$

Twice the Stephan-Boltzmann constant σ_s times temperature to power four, corrected for the relativity and quantum is the local velocity of light:

Velocity of light $1/3 \, c_s \equiv 2\sigma_s (6000-300)^4/(1.111 \ldots^{1/2})/1.04^{1/30} = 1.00012 \times 10^8$ m/s $\qquad (17.10)$

Equations with temperature to power four should be considered as wave function format of two pulses each $[\sqrt{(2^{1/2} \times 6 \times 5 \times 10^{-8})} \times T^2] \equiv 10^4$ identified with sun temperature of 10000

K produced by $\sim\sqrt{3}$ x 6000 x $1.04^{49/50}$.=10000.44 K, where 0.44 K is taken as 1/3 k_s x $1.111\ldots^3$.

The finding that Stephan-Boltzmann constant $\sigma_s = 5$ x 10^{-8} with T^4 term derive local velocity of light is considered a significant finding, since $2\sigma_s$ expresses atomic force F_{atomic}=1x10^{-7} N without corrections.

Dividing 5x10^{12} Hz by 3000 K derives 1.6666 . . . x 10^{10} Hz, which times 2/3 eliminates the effect of electron frequency and identifies it as the invert of gravitation constant $G_s \equiv 1$ x10^{-10}.

Integer cubic space model connection with high energy physics

Synonymous meaning of unsimilar parameters has lead to the formulation of several dozens of expressions for the gravitation force of the sun in Table 5. The force of 1 N/(K-kg) derived from the force of 10 N/(K-kg) generated from any planetary mass, atomic mass or sun mass generates radiation pulses displaying the mass density of 1x10^{-28} kg/m^3 and the gravitation constant G_s. Due to equivalence of dimensions 1x10^{28} kg/m^3 expresses also force 1x10^{28} N, energy 1x10^{28} J and mass 1x10^{28} kg called now integer muo10$_s$ parameters. In integer cubic model the relativity corrected value below represents at the atomic level gravitation energy of the proton-electron pair for three planes from the geometric mean mass with unity distance between the bodies:

$$\text{Int.muo10}_s \text{ force} = 3E_{g3} = 3\sqrt{(1\mathbf{x}10^{-30}\mathbf{x}1\mathbf{x}10^{-27})}\text{x}1.111\ldots^{1/2} = \mathbf{1.00\ldots x10^{-28}} \text{ J} \qquad (17.10a)$$

The muon mass of 1.881x10^{-28} kg appears to be the **integer mass density of radiation pulse 10^{-28} kg** called muo10$_s$ incremented by the not interacting flux and corrected for relativity and quantum:

Muon mass =1.881x10^{-28} kg $\equiv$1x10^{-28}x1.2x1.1546^2 x 1.111 . . .$^{3/2}$ x$1.04^{1/10}$=1.881 x10^{-28} J/m^3
$$(17.11)$$

The product of the force generated by 1 N/(kg-K) and the mass of muons expresses the radiation pulse 1x10^{-27} J with the correction, K_{rel}^3 x 1.111 . . .^{2} x $1.04^{1/10}$:

1 N/(kg-K)x1.881x10^{-28} kg $\equiv$ 1x10^{-27}x1/10x1.1547^3x1.111 . . .^{2}/$1.04^{1/4}$
= impulse of energy 1.882x10^{-28} J $\qquad (17.12)$

Expressing the radiation pulse 1x10^{-27} J in one dimension derives 10^{-9} J, which time invert of 1 step space quantization obtains the gravitation constant G_s:

$$G_s = 1/10 \text{ x }[1\text{x}10^{-27}]^{1/3} = 1 \text{ x}10^{-10} \text{ N-m}^2/\text{kg}^2 \qquad (17.13)$$

The other mass value quoted for muon is 105.656 MeV/c^2, which in the model integer SI units corrected for relativity and quantum expresses electron mass in kilogram:

$$105.656 \text{ x}10^6 \text{ x } 1 \text{ x}10^{-19}/9 \text{ x}10^{16} \text{ x}1/(1.1547 \text{ x } 1.04^{1/3}) = 1.0015 \text{ x}10^{-30} \text{ kg}$$

Energy 105.656 MeV/c² appears to be energy of the individual pulses in the shower of pulses produced, when proton becomes unstable and converts into electrons. Thus the muon does not contradict the integer model.

Equivalencies of energies of **tuons** and **quacks of the standard model** to the energies of the integer cubic model were dealt in the author's second book [46 pages 55-56]. **Higg's boson**s is discussed in Chapter 22. The present book inspects the sources of the physical constants looking within the atom witout breaking it, the process adopted in high energy physics.

Geometric mean of masses of sun and earth, m_{geo}

Geometric mean is the central tendency of a set of data. It expresses the value of an equivalent resistor of two resistors in parallel, with other examples given on WEB pages. For the earth-sun pair it is now the square root of the product of their masses expressed in integer values nominally 3.4641×10^{27}. After correction of $1/3 \times 1/1.1547$ it derives an exact value and pulse 1×10^{-27} J $\times 1 \times 10^{27}$ kg $\equiv 1.00$:

$$m_{geo} = (m_{earth} \times m_{sun})^{1/2}/\text{corrections} = 1/3 \times (2 \times 10^{30} \times 6 \times 10^{24})^{1/2}/(1.1547) = 1 \times 10^{27} \text{ kg} \quad (17.14)$$

For sun-earth pair/m$=\sqrt{(6 \times 10^{24} \times 1 \times 10^{30})}/(1.5 \times 10^{11} \times 1.111 \ldots \times 1.2^2 \times 1.04^{1/2} = 1.0008 \times 10^{16}$ kg/m.

Geometric mean of masses of sun and of earth, m_{geo}, have an obscure relation (square, invert or invert square with a simple multiplier) to the energy of radiation pulse of 1×10^{-27} J, to the number of electrons in the mass of earth 6×10^{54} and to the minimum energy 3.00257×10^{-54} J derived by the author in Equation 21.10. Wahlin [48] attributes to the minimum energy value 2.4495×10^{-54} J differing from the value proposed by the author by the relativity factor $1.11111^2 \times 1.04^{1/6}$.

Generally gravitation potential V_g (see WEB) is the gravitation energy E_g per unit mass, this being in SI units, so that $E_g = \text{mass (kg)} \times V_g$. At a distance L from the center of mass M the gravitation potential is $V_g = - G_s \times m_{geo}/L$. In this formula, at the atomic level, the integer natural atomic elementary gravitation value of G_s is assumed to be unity, $G_s = 1$. The gravitation energy E_g derived from the geometric mean of the sun-earth pair, at distance of sun-to earth separation, has been found to be twice the square of local velocity of protons or twice the square of the average local velocity. The added factor 3/2 accounts for electron energy in the total energy, while $\sqrt{2}$ indicate presence of two vectors added at 90^0. The same value energy appears in the proton oscillation.

Gravitation energy $E_g \equiv -\sqrt{2} \times 10^{27}$ kg$/(1.5 \times 10^{11}) \times 3/2 \times \sqrt{2} \equiv - 2 \times 10^{16}$ J $\qquad (17.15)$

$$\equiv -1 \text{ kg} \times 2 \times (1/3 \ c_s)^2 = -2.000 \times 10^{16} \text{ J} \qquad (17.16)$$

The factor of 2 in Equation (17.16) is attributed to the presence of 1/3 c_s being an oscillatory value (rate) and 1/3 c_s being also the forward velocity. For $(c_s)^2$ pathway factor 1/5 x 1.1111 has to be applied to obtain the correct value for E_g:

$$\text{Gravitation energy } E_g = 1 \text{ kg x } (c_s)^2 \text{ x1/5 x 1.111} \ldots = 2.00 \ldots \text{ x } 10^{16} \text{ J} \qquad (17.17)$$

The gravitation energy E_{g3} generated from 3 planes, 3×10^{16} J, time velocity of light and 2/3 factor, eliminating electron interaction in gravitation, derives:

$$\text{The integer earth mass} = 3 \times 10^{16} \text{ J x } 3 \times 10^8 \text{ x } 2/3 \equiv 6 \times 10^{24} \text{ kg} \qquad (17.18)$$

In the above Equations the 3×10^8 multiplier is the converter from integer atomic units to classical SI units indicating also that all the geometric mass values in the Table 4 below express such low values of mass because they are in atomic integer values. Gravitation energy 3×10^{16} J multiplied by the linear probability 1×10^{10} derives integer SI value 3×10^{26} J. Multiplied by 1.1547^2 x1000 x 1/1.111 $\ldots$ $^{1/4} \approx 3.8 \times 10^{29}$ W derives the sun power of Equations 5.1, 5.5-5.6. The geometric mean of masses of sun and of earth, $\sqrt{2}$ $\times 10^{27}$ kg multiplied by relativity and quantum corrections (1.111 $\ldots$ $^{1/2}$x1.04$^{1/6}$) referred to one quantum pathways by a factor 1/5 derives 3.001×10^{26} J identified with E_{g3} expressed in integer SI units:

$$E_{g3} = \sqrt{2} \text{ x}10^{27} \text{ kg } (1.111 \ldots ^{1/2}\text{x}1.04^{1/6}) \text{ x } 1/5 \equiv 3.001 \times 10^{26} \text{ J} \qquad (17.19)$$

Gravitation energy expresses gradient of mass, (kg/m), and this has been found to express the force of gravitation. The m_{geo} mass of 10^{27} kg is the input derived from ½ mass of the sun times $1/10^3$ probability and times linear probability 1/10 deriving the vector component of the gravitation energy 1×10^{16} J.

The geometric mass times square of the velocity of light drives energy $\sqrt{2}$ x1 $\times 10^{43}$ J, which can be also expressed as linear momentum of earth, of mass 6×10^{24} kg rotating at $v = 30\ 000$ m/s

$$m_{geo}(c_s)^2 = m_{earth} \, v^2 \text{ x } 1.11 \ldots \text{ x}\sqrt{2}/6 \text{ x}10^{10} \text{ J} \qquad (17.20)$$

The geometric mass of earth and sun 2×10^{27} kg with $L_{earth} = 1 \times 10^{11}$ m expresses the sun beam temperature in the following identity:

$$\text{Sun beam temperature} = \sqrt{2} \, m_{geo}/L_{earth} \text{ x}10^{-10}\text{x}1.111 \ldots ^{4/5}\text{x}1/10^3\text{x}1.04^{4/5} \equiv 1000.054 \text{ K} \qquad (17.21)$$

The geometric mean definition applies also to some relations, between the electron and proton, in which 1/6 of the gravitation energy corrected for relativity derives exactly the integer space value of the electron charge. The latter was in Equation 1.15 identified with the proton momentum, proton mass moving with local average velocity 1/3 c_s:

$$\text{Electron charge } e_s = 1/6\text{x}(1\text{x}10^{-30}\text{x}1\text{x}10^{-27})^{1/2}/0.5\text{x}10^{-10}\text{x}1/1.111 \ldots ^{1/2} = 1.000000000\text{x}10^{-19} \text{ C} \qquad (17.22)$$

The energy 1 J in two dimensional display of one plane per kg of radiation pulse of 10^{-27} kg is the geometric mean mass of the sun and earth, Equation 17.14:

$$m_{geo} \equiv 1 \times 10^{27} \, kg \equiv 1 \, J/1 \times 10^{-27} \tag{17.23}$$

The geometric mass of Bohr atom in integer cubic SI units is given by the proton-electron pair, corrected by relativity and times volume probability and derives exactly the 30 J ≡ 30 eV the Hartree energy of Bohr atom:

$$m_{geoH} = (1 \times 10^{-30} \times 1 \times 10^{-27})^{1/2}/1.111\ldots^{1/2} \times 1 \times 10^{30} \equiv 30.000\,000 \, J \equiv 30 \, eV \tag{17.24}$$

At the natural atomic level the geometric mass of Bohr atom in integer units is

$$m_{geoHatomic} = (1 \times 10^{-30} \times 1 \times 10^{-27})^{1/2}/1.111\ldots^{1/2} = 3.000\,000\,000 \times 10^{-29} = 30 \, m_{es} \tag{17.25}$$

Thus the gravitation energy E_{gl} for each planet from the sun with corrections has the same value of

$$E_{gl} = 1 \times 10^{16} \, J/K \tag{17.26}$$

The above energy 1×10^{16} J has to be vectorized to represent synchronized energy 1×10^{14} J of Equation 7.16:

$$1 \times 10^{16} \, J \times 1/10 \times 1/10 = 1 \times 10^{14} \, J \tag{17.27}$$

Table 4. *Gravitation energy of planets from geometric mean masses per metre-K of sun-planet pairs*

Gravitation energy E_{gl} of the sun-planet per plane $E_{gl} = (mplanetmsun)^{1/2}/L_{tosun}$ x cor.

Mercury $(3.39 \times 10^{23} \times 2 \times 10^{30})^{1/2}/57.91 \times 10^9 \times 1/(1.2 \times 1.1111^{3/2}) \times 1/(1.04^{1/3}) = 0.9985 \times 10^{16}$ J/m
Venus $(4.87 \times 10^{24} \times 2 \times 10^{30})^{1/2}/108.2 \times 10^9 \times 1/(1.2^3 \times 1.1111^5) \times (1.04^{1/3}) = 0.9984 \times 10^{16}$ J/m
Earth $(6 \times 10^{24} \times 2 \times 10^{30})^{1/2}/149.6 \times 10^9 \times 1/(1.2^3 \times 1.1111^3) \times (1.04^{3/5}) = 1.00026 \times 10^{16}$ J/m
Mars $(6.42 \times 10^{23} \times 2 \times 10^{30})^{1/2}/227.94 \times 1.2^2 \times 1.1111^3 \times (1.04^{9/20}) = 0.9994 \times 10^{16}$ J/m
Jupiter $(1.9 \times 10^{27} \times 2 \times 10^{30})^{1/2}/778.33 \times 10^9 \times 1/(1.1547^2) \times 1/6 \times (1.04^{1/4}) = 0.99976 \times 10^{16}$ J/m
Saturn $(5.68 \times 10^{26} \times 2 \times 10^{30})^{1/2}/1427.94 \times 10^9 \times 1/(1.2^3 \times 1.1111^3) \times 1.04^{1/10} = 0.99976 \times 10^{16}$ J/m
Uranus $(8.69 \times 10^{25} \times 2 \times 10^{30})^{1/2}/2870.99 \times 10^9 \times 1.2^2 \times 1.1111^4) = 1.0078 \times 10^{16}$ J/m
Neptune $(1.025 \times 10^{26} \times 2 \times 10^{30})^{1/2}/4564 \times 10^9 \times 1.2^3 \times 1.11111^6 \times 1/(1.04^{1/2}) = 1.0002 \times 10^{16}$ J/m
Pluto $(1.31 \times 10^{22} \times 2 \times 10^{30})^{1/2}/4913.52 \times 10^9 \times 1/(1.2^3 \times 1.1111^6) = 1.0093 \times 10^{16}$ J/m

$$\tag{17.28}$$

The power generated for each couple in three planes = **3 x 10^{16} W** $\tag{17.29}$

because for earth each mass of 1 kg generates power of 1 W. **Converted to atomic level times 10^{-30} it becomes the minimum energy 1 $\times 10^{16} \times 10^{-30} = 1 \times 10^{-14}$ J.**

The sun reacts individually with the planets only with one of his 30 beams at a time radiating radially as shown in Chapter 4-5. Planet equalizes energy of its three planes with energy supplied in the beam in one plane. Since two components are needed to generate a signal the probability for each beam decreases the energy per beam to 1×10^{30} kg by the factor of 30^2 to the equivalent of $1 \times 10^{30}/900 = 3 \times 10^{27}$ kg. The new mass times radiation parameter 1x10^{-27} J velocity of light and 2/3 factor eliminating electron from the interaction derives:

Unit of energy delivered from the sun = 3 x 10^{27} kg X 1x10^{-27} J x 2/3 ≡ 2 eV or J   (17.30)

Total energy from the 30 beams is 60 eV multiplied by space probability 1/10 and referred to one of three planes derives 2 J or eV per atomic event.

Gravitation power in one pathway in wave function format with velocity expressing the pulse rate:

$$E_{gl} = (1J \times 1/3 \text{ cs}) \times (1 \text{ J} \times 1/3 \text{ cs}) \equiv 1 \times 10^{16} \text{ W}$$ (17.31)

Converting ST to atomic terms for 10 pathways:

Sun pulse =10x10^{16} Wx1x10^{-30}/10^{-19}=1x10^{-13} J/10^{-19} ≡ 1 MeV pulse = 1/u_{os} (17.32)

The product of gravitation $2E_{gl}$ = 2 x10^{16} J and of the velocity of light define the earth mass:

$$2 \times 10^{16} \text{ J} \times 3 \times 10^8 \equiv 6 \times 10^{24} \text{ kg}$$ (17.33)

Data in Chapter 5 have not been changed to accommodate this new interpretation.

Jupiter data does not follow the pattern of typical relativity corrections by needing a factor of 1/6 induced possibly by very strong magnetic field inducing a Lenz's effect decreasing the sun's attraction force, or more likely by discharging energy through Kelvin-Helmholtz mechanism involving volume contraction (see WEB). Since all its interacting energy is directed on sun it has no factor 1.2 in the expression. All other planets have this factor with different power indexes. Neptune factor 1.2^3 is attributed to sun radiating to Neptune in three planes and by doing so increasing the force in N/kg about 3 times in comparison to earth as derived by the Appendix.

The product of gravitation energy 10^{16} J and of square of the local light velocity (1/3 x c_s)^2, and of the converter to SI units represented by another velocity 1/3 x c_s over minimum pulse energy 1x10^{-14} J derives the number of electrons in earth mass:

$$10^{16} \text{ J} \times (1/3 \times c_s)^2 \times (1/3 \times c_s)/1 \times 10^{-14} \equiv 1 \times 10^{54} \text{ electrons}$$ (17.34)

For six directions the number is 6 x10^{54} electrons. Integer SI current of earth is: 1 x10^{54} x 3000 m/s x 10^{-19}C x 2/3 x 10^{-1}= 2 x 10^{29} A.

That number times mass of electron over local velocity of light treated as a converter to nature atomic units derives the SI energy of gravitation of planets per pathway:

$$E_{gl} = 1 \times 10^{54} \times 1 \times 10^{-30}/(1/3 \times c_s) = 10^{16} \text{ J}$$ (17.35)

If 10^{16} expresses a number, 10^{16} x 10^{-14} J expresses the sun pulse of 100 J ≡100 eV.

Geometric gravitation forces for all planets except Jupiter require relativity correction for the local velocity of light 1/3 c_s giving strong support for the hypothesis of presence of local light velocities.

Gravitation energy of sun-planet pairs per second of space vector frequency $= 10\ E_{g3}$ x $v_{ovector} = 3$ x10^{16} J/K x 3x10^{21} Hz x$10=1$x10^{38} W$=1/e_s^2$, which converted to atomic level by volume probability, derives the gradient 1 J-s/m per unit local velocity of light:

$$1 \text{ x } 10^{38} \text{ W x } 1 \text{ x } 10^{-30} = 1 \text{ x } 10^{8} \text{ W} \equiv 1 \text{ J-s/m} \tag{17.36}$$

The geometric mean of electron and proton in the Bohr atom is:

Geometric mass of the Bohr atom $= (mps + mes)^{1/2}/1.111 \ldots = 3 \text{ x } 10^{-29}$ kg

Gravitation energy of the Bohr atom per cycle of space vector frequency converted to atomic terms in eV:

$$E_{g3(mps+mes)} \text{ x } E_{searthSI} \text{ x cor.} = 3\text{x}10^{-29} \text{ kg x } 3\text{x}10^{21} \text{ J-s x}10^{-30}/1\text{x}10^{-19} = 1\text{x}10^{-19} = e_s \tag{17.37}$$

Gravitation energy of earth-sun couple per plane expressed at 1000 K, believed to be generated by sun in each direction of space is $E_{gl}=1$ x 10^{16} x $1000 =1$ x 10^{19} J/m. Equation 1.15 expresses the electron charge e_s in mechanical units of [kg-m/s] making the product 1 x 10^{19} J/m x 1 x 10^{-19} kg-m/s = 1 W-kg $\equiv$ 1 J/m². **Conflict in integer SI units is immaterial** because at the atomic, molecular or wave packet kinetic formative stage the event has no units of time or space. Nevertheless units require further research.
Please note from Equation 2.2 that:

$$E_{sspaceSI} = E_{s1kgSI} =1/(\varepsilon_o \mu_o) \equiv 1 \text{ kg x } c_s^2\text{x}1.111 \ldots =(3\text{x}10^8)^2\text{x}1.11111=1.000 \ldots \text{x}10^{17}$$

$$E_{sspaceSI} = E_{s1kgSI} =3 \text{ x } 1.111 \ldots \text{ x } 3E_{gl}$$

Balancing probabilities

With pulses from the sun running at 6000 K over most of sun to planet separations the number degrees of freedom in three planes is $(3\text{x}10^{16}\text{x } 6000^4)^3\text{x}1.04^{1/2}=5.9937 \text{ x}10^{94} \approx$ 6 x10^{94}. With ε_{os} energy per cubic meter moving at c_s and relativity corrector 1.11111 . . , energy per unit (cubic meter) is 1.1111 . . . $\varepsilon_{os}\ c_s^2 =1x10^6$ J/unit. This expresses an interesting figure for total energy in six space orientations of the annual rotational volume of earth:

$$6 \text{ x}10^{94} \text{ units x}1 \text{ x}10^5 \text{ (J/unit) x } 1.2^3//1.04^{1/9} = 1.00084 \text{ x}10^{100} \text{ JSI} \equiv 1 \text{ x}10^{100} \text{ JSI} \tag{17.38}$$

The results of Chapter 21 indicate that the Universe runs for very periods like clockwork.
The balance of forces from the sun and from the earth requires probability to generate opposite, equal numerically and in phase, density momenta (forces) at their interactions. Probability of generating a resultant expressing one parameter directed in one specific direction from three eigen value unity vectors located in orthogonal axes is $1/10^3$. The same value applies to the second parameter, two parameters being V and I, E

and *B*. The probability of occurrence of combined parameters such as energy or momenta becomes then $(1/10^3)^3 = 1/10^9$. The same probability value applies to the pulse of sun as to the pulse of earth generating the total probability of the event $(1/10^9 \times 1/10^9)^9 = 1/10^{18}$ times linear probability creates a total of $1/10^{28}$ expressing the muons. The antiphase balance requirement provides another probability factor 10^{-2} making a total of 10^{-30}, so to 10^{30} W is required from the sun to generate balance 1 W vectors of the earth.

Otherwise the minimum particle energy in 3 planes of 3×10^{-14} J over the above gravity power of 3×10^{16} W represents the probability of 10^{-30}. With the earth temperature of 300 K:

$$\text{1 kg at 300 K}/1 \times 10^{-14} \text{ J} \equiv 3 \times 10^{16} \text{ units} \qquad (17.39)$$

The sun at 6000 K delivers 1/4 of its power density in each direction expressed by the gravitation energy E_{g3} for the sun-earth couple 3×10^{16} W distributed over the surface area of the sphere of earth annual rotation in which π is replaced by 3 and the invert of linear probability is applied: $\qquad (17.40)$
Power density $= 3 \times 10^{16}$ W/$[1.1111 \ldots \times 4 \times 3 \times (1.5 \times 10^{11})^2] \times 10^{10} \times 10^{-3} = 1.000000001$ W/m²

The sun constant is actually represented by 300 K generating 3 eV/(K- m²) with a major relativity and quantum correction $K_{rel}^3/1.04^{1/4} = 1.5396/1.04^{1/4}$ $\qquad (17.41)$

Sun constant from earth thermal radiation with a false assumption that earth comprises

electron radiation $= 300$ K $\times$ 3 eV/(K-m²) $\times K_{rel}^3/1.04^{1/4} = 1371.7$ W/m².

This is another example of energy equipartition expressing earth spreading thermal energy in all direction in three planes.

The surface area of the disk of earth annual rotation times the gravitation energy delivers 3×10^{16} W $\times 1.111 \ldots^2 \times 3 \times (1.5 \times 10^{11})^2 = 1/4 \times 10^{40}$ W-m² $\qquad (17.42)$

Formation of wave function

The function expresses two or three photon pulses in orthogonal orientations generating a third pulse. Such functions are separated by X. One type of wave function derive the product of components with corrections in which case the components represent probability of components' events. Single probability, < 1, applied to high rate of events imposed by velocity, temperature becomes a large number called **probability rate**. Other type of wave function expresses pulse progression at velocity of light or its large fraction, while the power of the input pulses is summed in the central beam. In vector addition relativity corrected two pulses increase velocity by $\sqrt{2}$; three pulses added vectorially from orthogonal positions and corrected by relativity increase their velocity by two. For pulses from earth a doubled stepped process is needed to increase velocity of protons from 1/4 c_s to c_s. The pulse in each of three orthogonal orientations is already vector sum of velocities ¼ c_s, which appear as 1/2 c_s. Second vector addition derives c_s.

Adding primary energies of 1 J (at 300 K) and correcting by 1.1111 . . . derives the input energy at 10 J. This satisfies the conservation law of energy and momentum. In the model laws apply to the product of pulses, not of waves. If the source of radation is clump of 9 protons, each clump has probability of 1/6 to radiate in a specific direction in above two step process discharging from initial $9 \times 1 \times 1.111$ . . .$= 10$ J energy 0.9922903×10^{-7} J $\times 1.0328^{1/3} = 1.00302 \times 10^{-7}$ J expressing 1×10^{-8} J output per joule input identified with $(1/3\ c_s)^{-1}$. The variance 0.003×10^{-7} J expresses 1 J per K of 300 K. Pulses from hydrogen around the sun are re-radiating other energy generated from the sun and the energy requirements to generate vector form may be smaller, with single process needed. The formation of the gravitation force cannot be described by a single mathematical formula.

Balance of the gravitation pulses is obtained if two equal antiphase pulses moving in the opposing directions interact generating no resultant magnetic flux.

Table 4 shows that the power index of the reduced radiation 1.2 varies from unity to three, meaning that some interactions occur in one plane, others in two or three planes. Jupiter directs 1/6 of its radiation in one orientation. Relativity corrections have power indexes ranging from 3/2 to 6. Power index of 6 indicates that pulses both from the sun and from the planet interact in three planes. The index for earth of three in Table 4 is consistent with model indicating in Equations 17.26 that sun interacts with one plane and earth with three planes.

Generation of a radiation pulse formed by the wave function format is well illustrated by a signal generated by a directional aerial formed by two rectangle shaped wires positioned at right angle to each other. The signal generated progresses mainly at 45° to the planes of the rectangles because other signals generated by one loop are cancelled by the other loop. The time constant of the signal formation following Equation 15.25 is 1.666 . . . $\times 10^{-8}$ s. Hence at local velocity $1/3\ c_s$ with relativity correction 1.111 . . . linear probability 1×10^{-10} the distance traveled per second by the signal is

Distance per second = $2/3 \times 1.666$ .. $\times 10^{-8} \times 1 \times 10^{8} \times 1 \times 10^{-10}/1.111$ . . . $= \frac{1}{2} \times 10^{-10}$ m (17.43)
in atomic units   = radius of Bohr atom a_{os}.

Thus at least two equal current pulses at right angle to each other are needed to generate a forward moving signal at local velocity of light. Since dissociation energy of Bohr atom is recorded this appears to be possible by converting atomic mass to energy. Three local velocities generate c_s, while two generate gravitation energy per plane.

Gravitation energy represents per plane kinetic energy of 1 kg moving at local velocity of $1/3\ c_s$ can be expressed by the square of $(1/3\ c_s)^2$ or as product of **probabilities-rates**

Energy $E_{g1} = 1$ kg $\times (10^8)^2 = 1 \times 10^{16}$ J $=$ **(1J x1/3c$_s$)X(1J x1/3c$_s$)** $\equiv$ **1x10^{16} J**   (17.44)

The force of repulsion F_{Grs} is the Special Relativity value force of acceleration affecting the mass of earth accelerated centrally during its rotation around the sun corrected by relativity and quantum: (17.45) $F_{Grs} = (6 \times 10^{24}$ kg$\times 3.7755 \times 10^{-3}m/s) \times 1.154700 \times 1.111$ . . .$^{3/2}/1.04^{27/50} = 3.0000 \times 10^{22}$ N $-1/2 \times 10^{19}$.

The variance converted to atomic units expresses $\frac{1}{2}$ electon volt.

Mass of earth in one orientation at atomic level 1×10^{24} kg $\times 1 \times 10^{-30} = 1 \times 10^{-6} = 1 \mu_{os}$ (17.46)
expresses energy of one unit of magnetic permeability. (17.47)

Hence $F_{Grs} \equiv 6\mu_{os} \times 10^{30} \times 3.7755 \times 10^{-3}$m/s $\times 1.154700 \times 1.111 \ldots^{3/2}/1.04^{27/50} = 2.9994 \times 10^{22}$ N.
The minimun energy in SI units is 1×10^{-14} J times volume probability 1×10^{-30} is 1×10^{-44} J at atomic level per direction; the earth mass converted to atomic level for six directions is $6 \times 10^{24} \times 1 \times 10^{-30} = 6 \times 10^{-6} = 6 \mu_{os}$.

The same numerical relation is also expressed by radiation particle of 1×10^{-27} J with probability 1/8 for 8 corners of the integer cube in 6 directions forming wave function by two such structures displaced by $90°$ and corrected for relativity and quantum:

(17.48)

$F_{Grs} \equiv 6(1 \times 10^{-27})^{1/8} \times 6(1 \times 10^{-27})^{1/8}/1.154700 \times 1.04^{1.4} \equiv 5.9874 \times 10^{-6} \approx 6 \times 10^{-6} = 6 \mu_{os}$.

Some classical SI constants appear to be already expressing atomic values, e.g. 1×10^{-27} J times $1 \times 10^{30} \equiv 1000$ K and $6(1000$ K$)^{1/8} \times 6(1000$ K$)^{1/8}$/cor. $= 202.44/(1+1/80) = 199.94 \approx 4 \times 50$ eV. When the product of two or three orthogonal impulses express correct result the components express the probability of components, $1 \times 10^{-27} = (10^{-9})^3$.

Formation of gravitation has not been considered from photon radiations of other elements than hydrogen atom or its molecule. Hydrogen surrounds the sun, hydrogen radiation comes from the outskirt of the Universe, and defects in NaCl have Bohr atom structures. These are the data. Combined effects of mass conversion in sun of other elements have been considered. Vector forces seem to arrive on earth re-radiated as elementary proton or electron signals of H or its conglobates from the gas surrounding the sun without clear vector contributors of other elements.

Reassessment of the gravitation balance

Rotation of a planet around the sun causing repulsion. Unless prevented by another force two masses attracting each other would collapse. Presence of the repulsive force is demonstrated by reversing the direction of one current in the Paris International Center newton force reference $F_r = 2 \times 10^{-7} I^2$ N/m expressing in the integer model force $F_s = 1/5 \mu_{os} I^2$ deriving with earth current 1.8×10^{29} A the exact gravitation force $2 \times 1.8 \times 10^{29} \times 1 \times 10^{-7} = 3.6000 \times 10^{22}$ N.

The particles of atoms do not collapse because they rotate, beside being subjected to the fundamental gravitation force of Equation 6.1. Similarly earth and planets rotation prevent their collapse to sun. The same rule applies to earth and to planets, which are rotating and do not collapse centrally. The suns of the galaxies also rotate around the center, which is usually the black hole. Galaxies treated as current carriers generate the centrally directed force of repulsion opposing Universe collapse. The most recent finding on WEB is that the whole Universe is rotating preventing its collapse.

It should be noted that the repulsive force of Equation 7.4:

$$5.38 \times 10^{14} \, \mathrm{J} \times 1 \times 10^{8} \, \mathrm{m/s} \times 1.1547/[\sqrt{3}(1 + 1/80)] = 3.5424 \times 10^{22} \qquad (17.49)$$

has to overcome the force of attraction of the sun to earth and the force of attraction of the earth to sun, which individually are one half of that value. In the final assessment repulsive forces were found for three planets to have equal components in the direction of the sun and against it producing null effect on gravitation.

Thermal balance of attraction not involving rotation considers static beams of thermal flux between the sun and earth generated by each body. Both fluxes are vectorized generating equal and opposite forces. Sun on earth and earth derives 2 J (Equations 16.6-7).

The total force F_{Gs} = 3.6 $\times 10^{22}$ N is divided equally between the sun and the earth. The proposed gravitation balance is essentially valid for all sun-planet couples. It is numerically explained below for the sun-earth couple. Because of equivalence of terms shown in Equation 2.52a imposed by kinetics interactions gravitation forces can be expressed in different integer and classical SI units. **Two different gravitation generation processes are clearly shown for rotating and stationary bodies. The sun generates per m^2 of circular surface per kelvin temperature to power four force of $1/(3 \times \sqrt{2}) \times G_s$ N $\approx 1/4 \times G_s \times 1/1.111 \ldots^{1/2}$ N with relativity and quantum corrections.** For sun reformatted reference force generated is expressed in natural atomic units by probability $1/10^3$ factor times 2/3 due to not interacting electrons, times 6_s, T^4 and A_{cirsun} with corrections. This is the gravitation force generated by nearly stationary bodies of high temperature, the stars of the planetary system. Planets temperature are generally low generating insignificant forces in such process. Planets generate gravitation forces by rotating. **The earth has no high source of energy and different factors are replaced by 2 N/m^2 vectorized by 1/1000 releasing proton energy by rotation increased by the square of the repetition rate expressed by the square of temperature (totally T^4).** The need to use different relativity corrections is the consequence of different earth and sun temperatures. The F_{Gsun} should be compared with Equation 19.10:

$$(17.50)$$

$F_{Gsun} = 1/3 \times G_s \times (6000-300) \mathrm{K}^4 \times 1.5 \times 10^{18} \mathrm{m}^2/(\sqrt{2} \times 1.111 \ldots^{1/2}) \times 1.04^{1/4} = 3.5441 \times 10^{22}$ N
$F_{Gsun} = 2/3 \times 1/10^3 \times 5 \times 10^{-8} \mathrm{W/m}^2 \times (6000-300) \mathrm{K}^4 \times 1.5 \times 10^{18} \mathrm{m}^2/(\sqrt{2} \times 1.111 \ldots^{1/2}) \times 1.0328^{1/34}$
$\qquad = 3.5440 \times 10^{22}$ N

$$F_{Gearth} = 2 \times 1/10^3 \times 600 \ \mathrm{K}^4 \times 1.278 \times 10^{14} \ \mathrm{m}^2 \times 1.154700^{1/2}/1.04^{1/9} = 3.5439 \times 10^{22} \ \mathrm{N} \qquad (17.51)$$

Above values are at 300 K derived from the classical Plank radiation 0.01 W. In integer SI units that value is 0.001 W. Conversion to integer atomic units by e_s and by linear probability 1/10 derives integer Boltzmann constant k_s = 0.001 x 1 x 10^{-19} x 1/10 = 1 x 10^{-23} J. Planck constant h_s= 1 x 10^{-33} J-s is k_s converted by linear probability 1 x 10^{-10} to integer atomic terms. In the integer model the energy per degree of freedom (DOF) is given by the same expression as in the classical SI system: ½ kT in SI joules or ½ kT/e_s in electron volts or joules at atomic level. This energy format applies to all radiation particles. It is slightly misleading because it suggests that temperature has been taken care of, while two pulse formation requires T^2, three pulse formation requires T^3, Stephan-Boltzmann constant requires T^4 and some expressions T^8. **Although the pulses in SI are 1 J, 1 eV or 2 J attributing temperature to pulse rate identifies the unit**

repeating itself as 1/2 k_s. In 1D (1/2 k_s)$^{1/3}$=(1/2 x 10^{-23} J)$^{1/3}$ ≈ 17 x 10^{-9} J which might be expressing 1 J energy of 17 DOF in space with probability 10^{-9} It is suggested that energy of each thermal emission, which has been converted into a vector force expressing atomic gravitation takes energy out of the mass of the body and expresses destruction of proton structure and turns it into radiation energy. However rotation generates a much more efficient process. Calculations in Chapter 21 and in Conclusions lead to this result.

It should be noted that Newtonian gravitation force of attraction 6.1 expressing sun and earth forces in terms of masses remains perfectly valid. The earth mass radiating 1N/kg delivers m_{earth}/L_{earth}^2 N on sun mass generating with the sun mass and G_s the reference value both in classical terms and integer cubic terms. The same argument applies to forces generated by the sun. Gravitation expressions discussed so far are equally valid. The research found the sources of their validity and constructive features of radiation.

18. *Earth as the satellite of the sun; special relativity modifies mass, distance, velocity and time; apparent inconsistencies.*

Satellite concept displays time dilation

Treating earth and planets as satellites of the sun attracted by the gravity of the sun is the simplest macroscopic explanation of the gravitation relation. In putting a satellite into the orbit a linear outward directed momentum must be supplied equal to the inward momentum of earth gravitation. The combined pair forms a gyroscope with fixed angular momentum, which is held constant for almost eternity until another external force is applied. Furthermore the Special Relativity Theory displaying time dilation and distance contraction or extension appears to be present at our doorsteps. Figure 11 shows a quarter of circular rotation pathway around the sun with earth moving with velocity 30000 m/s along the direction A→B, while the sun accelerates earth along the direction A→C. In a quarter of a year the earth linear velocity would cover in the direction AB the distance:

$$\tfrac{1}{4} \times 365 \text{ days} \times 24 \text{ hours} \times 3600 \text{ s} \times 30000 \text{ m/s} = \qquad (18.1)$$
$$\tfrac{1}{4} \times 31.557 \times 10^6 \times 30000 \text{ m/s} = 2.36677 \times 10^{11} \text{ m.}$$

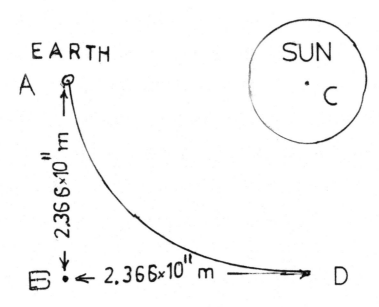

Figure 11. A quarter of circular earth pathway around the sun displays space dilation

At the same time as the sun gravity is pulling the earth towards itself with the earth moving in direction A→D. The two momenta added vectorially locate the earth for circular rotation at the position D. Hence the distance AC should be also 2.3667 x 10^{11}

168

m. The incremental factor over the earth to sun spacing 2.3667×10^{11} m/1.5000×10^{11} m = 1.5778 is identified to be the product of relativity correction attributed to be the relativity correction to power three for velocity ½ c_s and quantum corrector, because each of three local velocities is subjected to the correction:

$$1.1547^3 \times 1.04^{3/5} = 1.5768 \tag{18.2}$$

Relativity increases the mass of the particle and it decreases in the same ratio its velocity to keep the momentum constant. This view in another application is expressed also by Beiser [34]. Comparison of two other books [45, 46] and WEB pages shows that there is no uniformly accepted ways of dealing with relativity corrections. Now the decreased velocity induces space dilation increasing the sun to earth apparent spacing from 1.5000 $\times 10^{11}$ m to the value needed for circular motion 2.3667×10^{11} m.

The force of attraction decreases with the square of the separation so that the gravity force becomes now $(2.3667 \times 10^{11}$ m/1.5000×10^{11} m$)^2 = 2.4894$ smaller than the force of gravitation (3.5438×10^{22} N) time quantum corrector $1.04^{5/4}$:

One sided force of gravitation =
$$3.5438 \times 10^{22}/2.4894 \times 1.1111^{1/2} \approx 1.500 \times 10^{22} \text{ N} \tag{18.3}$$

This is the sun attraction of earth. Earth attracts the sun with the same force making the total relativity force 3.00×10^{22} N, the value derived in Equation 16.41.

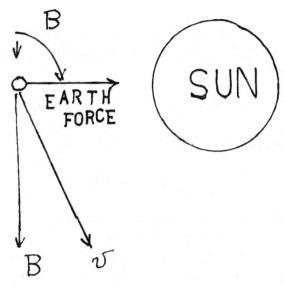

Figure 12. Rotation of earth induces current and rate of change of magnetic flux inducing a force opposing sun gravitation

Repulsive force derived from rate of change of magnetic flux created by the electrons of the mass of earth

The simplest macroscopic expression for the force of reflection created by the earth is expressed by the rate of magnetic flux generated by the earth. It is otherwise expressed by the linear earth velocity and by the flux density B in one plane with suitable corrections. The force created opposes, accurately, the force of gravitation along the sun earth axes. It is the vector product of the linear earth velocity v_{earth} and magnetic flux density B, created by the line current of earth I_{earth} with corrections remembering that for atomic units $u_o = 1$:

$$F_R \equiv - d\varphi/dt \equiv -v_{earth} \times B \equiv - (v_{earth} \times 1/2\pi) \times 10^{-10} \times 1/3 \times 1.1547 \times 1.1111/1.04^{1/2} \times 1.04^{1/24}$$
$$\equiv -29786 \times 1.7795 \times 10^{29}/2\pi \times 10^{-10} \times 1/3 \times 1.1547 \times 1.1111/1.04^{1/2} \times 1.04^{1/24} = -3.5439 \times 10^{22} \, \text{N}$$

$$(18.4)$$

The velocity vector v_{earth} is along the current I_{earth} while the downwards directed flux density B is spreading towards the sun. The three vectors could be considered to be those of the left hand rule applying to the motion of a conductor of an electron motor, which is obstructed from motion by the opposing force. Earth movement can be considered to be the expected reaction following the rules of Lenz's Law by the year long creation of a magnetic loop with current produced opposing sun attraction and flux B creating the opposing force. **This is not another force;** this is the vectorized force of the thermal flux generated by earth expressed in electro-magnetic terms.

Quantum radiation energies

With mass at the atomic level being unity the local velocity of 1×10^7 m/s derives the smallest energy 1×10^{-14} J, which, times one step quantization energy, is identified as local vectorized energy of Equations 2.36-7:

$$[1/(1 \times 10^7)]^2 \times 1/10 \equiv 10^{-14} \times 1/10 \, \text{J} = 10^{-15} \, \text{J} \qquad (18.5)$$

With electron mass being unity at atomic level, energy of 0.003 eV is derived from:

$$1 \times 10^{-15} \, \text{J} \times 5 \times 10^{12} \, \text{Hz} \equiv 0.003 \, \text{eV} \qquad (18.6)$$

Energy 0.003 eV is about 1/3 of quantum of 0.0083 eV deduced from the measured sequential time constants of author measurements [45].

The above 1×10^{-15} J $\times 10^{-30} = 1 \times 10^{-45}$ Jat at atomic level is considered to be equivalent electron energy, with electron oscillating together with proton also at 5×10^{12} Hz and generating $5h_s = 5 \times 10^{-33}$ J.

In the radiation of individual hydrogen atoms the pulse energy contribution of proton, of electron and of (proton+electron) pair is 1×10^{-15}.

The energy 1×10^{-15} J expresses rotational energy of the electron mass in Bohr atom rotating at $1/10 \, c_s$ in:

$$1.111 \ldots \times 1 \times 10^{-30} \times (3 \times 10^7)^2 = 1 \times 10^{-15} \, \text{J} \qquad (18.7)$$

Local energy started from unity values, times net temperature to power four corrected by relativity and quantum derives rare accuracy:

$$1 \text{ W/K} \times (6000-300)^4/[1.111 \ldots ^{1/2}(1+1/700)] \equiv 1.0000016 \times 10^{15} \text{ Hz} \times 1 \text{ W/Hz} \approx 10^{15} \text{ W} \quad (18.8)$$

The variance 16×10^8 is taken to be $5c_s$. Corrected square, of the atomic force in three planes, derives 10^{15} W with the calculator accuracy:

$$1 \text{ W} \times (3 \times 10^7)^2 \times 1.111 \ldots \equiv 0.999\ 999\ 999\ 9 \times 10^{15} \text{ W} \quad (18.8a)$$

Two pulses of $(6000-300)^4$ in wave function format generate sun power attracting earth with velocity of light

$$1 \text{ W/K} \times (6000-300)^4 \times (6000-300)^4 /(1.111 \ldots \times 1.04^{9/125}) \equiv (5700)^8/(1.111 \ldots \times 1.04^{9/125}) \equiv$$
$$\equiv 1.0000344 \times 10^{30} \text{ kg} \quad (18.8b)$$

The variance is $1/3 \times 10^{26}$. Variance times 300 of muo10_s mass 10^{-28} kg derives 1 W and 300 at the temperature 300 K.

Earth temperature cannot be neglected. The derivation gives the integer cubic exact value of the gravitation:

$$1 \text{ W/K} \times (6000-300) \times (6000-300)/(1 \times 10^{-15}) = 3.6000 \ldots \times 10^{22} \quad (18.9)$$

The square term of the wave function format expresses with two orthogonal vectors generation of the third resultant vector moving forward with $\frac{1}{2} c_s$, $c_s/\sqrt{2}$, or c_s.

Transferring a charge to a vacancy in a NaCl ionic crystal comprising only +ve and -ve charges can be largely expressed by the magical numbers 3 and 4 listed in the Conclussion 64. The excitation field has to be in a specific 1 of 3 directions and in 1 of 4 infrared 300 K rotational positions creating per event probability of $1/3 \times 1/4 = 1/12 = 0.083333$. Forward probability in three directions generates an interesting value $(1/3^2)^2 = 1/81 = 0.012345679$, which divided by $1.111 \ldots ^2$ generates 0.01000009 J. The event is the basic quantum unity or 1 eV per particle at the atomic level at 300 K. Relativity $1.111 \ldots ^{1/2}$, quantum $1.04^{1/3}$ and dielectric constant 5.62 are needed to increase the probability of two ions to reach 1 J energy level:

$$0.083333 \times 5.62 \times 1.04^{1/3} \times 1.111 \ldots ^{1/2} = 1.0003 \text{ eV} \quad (18.10)$$

Nearly 10 such steps create 10 eV needed to dissociate the Bohr atom structure formed earlier by prestressing. Actual lattice energy value corrected for strayed energy and for relativity is

$$7.91 \text{ eV} \times 1.2 \times 1.111 \ldots ^{1/2} = 10.0054 \text{ eV} \quad (18.11)$$

The value 7.9 eV is also quoted by Dekker [53, Table 5.2].

Photon radiation

The normal photon radiation was not discussed at all in the book. Photon radiation has no vector properties because photons have equal probability to oscillate in all directions. The linear velocity of light is generated by oscillations in three or two dimensions moving at local velocities and generating the velocity of light. Production of sun vector pulses requires at least 18-20 hydrogen atoms to be converted to helium to release energy needed for the pulse. The common statement that mass is a source of infinite chain of waves implies presence of a source of energy. Pulses and waves from the sun are no doubt supplied by energy obtained from mass to energy conversion. The equipartition of energy rule enforces the local gradients of energy on atomic scale in all orientations to be the same for radiations being in phase and for those being random in all directions. The photon energies are thermal in character and carry much higher power than vectorized and oriented derivatives generating electromagnetic forces and gravitation. This may lead to misjudgments; several expressions in Table 5 may incorrectly express not the gravitation force, but be derivative of the photon radiation having no resultant vector force moving in the specific direction. Wave **function is the requirement for the pulse signals to display momentum, a force and to proceed with velocity of light**. Individual gravitation forces are discussed in details in reference [45].

Further identities

The given below, difficult to explain, relation with two accurate quantum factors was developed displaying a high degree of accuracy. With integer value of earth linear velocity, $v_{earth}= 30000$ m/s, and $A_{cirearth}=1.26780 \times10^{14}$ m^2 and corrector 1.04 the derived gravitation force $F_{Gs}=3.5430 \times10^{22}$ N is within the reference limits of accuracy:

$$F_{Gs} \equiv 4 \times\sqrt{3} \times 1417.6 \times A_{cirearth} \times v_{earth}/(1.040685^{6/5} \times 1.040685^{1/20})= 3.5430 \times10^{22} \text{ N} \qquad (18.12)$$

The energy density 1417.6 W/m^2 is the sun constant value plus radiated thermal energy of earth in one direction $\sigma s \times 300^4 \times 1/10 \times 1.111 \ldots$ W generated by earth: $1373 + 45 = 1418$ W/m^2 on earth surface.

The gravitation force of the digital model generated by the sun is comprised of the following components: force of 1 newton per square meter of sun integer circular area 1.5×10^{18} m^2 times local $1/3 \, c_s$ light velocity, corrected for quantum $1.04^{5/4}$, relativities $1.333 \ldots$ and $1.111 \ldots$, space quantization 10^{-3}, applied only in one plane of earth rotation requiring a factor of 1/3. The expression derives a correct value of the model gravitation force in the expression:

$$F_{Gs}=1.000 \text{ N} \times A_{cirsun} \times 1/3 \times c_s \times 1/1.2 \times 1/1.111 \ldots^2 \times (1/10)^3 \times 1/3 \times /1.04^{9/10}=3.5470 \times10^{22} \text{ N}$$
$$(18.13)$$

Similar accuracy is obtained for the earth repulsive force using force of 1 N per square meter of earth total surface time local, $1/3 \, c_s$, light velocity corrected for relativity and quantum:

$F_{Rs} = 1$ N x $4A_{cirearth}$ x $1/3$ c_s x corrections
$= 4$ x 1.278 10^{14} m^2 x 1 x 10^8x$1/1.2$ x $1/1.111$ ...^{2}x$1.04^{5/16}$=3.5451x10^{22} N   (18.14)

The sun generates the power 1.0 W/m^2 per K per orientation of its surface area:

Sun power and mass in one direction:

$= 6000$ K x 1.0 W/m^{2}x1.5129x10^{18}x1 x10^8m/s x1.111 .../$1.04^{1/5}$
$= 1.0007$x10^{30} W $\equiv 1.0007$ x 10^{30} kg   (18.15)

Above value multiplied by volume probability 10^{-30} derives in atomic terms: 1.0007 J or eV.

The basic gravity model is definitely correct for the sun to earth couple, although there may be some minor numerical errors or unclear definitions in the two preceding books.

Is the number of orientations contributing to the total number degrees of freedom 10 or 30 ?

Equation 2.53 generates good agreement with 5 or 10 orientations expressed in majority of relations used in the book. Several accurate expressions were obtained with 30 orientations. In view of equality of gravitation energy for each nine of sun-planet couples derived in Equations 17.28 it is suggested that the rate of gravity pulses delivered individually by the sun to all planets is the same; Pluto is included as a planet. Sun distributes its energy equally to all planets. With the sun included there are ten bodies with 6 x $(10^{10})^{10} = 6$x10^{100} degrees of orientation freedom. This figure uncorrected for atomic scale is in agreement with the total integer SI overstated power of the sun of 6x10^{100} W generating 1 W per degree of freedom. These bodies circulate in one plane, so number of planes does enter the equation. Gravitation is distributed from the sun in three planes, while interacting with planets only in one dimension. It is generated from the planets in three planes. This hypothesis affects the whole sun system and it should apply to all other sun systems in the Universe. However response of the planets to sun attraction turned out to be unexpected as explained in Chapter 20.

Numbers of degrees of freedom; the converter defines relations between parameters

The sun interacts with earth with one half of its mass and ½ or 1/6 of its temperature. The relations quoted below are expressed in atomic units. The product of average velocity 1/3 c_s squared, converted to joules, is one tenth of the 1st term of Planck radiation law and expresses energy per quantum pathway:

Energy per quantum pathway = $(1/3$ $c_s)^2$**x** (unity atomic unit of mass)x10^{-19} = 0.001 J (18.16)

One tenth of the average velocity **1/3 x c_s**, expressing the velocity of light in local pathway, is equal to temperature squared, corrected by relativity:

$$1/3 \text{ x } c_s \text{ [m/s]x } 1/10 \equiv T_{3000}{}^2 \text{x} 1.111 \ldots \equiv 1 \text{ x } 10^7 \text{ m/s} \tag{18.17}$$

The product of the atomic force and of the average velocity reduced by 1/10 in each parthway derives power 1 W per pathway and 10 W for ten pathways:

$$\textbf{1/3 x } c_s \text{ [m/s]x } \textbf{1/10 x } F_{as} \text{ [N] = 1 x } 10^8 \text{ x 1/10 x 1 x } 10^7 = 1 \text{ [W]} \tag{18.18}$$

Orientational probability applied to temperature and mass derives correct results

Gravitation is postulated to follow the conversion of mass to energy within the body of the sun. Mass generates a force of 10 N/kg per K converted to linear value 1 N/kg per K. Mass of proton in one direction, 10^{-27} kg, moving at local velocity 1×10^8 m/s becomes in atomic terms momentum expressing electron charge $e_s = 10^{-27} \times 1 \times 10^8 = 1 \times 10^{-19}$ C. In the far past the momenta emitted by electrons have formed protons, congregated them into the mass of the sun and have formed the planetary system. Protons, generating thermal energy were converted to vectors and they are responsible for the gravitation forces between large masses and between atomic particles. Both moments and spin of electrons creates momentum and proton mass. Hence postulating presence of both these forces is not inconsistent.

Calculations derive equal gravitation energies or moments in atomic terms. Although the distance between the planet and the sun for the circular model remains constant, gravitation energies are dissipated by the sun and by the planet. Atomic terms expressing true relations display the annulled sum of two momenta expressing in SI terms a loss of mass.

Creation of radiation by mass is generally acknowledged in science. The process is likely to be reversible with radiation creating momentum and mass.

The following expressions illustrate further the integer model findings. The corrected mass of proton to power 1/30 expressing quantom orientation probability in one plane is taken to derive 1st term of Planck radiation law:

$$(2 \text{ x } 10^{-27})^{1/30}/(1.1111^2 \text{ x } 1.04) \equiv 0.01003 \text{ J} \tag{18.19}$$

Sun mass of 1 kg at temperature of 6000 K to power 1/30 with corrections and in three planes derives the energy of pulse of 3 J:

$$1 \text{ kg x 3 x } (6000)^{1/30}/(1.1547^2 \text{ x} 1.04^{1/18}) \equiv 3.0004 \text{ J} \tag{18.20}$$

The mass of earth in kilograms in 10 pathways of one plane generates the temperature 300 K:
$$(6 \text{ x} 10^{24})^{1/10} \equiv 300.48 \text{ K} \tag{18.21}$$

The mass of sun to power 1/30 derives the ionization energy of hydrogen atom 10.23 x eV x 1.1547^2 =13.64 eV:

$$(2 \times 10^{30} \text{ [kg]})^{1/30} \equiv 10.23 \text{ eV} \qquad (18.22)$$

The earth mass of 1 kg at temperature 300 K generates power of 3 W:

$$1 \text{ kg} \times 3 \times (300 \text{ K})^{1/30} \times 1/(1.2 \times 1.04^{1/5}) \equiv 2.9988 \approx 3 \text{ W} \qquad (18.23)$$

Does the emission from the mass of 1N/kg establish gravitation balance?

Eight calculations of moon gravitation forces using different classical formulae derive an average value of 2.0103×10^{20} N [46]. To confirm this value using the integer model data and rounded up mass values of the earth and of the moon (integer 8×10^{22} kg) the gravitation constant G used had to be equal $2/3 \times 10^{-10}$ Nm^2kg^{-2} value generating at the distance 3.844×10^8 m the integer model force of:

$$F_{Gsmoon} = 2/3 \times 10^{-10} \times 6 \times 10^{24} \times 8 \times 10^{22}/(3.844 \times 10^8)^2 \times 1.111 \ldots^2 \times 1.04^{7/24} = 2.0067 \times 10^{20} \text{ N} \quad (18.24)$$

The gravitation force of the moon per its unit mass 7.3477×10^{22} kg subjected to one $1.111 \ldots^{1/2}$ correction is 2×10^{20} N/($7.3477 \times 10^{22} \times 1.111 \ldots^{1/2}$) = 0.002583 N/kg. That value varies significantly from 0.006 N/kg for the earth to sun calculation. The WEB pages quote 107 K and 153 K as the day and night moon temperatures deriving now an average value of 130 K, this being a fraction of 130/300 = 0.433 of mean earth temperature. If the force of moon gravitational attraction inducing repulsion is proportional to its temperature it would at 300 K amount to:

$$0.002583 \text{ N/kg}/0.4333 = 0.00596 \text{ N/kg} \qquad (18.25)$$

This suggests that the gravitation balance expresses the ability of the smaller body to generate sufficient force. This view is confirmed in calculations in Chapter 20 for Jupiter, which generates other than thermal energy to balance the sun attraction. While the force 1 N/kg appears to be specific to the inside of the Bohr atom, it does not impose a gravitational model valid all-over the Universe. It balances the sun beam of gravitation locally around its spherical surface.

Does our time perception need correction by Special Relativity?

The classical way of deriving basic laws in physics is to obtain a dimensional agreement between physical parameters using dimension analyses and then finding a numerical factor logically fitting the relation. The procedure adopted most often in the research generating the three books of the author was just the opposite. Numerical agreement was first searched between parameters including likely relativity and quantum corrections and then missing dimensional terms were inserted in the relation.

A simple relation in integer units finds that in integer cubic model SI units three velocity of light are derivative of ionization energy of hydrogen and equal the squared linear earth velocity: (18.26)

$$3\,c_s \equiv 3(13.62\ \text{eV}/1.04^{3/4}\text{x}10)^4/1.04^{1/2} \equiv (\text{squared linear earth velocity, } v_s^{\ 2}) \equiv 30000^2\,(\text{m/s})^2$$

Long analyses discovered that this identity was the consequence of the duration of one solar year, which is found now in terms of the model to have undilated value $3.0\text{x}10^7$ s. The undilated duration of the solar year t_{ss} is derived by correcting the classical SI measured value $31.558\text{x}10^6$ s by arbitrary relativity and quantum correctors:

$$t_{ss} = \textbf{31.5581x } 10^6 \textbf{ s}/(\textbf{1.111} \ldots ^{1/2}\textbf{x1.04}^{1/20}) = \textbf{29.9973 x}10^6 \equiv \textbf{3.0 x}10^7\,\textbf{s} \equiv \textbf{1/10}\,c_s\,\text{m/s} \quad (18.27)$$

The identity states that velocity of light is formed in 10 quantum orientations of space plane and the probabilty of it being formed in the specific direction is 1/10 so although time is velocity it takes ten times longer for the earth to rotate around the sun. The variance $2700\text{x}1.111 \ldots \approx 3000$ K is 1 m/s per kelvin. **Above equality confirms identity Time $\equiv$ Velocity of Equation 2.52b arising out of the absence of units of time and of length in kinetic interactions.** This statement was considered trivial until Equation 18.27 established its validity coinciding with kinetic evidence.

The undilated duration of solar year derives, in meters of the circular circumference, earth path of rotation around the sun in the following identities:

Earth circular circumference of rotation around sun =
$$v_s t_{ss} \text{ x } 1.111 \ldots = 30000 \text{ x } 3.0\text{x}10^7 \text{ x } 1.111 \ldots = 1 \text{ x } 10^{12}\,\text{m} \quad (18.28)$$

It is surprising that starting with distance to sun L_{earth}, earth rotation circumference is

$$1.5 \text{ x}10^{11} \text{ m x } 2\pi \text{ x } 1.111 \ldots /1.04^{1.2} = 0.99905 \text{ x } 10^{12} \approx 1 \text{ x } 10^{12}\,\text{m} \quad (18.29)$$

It is also surprising that starting with L_{earth} and replacing π by 3 the identity becomes simpler $1.5 \text{ x}10^{11}$ m x 2 x 3 x 1.111 $\ldots = 1 \text{ x } 10^{12}$ m (18.30)

The distance covered per infrared cycle of local velocity is $1\text{x}10^{12}$ m$/1\text{x}10^{12} = 1$ m/cycle. To express the length of 1 m in nature atomic terms requires multiplication by linear probability of $1\text{x}10^{-10}$ m generating Bohr atom diameter per cycle. Hence the observed period of the earth rotation may be dilated, meaning that the period of the time unit of one second is dilated time by

$$31.5581 \text{ x } 10^6/3.0 \text{ x } 10^7 = 1.05194 \text{ factor or about } 5.2\% \quad (18.31)$$

The deviation can be treated more likely as quantum and relativity correction for $1/3\,c_s$:

$$1.111 \ldots ^{1/2}/1.04^{1/19} = 1.05192.$$

The Global Position System allows only a minute value for time dilation, but it does not recognize that an average particle moves with a larger velocity.

Momentum of earth at unit velocity in three orientations converted to nature atomic units by multiplying by c_s derives:

(Mass of earth x light velocity) x 3 = 6 x 10^{24} x 3 x 10^8 x 3 $\equiv$ 54 x 10^{32} J/m (18.32)

Above value time L_{earth} derives energy or force in the sun to earth beam as calculated in Table 2 in Planck units

$$54 \times 10^{32} \text{ J} \times 1.5 \times 10^{11} \text{ m} \times 1/10 = 8.1 \times 10^{43} \text{ J-m} \qquad (18.33)$$

And in three planes $3c_s$ x m_{earth} = 5.4 x 10^{33} kg-m/s, converted to nature atomic units by multiplying by c_s corrected by relativity and applied to one of 5 pathways derives

$$1/5 \times 5.4 \times 10^{33} \times 1.333\ldots/1.111\ldots \times 3 \times 10^8 = 3 \times 10^{42} \text{ N} \qquad (18.33a)$$

again that value is derived in Table 2 for the force in Planck units.

Example of energy conversion

Over the span of human species the interactions between the large masses of our solar system express rules of bodies in thermal equilibrium. True events are expressed by their probability derived from atomic radiation interactions. Energies derived in integer and classical SI systems have essentially the same values with the classical system considering essentially one particle velocity, the measurable velocity of light c. The integer model considers velocities: $1/10\ c_s$, $\frac{1}{2}\ c_s$, $1/3\ c_s$, $\frac{1}{4}\ c_s$ and c_s. Relativity and quantum correction for these value are relatively small. Linear, surface and volume probability, velocities of light and charge effect greatly numerical values.

A 300 K exponent of the Planck radiation equation expressed in the model SI values converts radiated energy to;

$$E(\lambda,\ T) = 8\pi h_s/(c_s)^3\ v^2 dv/(e^{vhs}/k_s T300\ -1) = 8\pi h_s/(c_s)^3\ T300 = 2.7888 \times 10^{-55} \text{ J} \quad (18.34)$$

indicating that energy remains proportional to temperature. Relativity and quantum corrections derive the minimum energy of 3 components of Equations 2.36-7:

$$\text{Minimum energy} = 2.7888 \times 10^{-55} \times 10^{10} \times 1.11111^{1/2} \times 1.04^{1/2} \approx 3 \times 10^{-45} \text{ J} \qquad (18.35)$$

Planck radiation per kelvin in classical SI units of 0.01 J as in Equation 3.14 times linear converter 1×10^{-10} and temperature divided by the mean particle velocity $1/3\ c_s$ and electron charge e_s:

Energy at 300 K in eV in SI terms = 0.01x300x3 x 1.11111 = 10 eV (18.36)

Reassessment of Planck's units in view of probabilistic derivation of the model physical constants and geometry of space

An event has unity probability and for Bohr atom it is related with energy value of unity. An event with lower probability is a misnomer. It is an event of unit probability together with several non-events. Presently one degree of freedom is identified with an event and so number of degrees of freedom, DOF, represents the number of events of unity value. If the end output is unity and since probability cannot exceed unity, input energy values in excess of unity express number of events required to generate a unity output inclusive of converter factors, quantum and relativity correctors and temperature correctors.

Unity value, 1 J, 1 W or 1 eV events are derived in Equations 1.2, 2.27, 2.29, 2.66 and others. Studying them is important in understanding the model proposed. Energy of 0.78 eV in [12] is a derivative of unity, and the average of 30 research studies, so it presents an overwhelming evidence for the validity of the model postulates.

Earth mass 6×10^{24} kg is attributed with 6×10^{24} DOF. Each kilogram of mass is free to move in 6 directions at local velocity of 10^8 m/s which in 3 D expresses value $(10^8)^3$ $= 10^{24}$ deriving 6×10^{24} DOF. Sun pulse of 1×10^6 eV (or DOF) and its mass in one local direction expresses balancing DOF in one plane $1/3 \times 6 \times 10^{24} \times 1 \times 10^6 = 2 \times 10^{30}$ DOF. Note that in SI units

$$\text{Pulse power} = (1 \text{ W} \times 1000 \text{ K})X(1 \text{ W} \times 1000 \text{ K}) = 1 \times 10^6 \text{ W} \qquad (18.36a)$$

Physics based on probability requires all Planck units to be integer values needing following three definitions for Table 2. Planck length is taken as $l_{ps} \equiv 1/10 \ h_s \equiv (1/3 \ v_{os})^2 \equiv 1 \times 10^{-34}$ m. Planck mass $m_{ps} \equiv [10 h_s \times 1/3 \ c_s/G_s]^{1/2} \equiv 1 \times 10^{-8}$ kg. Time $l_{ps}/c_s = 1 \times 10^{-42}$ s.

It is suggested that for the first time all integer cubic model SI physical constants are derived directly using Planck units and with minor correction derive also atomic units. Referring to Table 2:

velocity is 10^8 m/s; energy 10^{10} J times linear probability 1×10^{-10} derives 1 J; mass 1×10^{-8} kg times average velocity derives unity momentum; **actually unity reference is momentum expressed by Planck atom mass of proton 10^{-8} times Planck velocity 10^8 m/s with Planck being 3D display of $(1 \times 10^{-27})^{1/3}$kg $\times 10$.** Charge 2×10^{-18} C in one direction of 10 pathways derives $e_s = 1 \times 10^{-19}$ C; Planck constant is derived directly 1×10^{-33} J-s; force 3×10^{42} N derives atomic force as $(1/3 \times 3 \times 10^{42})^{1/6} = 1 \times 10^{-7}$ N; current 2×10^{24} A $\times$ 30000m/s $\times 1.1547^{3.5} \times 1.0328^{1/4}$ appears to be 2×10^{29} A that displayed by the earth mass of 6×10^{24} kg generates force inclusive of electrons of 3 N/kg; proton energy, 10^9 J, derives energy of radiation pulse $(1/10^9)^3 = 1 \times 10^{-27}$ J; time 1×10^{-42} s times force 3×10^{42} derives impulse of the field 3 N-s.

The product of 10^{10} J and volume probability10^{-30} derives 10^{-20} J or eV $\equiv 1/10 \ e_s$ re-presenting, according to Equation 1.15, 1/10 of proton local momentum suggesting that all 10 local momenta are discharged in one orientation during the radiation process. Electron charge is local proton proton momentum $e_s \equiv m_{ps1D} \times 1/3 \times c_s$., which times above surface (2D) converter $10^{20}/10$ to SI becomes unity equal $10^8 \times 10^{-8}$. Hence electron charge in SI terms can be also considered to be the reference unity pulse.

Further analysis of the integer Planck units length l_{ps} has uncovered direct relations between l_{ps} and other integer cubic constants:

$$l_{ps} \equiv h_s \varepsilon_{os} \equiv 10^{-33} \times 10^{-11} \equiv v_{os}^2 \equiv (1 \times 10^{-17})^2 \equiv 1 \times 10^{-34} \, \text{m} \qquad (18.37)$$

$$2\pi c_s^3 \times 6000 \, \text{K} \times l_{ps}^2 \equiv e_s^2 = 1 \times 10^{-38} \, \text{C}^2 \qquad (18.38)$$

$$2\pi c_s^3 \times 6000 \, \text{K} / 1.04^{9/20} \equiv 1.00007 \times 10^{30} \, \text{W or kg} \qquad (18.39)$$

Radiation in six orientations is generated by three sets of oscillations in any two planes 1-2, 2-3 and 1-3 of the orthogonal geometry of space. Any two planes of that geometry at right angle to each other generate a signals at 45^0 to the two planes a total of six directions. Thus the radiation can be considered to be generated by the configuration of the radiation direction receiving antenna comprising two circular loops at right angle to each other. Currents flowing in such antenna generate signals in four directions only because signals generated by one plane are partially cancelled by signals generated by the other plane. With ten optional quantum directions per plane the probability per plane is 1/10 and it decreases to $1/10^{10}$ if all signals from the radiation need to be directed in one direction to produce a signal. To produce a signal produced by two planes the total probability becomes $1/10^{10} \times 1/10^{10} = 1/10^{20}$ deriving 1/10 e_s on division by 1×10^{-19} C. Such pulses is generated in three planes with the total probability $1/10^{20} \times 1/10^3$ as in Figure 5 producing 1×10^{-23} J identified as Boltzmann constant k_s J. Thus an input of 1 J in three planes and in six directions in each generates 1×10^{-23} J in the model SI units and treated now as H. u. per kelvin.

Energy of 0.001 J $= (\mu_{os})^{1/2}$ and 0.001 J$=0.01$ J (of Planck radiation) $\times 1/10$ (18.40)

Radiation takes place in six orientations with the observer relating it to the pulse along the source-receiver axis:

Permittivity $\varepsilon_{os} = 10 \times (0.001 \times 1/10^3)^2 = 10 \times [(\mu_{os})^{1/2} \times (1/10^3)^2 = 1 \times 10^{-11}$ F/m (18.41)

Above relation would be generally expressed as μ_{os} with two sequential probabilities $1/10^3$:

$\varepsilon_{os} = 10 \times (\mu_{os} \times 1/10^3 \times 1/10^3) = (\sqrt{10} \times \mu_{os}) X (\sqrt{10} \times \mu_{os}) = 1 \times 10^{-11}$ F/m (18.42)

$\sqrt{10} = 3.16227766 \rightarrow \times 1/111 \ldots^{1/2} = 3.00000000 = 3$ appears also in other expressions.

Please note that is not a wave. It is a pulse moving at the local velocity $1/3 \, c_s$ and generating per second in each of ten local pathways:

$$\text{Power} = \varepsilon_{os} \times 1/3 \, c_s = 1 \times 10^{-11} \times 1 \times 10^8 = 0.001 \, \text{W} \qquad (18.43)$$

Above energy needs for ten quantum states input of 0.01 W as specified by Planck radiation law and three times the value for three planes per kelvin. Permittivity ε_{os} times invert of linear conversion 10^{10} and linear probability 10 derive 1 J/m^3.

The integer model is not concerned with radiations, frequencies and energies of photons generated by Bohr atom, the earth or the sun unless they lead to generation of vectors.

Apparent inconsistencies due to different positions of the observer, varying relativity corrections

Changing viewing position of the observer and his velocity modifies the value of some result and of the formative expression. Different values of gravitational forces calculated, 3.00×10^{22} N, 3.50×10^{22} N, 3.60×10^{22} N, 4.00×10^{22} N, 3.54×10^{22} N and 1.7719×10^{22} N, were attributed to the Rules of Special Relativity without justifying all values. Value 1.7719×10^{22} N is obtained for the observer being at the mass centre of the sun and the earth. Value 3.54×10^{22} N is obtained for the sun and the earth, but forces change directions with reference. 3.50×10^{22} N is the quantum corrected value of 3.54×10^{22} N. The ratio $4.00 \times 10^{22}/3.60 \times 10^{22} = 1.1111$, the relativity correction. $3.60 \times 10^{22}/3.00 \times 10^{22} = 1.2$ expresses the energy loss on radiation so all values can be fitted with veritable explanations.

The sun constant 1372.15 W/m² requires adding to its radiation $\sigma_s T_{300}{}^4 = 405$ W/m² from earth subjected to linear probability 1/10 deriving 40.5 and 1412.65 W/m². This value varies only 2 W/m² from the basic sun pulse $\sqrt{2} \times 1/6 \times 6000$ K $\equiv 1414.71$ W/m² the variance expressing the quantum value of the sun pulse 2 J. The pulse of 2 J is consequence of the sun attraction being only in one plane of earth rotation around the sun. The pulse of 2 J is generated by three 1 J pulses as in Figure 4 with the resultant being $\sqrt{3} \times 1.1547 = 2.0000$ J.

One half of the sun pulse 1412.65 W/m² = 707.325 W/m² is observed from the centre of mass towards the sun and 707.325 W/m² towards the earth. Some values may express the observer being proton or the electron requiring different corrections. The large number of expressions in Table 5 is likely to express the presence of oscillations and not due to unidirectional vector forces.

Uncertainties of measurements and accuracies in the probabilistic integer model

This topic requires further research; preliminary analyses given here creates interesting and unexpected results. Re-examination of the author's old experimental data deduced the presence of local light velocity components $\frac{1}{2} c_s$ from the incremental energy quanta $0.6926kT$ exceeding the classical value $\frac{1}{2} kT$, as described in Chapter 1. Presently the quantum q is re-defined by:

$$q = 1/3 \, h_s v_{Lms} = \frac{1}{2} \, k_s T_{300} \times 1.1111 \qquad (18.44)$$

Since numerically $h_s = k_s \times 10^{-10}$ Planck constant can be considered Boltzmann constant expressed in the natural atomic units, with both terms expressing per kelvin value. Hence expressed in the natural atomic units the energy for two degrees of freedom at 300 K:

$h_s T_{300}/e_s = 3 \times 10^{-12}$ J or eV becoming on multiplication by c_s and by relativity correction

$$3 \times 10^{-12} \times 3 \times 10^8 \times 1.1111 = 0.001 \text{ W-m} \qquad (18.45)$$

In 10 local orientations the above power totals in 10 local orientations the Planck radiation law value 0.01 J/s.

The graviton energy G_s expresses energy of 30 orientations relativity corrected:

$$G_s = 3 \times 10^{-12} \times 30 \times 1.1111 = 1 \times 10^{-10} \text{ J} \qquad (18.46)$$

The wave function energy at the natural atomic level at 300 K is

$$(300 \, h_s) \text{X} (300 \, h_s)/e_s \times 1.1111 = 1 \times 10^{-42} \text{ J} \qquad (18.47)$$

Above value times 10^{30} on converting to SI units becomes again 1×10^{-12} J per plane $= 1/3$ $h_s T_{300}/e_s$

The minimum uncertainty values $h/4\pi$ or $h/2\pi$ apply probably to signals expressed in 2 or 3 dimensions. In NaCl impurity Bohr atom model the quantum is $\frac{1}{2} k_s T_{300} \times 1.111$ in SI units; in atomic natural units it would take value $\frac{1}{2} h T_{300} \times 1.111$. Since $\frac{1}{2} h \approx 1/3 \, h_s$ per kelvin value is $1/3 \, h_s \times 1.111 \ldots = 3.7037037 \ldots \times 10^{-34}$ J-s. Multiplied by the vectorized power pulse $E_{searth SI 3} = 3 \times 10^{21}$ Hz (Equation 21.3) derives $1.23456790 \times 10^{-13}$ J.
Divided by relativity corrector $1.111 \ldots^2$ derives:

$0.999999999 \times 10^{-13}$ J $= 1 \times 10^{-13}$ J obtained earlier as 10 minimum energy pulses 1×10^{-14} J.

Above energy referred to 300 K derives $3 \, \varepsilon_{os} = 1.00 \times 10^{-13} \times 300 = 3 \times 10^{-11}$ F/m $\qquad (18.48)$

Please note $1/3 \, h_s/1.1111 \ldots = 3.000000000 \times 10^{-34}$ J-s $= 3 \, l_{ps}$ of Table 2.

The model is postulating that Boltzmann constant expresses limit of the accuracy of measurement per kelvin or $\frac{1}{2} kT$ per degree of freedom. Hence the SI uncertainty per one unit of space wave at 300 K is

Plank radiation term $= k_s \times E_{searthSI} = 1 \times 10^{-23}$ J $\times 1 \times 10^{21}$ Hz $= 0.01$ J $\qquad (18.49)$

The values of Boltzmann constant, Planck constant and Heisenberg uncertainty remain unaffected by the velocity of the observer. So is the space $E_{sspaceSI}$.

$$E_{sspaceSI} = (\mu_{os}\varepsilon_{os})^{-1} \equiv 1 \times 10^{17} \text{ J} \equiv (c_s)^2 \times cor. = (3 \times 10^8)^2 \times 1.1111 \ldots \equiv 1 \times 10^{17} \text{ J} \qquad (18.50)$$

Boltzmann constant is presently derived to be the product of the space frequency and the radiated (converted) electron mass moving at the local average squared velocity of particles multiplied by $1/3 \, c_s$ to convert atomic units to model integer SI units:

Boltzmann constant $k_s \equiv m_{es}\, xv_{os} = 1\times10^{-30}\, x\, 1\, x\, 10^{17} = 1\, \times10^{-23}\, J$ \hfill (18.51)

Inverted ε_{os} expressed in the wave function format is:

$$1/\varepsilon_{os} \equiv [m_{es}{}^{1/2}\, x(1/3\ c_s)]X[m_{es}{}^{1/2}\, x(1/3\ c_s)]\, x1/3\ c_s\, xv_{os} = 1\ x10^{11}\, J \hfill (18.52)$$

where $m_{es}{}^{1/2} \equiv 1/(\text{electron frequency per pathway}) \equiv 1/(1\ x10^{15}\ Hz)$ and $1x10^8/1x10^{15} \equiv 1/10^7$
The value $1/10^7$ is taken to be the energy $10^{-7}\, J$ per N, which times the Bohr atomic force $10^7\, V/m$ derives energy of 1 J or 1 eV. For atomic units h_s replaces k_s and uncertainty takes value $\frac{1}{2}\, h_s T$ per degree of freedom.

Boltzmann constant k_s at the atomic level decreases by surface probability $1\ x\ 10^{-20}$ so that resultant energy becomes $1\ x10^{-23}\, J\ x\ 1x10^{-20} = 1x10^{-43}\, J$. This sums energy of ten quantum position of the minimum energy $1\ x10^{-14}\ x\ 1x10^{-30} = 1x10^{-44}\, J$.

Increasing temperature to 6000 K and correcting for relativity and quantum derive energy of Equations 2.5:

$$1x10^{-43}Jx6000^4\ K/(1.11111^2 x1.04^{1.2}) \equiv muo10_s\ \text{energy} \equiv 1.0015x10^{-28}J = (h_s/e_s)^2$$
$$\equiv (1x10^{-14}\ J)^2 \equiv 1/10xh_s/\mu_{os} = 1x10^{-11}\ //1x10^{17} = \varepsilon_{os}/E_{sspaceSI} \hfill (18.53)$$

The variance from the integer 10^{-28} is $0.0015\ x10^{-28}\ J \equiv 1/6\ x\ m_{es}/1.11111$

Extending analyses generates new relations between integer physical constants suggesting strongly that some constants may be superfluous. Uncovering the kinetic aspects of radiation interactions makes this suggesrion a fact and the cause for the huge number of equations expressing the earth gravity force in Chapter 19.

Do observers moving at various velocities relative to space particle generate each a different set of physical constants or just different spacings and time intervals?

Analyses of above seemingly frivolous question lead to improved understanding of the integer cubic space model postulated. The observer on earth is looking at particles in space that are smaller than stationary particles on earth and these particles are moving at a local velocity of $1/3\ c_s$ requiring a quantum correction and a relativity correction of $1.11111 \ldots$

Some value obtained by the author may correspond to the observer moving with velocity of protons, $1/4\ c_s$, or electrons, $1/2\ c_s$.
Millikan experiment measured on earth neglected particle velocity and derived the classic SI charge $1.602\ x10^{-19}\, C$. Equation 1.10 derives value $1x10^{-19}\, C$ with particle velocity $1/2\ c_s$ because beside relativity factor $1.15471^{1/2}$ the charge is reduced by factor $1/1.2^3$. The velocity portion generating relativity correction 1.0328 derives also charge $1x10^{-19}\, C$:

$$1.6021\ x10^{-19}/1.2^3\ x\ 1.111\ldots/1.04^{3/4} = 1.00029\ x\ 10^{-19}\ C \hfill (18.54)$$

The variance from the integer value is identified as $1/3$ k_s = $1/3$ x 10^{-23} J, which time 300 K x1.111 . . ./1 x 10^{-19}= 0.01 J, the value of Planck radiation law.

Thus Milliken experiment derives two sets of physical constants. Observers moving with relative velocity other than $1/3$ c_s will see variations in force, distance and time interval, but most physical constants will remain represented by the two sets of values. The values of velocity of light and the space frequency v_{os} would be observed to have the same values because relativity correction changes ε_{os} and μ_{os} in inverse ratio. The electron and proton masses would remain the same, so would their frequencies, σ_s, G_s and the parameters expressing momentum and energy. Other parameters would have to be considered individually. Since our observations are from earth this topic will not be pursued.

Inconsitentcies in dimentions

Dimensional inconsistencies are unacceptable in Science. The author considered that a good numerical agreement is equally important and suggested fitting the discovered identifies with unity value having dimensions correcting the errors. This procedure was used for ten years until it was realized that proposed identities express interactions between particles, which obey the rules of kinetic elastic collisions. Such interactions are excellently described by Kittel and ales [64]. They have no units of time and distance (length).

The author has observed that SI equation $F = m$ x a applied to within the atom reads:

$$\text{Mass} \equiv \text{Force and Time} \equiv \text{Velocity follows} \tag{18.55}$$

Since mass generates 1 N/kg per kelvin, mass in kilograms becomes force and $F = T^4 m_{kg}$ newtons. With velocity c_s supplying momentum (Equation 1.15) Special Relativity gravitation force becomes

$$F_{Gs}=1 \text{ N } T^4 c_s \text{ x10 x 10} = 1000^4 \text{ x100 x 3 x } 10^8 = 3 \text{ x1/}\varepsilon_{os}^2 = 3\text{x}10^{22} \text{ N} \tag{18.56}$$

The force within an atom represents the rate of 1 J (eV, W) pulses generated.
Earlier text had to be revised and corrected.

Possibility of sun true electron moments being absorbed by the earth

Some texts quote that $1/3$ of sun electron radiations is absorbed by earth. Only $1/2000$ of single electron moment is absorbed by the proton with multiple reflections being possibly able to increase this factor to $1/6$ =0.1666 . . . This was observed by the author in epoxy experiments and that moment might increase the force of gravitation by 1.1666 . . . Reviewed calculation of the sun gravitation of the Equation 17.50 derives:

$$F_{Gs}= 1/3 \ G_s \text{ x } 5700^4 \text{ x 1.5 x } 10^{18}/(\sqrt{2} \text{ x 1.111 } \ldots^{1/2}) \text{ x } 1.0328^{1/32} = 3.5442 \text{ x } 10^{22} \text{ N} \tag{18.57}$$

Please note that quantum correction has identified with relativity correction for velocity of proton, $\frac{1}{4}\,c_s$, and its index 32 expresses number of interacting DOF.

Increasing gravitation by $1.1666\ldots\approx1.1547\text{x}1.0328^{1/3}$ might be hidden, if it is expressed by arbitrary use of above relativity correction in the expression:

$$F_{Gs}=1/3G_s\text{x}5700^4\text{x}1.5\text{x}10^{18}/(\sqrt{2}\text{ x }1.111\ldots^{1/2})\text{ x}1.1666\ldots/(1.1547\text{ x}1.0328^{1/3})\text{ x}1.0328^{1/32\cdots}$$
$$=3.5458\text{x}10^{22}\text{ N}$$

The resultant error 47 parts per 10^5 is excessive, although such accuracy was acceptable during the investigating period. The possibility of electron absorption is refuted. Reflections diffuse on earth electron discharge moments equally in six orientations.

19. *Apparent total gravitation forces of earth*

The equivalence of physical parameters explained in previous Chapters with kinetic interactions having no dimensions of time or distance is the cause of very large number of expressions of forces of sun attraction and of earth repulsion listed in Table 5. They comprise also many oscillations of equal magnitude without vector properties. Uncovering these relations for earth and planets look several years followed by analyses of the constants in the expressions leading to the integer model. With the model described these data add few undiscussed topics. Using standard SI units or integer SI units produced equally accurate results. Several expressions were discussed earlier [45, 46] more relations were added and only a few items will be covered now.

To bring the values closer to the reference range the quantum corrector has been applied to some, but not to all relations.

Table 5. Apparent total gravitation forces of earth   **Equation No**

$F_G = G \times m_{earth} \times m_{sun} / L_{earth}^2 = 3.5438 \times 10^{22}$ N   (19.1)

$F_R = m_{earth} \times v^2 / L_{earth} = 3.5430 \times 10^{22}$ N   (19.2)

$F_{Gs} \equiv (h \times v_{Lms} / a_{os} \times m_{earth} \times m_{sun}) / (X_{earth}^2) = 3.5438 \times 10^{22}$ N   (19.3)

$F_{Gs} \equiv 1/3 \times g_{sJ}^2 \times m_{earth} \times m_{sun} / m_{es}^2 \times 1 / (X_{earth}^2 \times K_{rel} \times 1.0357) = 3.5452 \times 10^{22}$ N   (19.4)

$F_{Gs} \equiv 1/3 \, m_{earth} / m_e \times g_{eV} \times 1 \times 10^{-30} \times K_{rel}^{1/2} / 1.04^{1/6} = 3.5432 \times 10^{22}$ N   (19.5)

$F_{Rs} \equiv \sqrt{2} \times L_{earth}^2 \times 1.1111 \times 1.04^{1/20} = 3.5434 \times 10^{22}$ N   (19.6)

$F_{Rs} \equiv 4 A_{cirearth} \times 1/4 \times c_s \times 1 / K_{rel}^{1/2} \times 1 \, N / 1.04^{1/6} = 3.5447 \times 10^{22}$ N   (19.7)

$F_{Rs} \equiv 1/10 \times m_{earth} / X_{earth} \times 10^{10} / K_{rel} \times K_{rel} \times 1.04^{2/3} = 3.5503 \times 10^{22}$ N   (19.8)

$F_{Rs} \equiv 1/2 \times 10 (m_{earth} / m_{es}) \times h_s \times 1 / K_{rel} \times 1.111^{1/4} = 3.5413 \times 10^{22}$ N   (19.9)

$F_{Gs} = 1/3 \times 1/10^3 \times \sigma_s (6000 - 300)^4 \times A_{cirsun} \times K_{rel}^2 \times 1.04^{1/3} = 3.5446 \times 10^{22}$ N   (19.10)

$F_G = 1/3 \times 1/10^3 \times \sigma (5780 - T_{earth})^4 \times A_{cirsun} \times K_{rel}^2 \times 1.04^{1/6} = 3.5346 \times 10^{22}$ N   (19.11)

$\mathbf{F_{Gs} = I_{earth} \times v_{earth} / L_{earth}} = 1.7795 \times 10^{29} \times 29786 / 1.4961 \times 10^{11} = 3.5428 \times 10^{22}$ N   (19.13)

$\mathbf{F_{Gs} = 1/2 \times I_{earth} / L_{earth} \times 10^{-30} / (h_s / 60)] / 1.04^{1/6}} = 3.5419 \times 10^{22}$ N   (19.15)

FGs=2/3 xGsxmearthxmsun/Xearth^2 (19.17)

$\equiv 2/3 \times 1/10 \times h_s^{1/3} \times m_{earth} \times m_{sun} / X_{earth}^2 = 3.5398 \times 10^{22}$ N   (19.18)

$F_G \equiv 1/8 \times 1/10^3 \times m_{earth} \times m_{sun} / m_e^2 \times h^2 / X_{earth}^2 \times ([1 + 1/10^2 + 22 \times 1/10^5) = 3.5438 \times 10^{22}$ N

$F_R \equiv 2/3 \times 1/10^2 \, m_{earth} / K_{rel} \times 1 / (1 - 1/10)^{1/4} = 3.5413 \times 10^{22}$ N   (19.20)

$F_G \equiv 1/10^3 \times m_{sun} / m_e \times h \times 1/50 \times \sqrt{2} / K_{rel} = 3.5439 \times 10^{22}$ N   (19.21)

$F_{Gs} \equiv 1/10^3 \times m_{sun} / m_{es} \times \sqrt{2} / (1 + 2/3) \times h_s \times 1/50 \times \sqrt{2} / K_{rel} = 3.5805 \times 10^{22}$ N   (19.21)

$\equiv (1/10^3)^2 N_e \times e \times \sqrt{2} / (1 + 2/3) \times 10^{-30} \times 1/50 \times \sqrt{2} / K_{rel} = 3.5805 \times 10^{22}$ N   (19.22)

$\equiv (1/10^3)^2 Q_{sun} (C) \sqrt{2} / (1 + 2/3) \times 10^{-30} \times 1/50 \times \sqrt{2} / K_{rel} = 3.5805 \times 10^{22}$ N   (19.23)

$F_{Gs} \equiv 1/10 \times 1/4 \times I_{os} \times 4 A_{cirsun} \times v_{Hms} / K_{rel}^{1/2} \times 1 \, N = 3.5437 \times 10^{22}$ N   (19.24)

$F_{Gs} \equiv 1/2 \times \varepsilon_{os} \times A_{cirsun} \times v_{Hms} / K_{rel}^{1/2} \times 1 \, N = 3.5437 \times 10^{22}$ N   (19.25)

$F_{Gs} = 4 \times 10^{10} \times (B_s \times A_{cirsun}) / dt \times K_{rel}^2 \times 1 \, N = 3.5478 \times 10^{22}$ N   (19.26)

$F_{Gs} = 8 \times 1/4 \times I_{os} \times A_{cirsun} \times v_{Lms} \times 50 / K_{rel}^{1/2} \times 1 \, N = 3.5375 \times 10^{22}$ N   (19.27)

$F_{Gs} \equiv (1/2 \times 1/48)^2 \times 4 A_{cirsun} \times 1/4 \times c_s / (K_{rel}^2 \times K_{rel}^{1/3}) \times 1 eV = 3.5414 \times 10^{22}$ N   (19.28)

$F_{Rs} \equiv 1/3 x 10^3 m_{earth}/m_{es} x g_{sJ} x K_{rel}^{1/2} = 3.5651 \times 10^{22}$ N   (19.29)

$F_{Gs} \equiv m_{earth}[C] x m_{sun}[C]/(4\pi\varepsilon_{os} X_{earth}^2) x 1 x 10^{-20}/K_{rel}^2 x 1.1111^{1/2} = 3.5210 \times 10^{22}$ (19.30)

$F_{Gs} \equiv (m_{earth}/m_{es}[C])/[4\pi\varepsilon_{os} X_{earth}^2] x 1 x 10^{-20} x 1/K_{rel}^2 x 1.1111 = 3.605 \times 10^{22}$ N   (19.31)

$F_{Gs} = 2 \ x 50 \ m_{earth} \ x m_{sun} x p_{vol} J \ x \ K_{rel}^4 \ (1/1.1111) = 2 \times 50 \times 6 \times 10^{24} x 2 x 10^{30} \times 1 \times 10^{-30} x$
$1.7777 \ldots = 3.5555 \times 10^{22}$ N   (19.32)

$F_{Rs} = \mathbf{1/6 \ x \mu_{os} \ I_{earth} \ K_{rel}^{1/2}} = 1/6 \ \mu_{on} m_{earth} \ v_{earth} K_{rel}^{1/2}$
$= 1/6 \ x 1 \ x 10^{-6} x 5.9742 \ x 10^{24} x 29786 \ x 1.07564/1.11111 = 3.5445 \times 10^{22}$ N   (19.33)

$F_{Rs} = 6 A_{cirearth} \ x \ 50 \ x K_{rel}^{1/2} x 10^6 = 3.5679 \times 10^{22}$ N   (19.34)

$F_{Gs} \equiv 1/8 (\text{Mass conversion rate of sun}) x c_s^2 x 1/(1.02 K_{rel})^2 = 3.5565 \times 10^{22}$ N   (19.35)

$F_G = 1/2 P_{sun}/c \ x \ p_{1Dv} \ x A_{cir}/C_r = 3.5429 \times 10^{22}$ N   (19.36)

$F_G = 1/2 \ x \ |P_{sun}| \ x \varepsilon \ o \ x \ vLmxAcirsun \ xCr$ (19.37)

$= 1/2 \ x \ |P_{sun}| \ x\varepsilon_o \ x \ (kT \ x \ ln \ 2)/h \ x \ A_{cirsun} x C_r = 3.5443 \times 10^{22}$ N   (19.38)

$F_R = [(3 m_{earth} \ x \ v_{earth})^2 x Z_{space}]^{1/2}/c = 3.4557 \times 10^{22}$ N   (19.39)

$F_{Rs} = m_{sun}/m_{es} \ x 1 \ x 10^{-27} x 1/8 \ x 1/50 \ x 1/5 x K_{rel} = 3.574 \ x \ 10^{22}$ N   (19.40)

$F_{Gs} = 1 N \ x \ 31.557 \ x 10^6 \ x \ 6000^4/K_{rel} = 3.5412 \ x \ 10^{22}$ N   (19.41)

$F_{Gs} = 1/8 \ x \ [m_{sun}/k_s]^{1/2} = 3.5975 \times 10^{22}$ N   (19.42)

$F_R = I_{os} \ x [1/6 \ x \ m_{earth}/m_e]^{1/2} x 10^3 (v_{earth}/c_s)^2 x m_{sun}/(X_{earth}^2)^2 x (1 + 1/10 \ x 1/10)$ (19.43)
$= 3.5446 \times 10^{22}$ N

$F_G = 1.06483 \ x 10^{17} x 6000 \times 50 \ x \ 1.1111 = 3.5494 \ x \ 10^{22}$ N   (19.44)

$F_{Gs} = 4 A_{cirsun} \ x 1/4 \ \mathbf{x} \ c_s \ x T_{sun} x 10^{-6} x K_{rel}^2 = 3.6521 \ x \ 10^{22}$ N   (19.45)

$F_{Gs} = 4 A_{cirsun} \ x 1/4 \ \mathbf{x} \ I_o \ x \ a_o \ x \ v_{Hms} /K_{rel}^{1/2} \ x \ 10^{10} = 3.5405 \ x \ 10^{22}$ N   (19.46)

$F_{Rs} = [1 N/ c_s]^{1/2} \ x 1./1000 \ x 1./(\sqrt{3} K_{rel}^2) = 3.5267 \ x \ 10^{22}$ N   (19.47)

$F_{Gs} = m_{sun}/m_{es} \ x 1 \ x 10^{-27} x 1/8 \ x 1/50 \ x 1/5 x K_{rel} = 3.5740 \ x \ 10^{22}$ N   (19.48)

$F_{Rs} \equiv m_{earth}/m_{es} \ x 10^{-30} \ x \ 1/10 \ x 1/10 \ x \ 1.02^{3/2}/\sqrt{3} = 3.5523 \ x \ 10^{22}$ N   (19.59)

$F_{Rs} = 1/4 \ x \ 29786 \ m_{sun}/(L_{earth}) \ x 1/3 x 1/1.111 x K_{rel} \ x 1.04 = 3.5670 x 10^{22}$ N   (19.50)

$F_{Gs} = P_s \ x c_s x A_{cirearth}/(10^3 \ x \ \sqrt{2} \ x \ K_{rel}^{1/2}) \ x \ 1.02 = 3.5332 \ x \ 10^{22}$ N   (19.51)

$F_{Gs} \equiv h_s^2/m_{ps1D} \ x \ m_{earth} \ x 3 \ x 10^3/K_{rel}^{1/2} = 3.5679 10^{22}$ N   (19.52)

$F_R \equiv [I_{earth}^2 Z_{space}]^{1/2} \ x \ c \ x \ 1/K_{rel}^2 x \ 1/(1 - 1/10)^{1/2} = 3.5436 \ x 10^{22}$ N   (19.53)

$F_R \equiv [(3 m_{earth} \ x \ v_{earth})^2 x Z_{space}]^{1/2}/c = 3.4557 \ x 10^{22}$ N   (19.54)

$F_G = m_{earth} \ x \ v_{earth}/(1/4 \ \text{seconds per year}) x 1.0202 \ x \ K_{rel}^3 = 3.5470 \ x \ 10^{22}$ N   (19.55)

$F_{Rs} = 4 A_{cirearth} \ [m^2 \ x \ 1N \ m^{-2}] \ x 1/4 \ \mathbf{x} \ c_s \ x \ K_{rel}^{1/2}$
$= 4 \ \mathbf{x} \ 1.278 \ x \ 10^{14} \ x 1/4 \ x 3 \ 10^8/1.07547 = 3.5650 \ x \ 10^{22}$ N   (19.56)

$F_{Rs} \equiv A_{cirearth} \ x 1 N-m^2 x c_s/(1.043/2 K_{rel}^{1/2})$
$= 1.278 x 10^{14} x 3 x 10^8/(1.04^{3/2} x 1.07547) = \mathbf{3.5437 x 10^{22}}$ **N**   (19.57)

Using integer values for masses and distance derives often reference value

$F_{Gs} \equiv 1 N-m^2/kg^2 x \ 2 x 10^{30} x 6 x 10^{14}/(1.5 x 10^{11})^2 x 1.1547/(\sqrt{3} x 1.04^{1/12}) =$
3.5439x10²²N.

$F_{Gs} \equiv 1/2 x 2 x 10^{30} kg x 1/10^6 x K_{rel}^{1/2}/(30 eV x 1.04^{1/4}) = \mathbf{3.5472 x 10^{22} N}$ (19.59)

$F_G \equiv 1/2 x 1.9891 x 10^{30} kg x 1/10^6/(27.2138 eV x K_{rel}^{1/2} x 1.04) = \mathbf{3.5370 x 10^{22} N}$ (19.60)

$F_G \equiv v_{Lms} \ x 10^{10}/ K_{rel}^2 x 1.04^{13/9} = 5 \ x 10^{12} x 10^{10}/ K_{rel}^2 x 1.04^{13/9} = \mathbf{3.5435 x 10^{22}}$ **N**   (19.61)

$F_G \equiv 1/2 \; x1.9891x10^{30}/(3 \; x10^8)2x \; 10 \; x1.04^{3/2} \; x \; 1.04^{1/5} = \textbf{3.5437 x10}^{22} \textbf{ N}$ (19.62)

$F_{Gs} \equiv 2\sqrt{\pi} \; x1/\varepsilon_o^2 = 3.5449 \; x10^{22} \, N$ (19.63)

$F_{Gs} \equiv e_s \; xc_s/h_s \; x1.1547x1.04^{3/5} = 3.5466 \; x \; 10^{22} \, N$ (19.63)

$F_G \equiv exc/h \, x\text{correctors}$ (19.64)

$=1.6022x10^{-19}x2.99792x10^8/6.62607x10^{-34}/(1.1111^2x1.1547) \; x1.04^{1/10}=3.5470x10^{22} \, N$

$F_{Gs} \equiv 10 \; Nx \; 6000^4 \; x \; 3 \times 10^8 \; x \; 1/10 \; x \; 1/10/1.1547 \; x \; 1.04^{7/5} \equiv 3.5440 \; x \; 10^{22} \, N$ (19.65)

$F_{Gs} = m_{earth} \; x \; \text{circular velocity}/L_{earth} = 3.5438 \; x \; 10^{22} \, N$

$\quad = m_{earth} \; x \; (\text{ velocity at 2/3 earth radius})/ (\text{2/3 earth radius})$

$\quad = 6 \; x10^{24} \, kg \; x \; 196.62/8.5x10^6 = 3.5438 \; x \; 10^{22} \, N$ (19.66)

Earth repulsive force = 1 N/m^2 x Circular earth area x velocity of light /1.04^2

$\quad\quad = 1 \; x \; 1.278 \; x \; 10^{14} \; m^3 \; x \; 3 \; x \; 10^8 \; m/s \; /1.04^2 = 3.5447 \; x \; 10^{22} N \; (19.67).$ (19.68)

$F_{Gs} = 3.8900x10^{29} \, W \; x \; 10^{-10} \; x \; 10^3 \; x \; 1/1.07457x1/1.04^{1/2}=3.5497 \; x10^{22} \, N$

$F_{Rs} = -F_{Gs} = 6x10^{24}x30000^2/1.5x10^{11}x1/1.04^{3/8}=3.5436 \; x10^{22} \, N$ (19.69}

$F_{Gs} = 6 \; \boldsymbol{x} \; 10^{24} \, x10N/kg \; x1/(3 \; \boldsymbol{x}300)x1/1.3333^2x1.04^{5/4} = 3.5446 \; x \; 10^{22} \, N$ (19.70)

Force of earth repulsion = 6x10^{24} kg x10 N/ kg x1/10 /$\sqrt{3}$x1.04$^{10/17}$=3.5448 x10^{22} N

$F_{Gs} \equiv /\sqrt{3} \; (\sqrt{2} \; x \; 1/5 \; x \; 5 \; x \; 10^{12})^2 \equiv 3.5438 \; x \; 10^{22} \, N$ (19.71)

$F_{Gs} = 1.5x10^{18}x3x10^8x6000x1x10^{-8}x1.3333/1.04^{3/4}=3.6000$
$x10^{22}/1.04^{3/4}=3.4957x10^{22}$

$\quad = 3.500x10^{22} \, N$ (19.72)

$F_{Gs} = 1 \; N/m^2 \; \boldsymbol{x} \; \sqrt{3} \; (L_{earth})^2/K_{rel1/3cs} = \sqrt{3}(1.5x10^{11})^2/1.1111=3.5074x10^{22} \, N$ (19.73).

$F_G = 3.500x10^{22} \, N \; x1.04^{1/4}=3.5420x10^{22} \, N$ (19.74)

$F_{Gs} = 1 \; N/m^2 \; \boldsymbol{x} \; 3 \; \boldsymbol{x} \; 2/3 \; \boldsymbol{x} \; X_{earth}^2/2 \; \boldsymbol{x}K_{rel}^3 = 3 \; \boldsymbol{x} \; 2/3 \; \boldsymbol{x} \; (1.5x10^{11})^2/2x1.3333 \; \boldsymbol{x} \; 1.1547$
$1.04^{1/4}$

$\quad = 3.500009 \; \boldsymbol{x}10^{22} \, N$ (19.75)

$F_{Gs} = \sqrt{3} \; \boldsymbol{x}13.6401 \; eV \; \boldsymbol{x}1.5x10^{11}m/1x10^{10} \; m \equiv 3.5438 \; \boldsymbol{x} \; 10^{22} \, N$ (19.76)

Gravitation force per meters is ionization energy of hydrogen atom with corrections:

$3.5438 \; \boldsymbol{x}10^{22} \, N/1.5 \; \boldsymbol{x}10^{11}m \equiv 23.625/(1\boldsymbol{x}10^{-10})$

$\quad\quad =30 \; eV \; \boldsymbol{x} \; (1.1111/1.3333) \; \boldsymbol{x} \; 1.04^{1.45}/(1\boldsymbol{x}10^{-10}) =23.618/(1\boldsymbol{x}10^{-10})$ (19.77)

Atomic $F_{Gsa} = \sqrt{3} \; x10 \; N/kg \; x \; m_{es} m_{ps} /(1/2a_{os})^2x3 \; x300 \; K \; x1x10^{30}x \; 1.111^3/$
$(1.3333x1.04^{1.7})$

$\quad\quad \equiv 0.006001 \; N \approx 3.5438x10^{22} \, N/6x10^{24} \, kg \; \text{of earth}$ (19.78)

Atomic $F_{Gsa} =$
$3x6x\sqrt{3} \; x \; 0.006 \; x \; 300^2 \; x \; m_{es} m_{ps} /(1/2a_{os})^2x1x10^{30}/(1.11111x1.04^{1/5})=0.0060007 \; N$ (19.79)

$F_{Gs} = G_s x\text{msun}x\text{mearth}/(\text{distance})^2x1/K_{rel}^2$

$\quad\quad =10^{-10}x2x10^{30}x6 \; x10^{24}/(1.5 \; x10^{11})^2x1/1.1547^2=4.0000035 \; x \; 10^{22} \, N$ (19.80)

$F_{Gs} = G_s x\text{msun}x\text{mearth}/(\text{distance})^2x1/K_{rel}^2$

$\quad\quad =10^{-10}x2x10^{30}x6 \; x10^{24}/(1.5 \; x10^{11})^2x1/1.1547^2=4.0000035 \; x \; 10^{22} \, N$ (19.81)

$4.0000035 \; x \; 10^{22} \, N/(1.1111111)^{3/2} \; x1.04^{1/10} = 3.5438 \; x \; 10^{22} \, N$ (19.82)

$\textbf{3}. \; 00000 \; \boldsymbol{x} \; 10^{22} \; \boldsymbol{x} \; 1.1547 \; \boldsymbol{x} \; 1.03972 \; \boldsymbol{x} \; 3/2 = 3.5460 \; x \; 10^{22} \, N$ (19.83)

The term $4A_{cirsun}$ in Equation 19.56 is taken to indicate that sun radiates gravitation along the plane of its rotation. Equation 19.56 shows that earth replicates with proton velocities at 1/4 c_s also generating 1 N/m² in the same plane. Also

$$F_{Rs} = \mu_{os} I_{earth} \times 10 \times 10^{10} \sqrt{3}/(2\pi L_{earth} \times 1.04^{3/4} \times 1.11111 \times 1.04^{1/20}) = 3.5450 \times 10^{22} \text{ N} \qquad (19.84)$$

Local velocities convert to velocity of light c_s and vectorized thermal flux generated by proton infrared oscillations of earth $1/10^3 x \sigma_s T_{earthrmss}{}^4 = 0.436$ W/m² generates in 3 planes the repulsive force requiring quantum corrector $1.04^{9/10}$ in

$$F_{Rs} = 3A_{cirearth} \times 1/10^3 x \sigma_s T_{earthrmss}{}^4 \times c_s \times K_{rel}{}^{1/2}/1.04^{9/10} = 3.5427 \times 10^{22} \text{ N} \qquad (19.85)$$

$F_{Gs}{}' = 2/3 \times$ msun $\times$ mearth/(separation squared) $\times$ corrections

$$= 2/3 \times 1 \times 10^{-10} \times 2 \times 10^{30} \times 6 \times 10^{24}/(1.5 \times 10^{11})^2/1.11111 \times 1.04^{3/2}/1.04^{1/7}$$

$$= 3.00003 \times 10^{22} \text{ N} = 1.33333 \times (1.5 \times 10^{11})^2 \text{ N} \qquad (19.86)$$

$$= (1.73205 \times 10^{11})^2 = (\sqrt{3} \times 1.0000 \times 10^{11})^2 \text{ N}$$

$$5.9742 x 10^{24} \text{kg}/[\pi(1.4961 x 10^{11})^2] x 1/10 \times 1/10 \times 1.1111^2 \times 1.04^{3/2} = 1.0012 \text{ N/m}^2 \qquad (19.87)$$

$$F_{Gs} = 1 \, x 10^{-10} \times 6 \, x 10^{24} \times 2 \, x 10^{30}/[(1.5 \, x 10^{11})^2 \, x 133333 \ldots = 4.0000 \ldots x 10^{22} \text{ N} \qquad (19.88)$$

$$F_{Gs} = (1 \, x 10^{30})^{3/4}/1.1111 \ldots {}^{1/2} = 3.000000001 \, x 10^{22} \text{ N} \qquad (19.89)$$

$$F_{Gs} = 1/10 x 6 \, x 10^{24}/1.5 x 10^{11} x 10^{10}/[1.11111 \, x 1/(2\ln2^2)^{2/5}] = 4.0000 \ldots x \, 10^{22} \qquad (19.90)$$
N/1.12873

$$F_G/\text{mass}_{earth} = 3.5438 x 10^{22} \text{ N}/5.9742 x 10^2 \text{kg} x 1.04^{1/20} = 0.00599 \approx 0.006 \text{ N/kg} \qquad (19.91)$$

$$3.5438 \, x 10^{22}/[6 \, \sigma_s(6000-300)^4]/1.04^{1/5} = 4.9964 \, x 10^{12} \text{ Hz} \qquad (19.92)$$

$$F_R \equiv -d\varphi/dt \equiv \upsilon_{earth} \text{ X } B \equiv (\upsilon_{earth} \text{ X } I/2\pi) \, x 10^{-10} \, x 1/3 \times 1.1547 \times 1.1111/1.04^{1/2} \qquad (19.93)$$

$$\equiv 29786 \text{ X } 1.7795 \, x 10^{29}/2\pi \, x 10^{-10} \, x 1/3 \, x 1.1547 \, x 1.1111/1.041/2 = 3.5376 \qquad (19.94)$$
$$x 10^{22} \text{ N}$$

$$F_{Gs} = 1000 \text{ K} \times (1/10 x 1/10)^3 x 1/3 c_s x 1/1.3333 \, x 1/1.11111 x A_{cirsun} \, x 1/3 \qquad (19.95)$$
$$/1.04^{9/10} = 3.5477 x 10^{22} \text{ N}$$

$F_{Rs} = 1$ N x 4Acirearth x $1/3c_s$ x corrections

$$= 4 \times 1.278 \, 10^{14} \text{ m}^2 \times 1 \times 10^8 x 1/1.3333 \, x 1/1.11111 x 1.04^{5/16} = 3.5451 x 10^{22} \text{ N} \qquad (19.96)$$

With the mass of earth comprising 6 x 10^{54} charges and 32 quantum orientation positions on the spherical surface the probability of generating a quanta per charge is 10^{-32} generating total Special Relativity attractive force of:

$$F_{Gs} = 1/2 \times 6 \times 10^{54} \times 10^{-32} = 3 \times 10^{22} \text{ N} \qquad (19.97)$$

The correctness of representing the earth by the current, and so earth mass by number of electrons, is demonstrated by Equations 19.33, 19.13 and 19.15.

Square root is taken to indicate that force of gravity is applied only along one plane. To make it dimensionally correct the time in atomic units is taken to be 1 year and the average magnetic flux density in atomic units within the area of earth circular rotation around the sun is taken to be 1 Weber/m². Then the relation represents the well known

Gravitation force = Rate of change of magnetic flux.

All gravity forces for earth and for the planets in Tables of Chapters 19 and 20 are rough data derived and accumulated for further analyses; they may contain early misconceptions, which were not corrected.

Please note that **the gravitation force in Equation 19.96 of 3.5451×10^{22} N is in SI units, which increases its real value by linear factor 1×10^{10}. So in real atomic terms the force is 3.54381×10^{12} N$_{at}$, which expresses $3.5438 \times 10^{12}/(1.1547^2 \times 1.1111^3) \times 1.0328 \equiv$ 2.00113 J/m ≈ 2 J/m times frequency 1×10^{12} Hz.**

During earth rotation around the sun the gravitation forces act essentially in one plane only. The Sun attracts earth with the force of $3.5438(05) \times 10^{22}$ N as given by Equation 6.1. The earth rotation generates an outward centrifugal force in Equation 6.6 of 3.5430×10^{22} N opposing sun attraction.

Independently of the above forces, vectorized thermal force of the sun is balanced by the vectorized forces of the earth, which for earth take the energy from instability of earth's protons.

The large number of equations expressing the gravitation forces is the consequence of physical constants not being independent and overabundant. For example relativity corrected square root of Boltzmann constant in integer space constants can be expressed by 300 K, Planck radiation energy and 10^{12} Hz frequency:

$$k_s^{1/2}/1.111\ldots^{1/2} = 300 \text{ K} \times 0.01/10^{12} \text{ Hz} = 3 \times 10^{-12} \qquad (19.98)$$

20. Apparent total gravitation forces of planets

To test the accuracy and to extend of validity of gravitation concepts proposed for the sun-to-earth system several relations used for earth were applied to all planets. The sun-to-earth gravitation appeared simpler, than sun-to-other planet gravitation. The sun to planets classical gravitation force of attraction is expressed by the same Newton product of their two masses over the square of the separation, times the International gravitation constant G for all the planets except Pluto, which requires a factor of $1/\sqrt{2}$. The classical centrifugal force of repulsion F_R is given by the same formula for Mars, Mercury, Neptune, Uranus and Jupiter as for earth. Comparison with original equations identifies changes needed to get the correct values. F_R needs to be corrected for Pluto and Saturn essentially $F_R \times 1/(1.1547 \times 1.1111/1.0328^2)$ and for Venus $F_R \times 1.1111^{1/2}$. Thus although the three planets require in the classical centrifugal forces specific relativity corrections their repulsion is produced by acceleration induced by that macroscopic force. The sun and the planets supply each one half of the forces listed in Tables below.

Both SI and integral space constants were used in evaluations. Table 6 gives values of parameters of planets used in the calculations including the currents generated by the moving masses and hence charges. The currents are accurate to the 3rd decimal point. The accuracy of the reference values of the gravitation force, limited mostly by the accuracy of its mass, is to the 3rd or the 4th decimal point for the closest bodies. The current values in amperes are masses of planets in kilograms times their linear velocities in meters per second, thus their mechanical momenta. This statement makes a very clear physical connection between the electrical and mechanical terms. Current produced by any planet is originally expressed in atomic units with e_s being unity, converted to SI units with volume probability P_v over electron mass being unity:

$$I_{planet} = e_s \times m_{planet} \times v_{planet} \times P_v/m_{es} = m_{planet} \times v_{planet.} \tag{20.1}$$

I_{planet} creates a force opposing the gravitation of the sun, in the same way as currents flowing in the opposite directions in parallel conductors. For planet closer to sun I_{planet} forms a current loop around the sun inducing the same force expressed by the Ampere force directing against the sun. Similarly the whole galaxy rotating around its center creates such Ampere force or its fraction. It is believed that the back hole is also rotating in one plane. Consequently, while the gravity forces act centrally **the Ampere force** having axial direction opposes the gravity force and **creates a leak in the black hole leading to generation of a very powerful beam of radiation in both directions at right angle to the black hole rotation axes. These leaks form powerful jets of reported radiation.**

The planets generate equal and opposite currents to those generated by the sun, so that the sum of moments for each pair is zero. The sun temperatures T_{sun} and T_{suns} have been replaced by their values of 5780 K and 6000 K to improve clarity. Force of gravitation of Equation 6.1 applies both to earth and to sun.

Replacing in that Equation sun mass by planet mass and earth distance by planet distance to sun derives gravitation force F_{planet} for the planets not emitting their own energy:

$$F_{planet} = 3.5438 \times 10^{22} \times [X_{earth}/X_{planet}]^2 \times (m_{planet}/m_{earth}) \tag{20.2}$$

Table 6. Parameters of planets and currents generated by them

Planet	Distance to sun	Mass	Mean temperature	Linear velocity	Generated current (A) x10[19]
	(10^9 m)	m (kg)	(K approx)	v(m/s)	mass x velocity
Mercury	57.91	3.39e23	200-340	47868	1.6227×10^{28}
Venus	108.20	4.87e24	735	34978.51	1.7035×10^{29}
Earth	149.60	5.98e24	330	29786	1.7795×10^{29}
Mars	227.94	6.42e23	227	24117.3	1.5483×10^{28}
Jupiter	778.33	1.90e27	165	13070.4	2.4834×10^{31}
Saturn	1426.94	5.68e26	134	8783	4.9887×10^{30}
Uranus	2870.99	8.69e25	76	68048	5.9134×10^{30}
Neptune	4504	1.025e26	72	95437	9.7830×10^{30}
Pluto	4913.52	1.31e22	44	4740.5	6.2101×10^{25}

Table 7. Common Equations for gravitation force for planets with corrections

F_{Gs}=1/3 x 1 N x(sun distances earth/planet)^2 x m_{planet} x g_{eV} x corrections; g_{eV}=1/60 eV/K

Planet	Force F_{Gs} with arbitrary chosen corrections
Earth	F_{Gs} =1/3 x1/60 x1.1547$^{1/2}$ x(149.6/149.6)^2 x5.98 x10^{24}/1.04$^{1/4}$=3.5470 x10^{22} N
Jupiter	F_{Gs} =1/3 x 1/60 x 1.1111x (149.6/778.33)^2 x1.9 x10^{27}=4.3328 x10^{22} N
Mars	F_{Gs} =1/3 x 1/60 x1.1547x(149.6/227.94)^2 x6.42 x10^{23}/1.0328/1.04=1.619 x 10^{21} N
Mercury	F_{Gs} =1/3 x 1/60 x 1.1111$^{1/2}$ x(149.6/57.9)^2 x3.39 x10^{23}=1.291 x10^{22} N
Neptune	F_{Gs} =1/3 x 1/60 x x1.1547$^{1/2}$ x (149.6/4504)^2 x1.025 x10^{26} = 6.7481 x 10^{20} N
Pluto	F_{Gs} =1/3 x 1/60 x1/1.3333 x (149.6/4913.5)^2 x1.31 x10^{22}= 5.0598 x 10^{16} N
Saturn	F_{Gs} =1/3 x 1/60 x 1.1547$^{1/2}$ x(149.6/1426.94)^2 x5.68 x10^{26} = 3.7270 x 10^{22} N
Uranus	F_{Gs} =1/3 x 1/60 x1.1547$^{1/2}$x(149.6/2871)^2 x8.69 x10^{25}= 1.4084 x 10^{21} N
Venus	F_{Gs} =1/3 x 1/60 x(149.6/7108.2)^2 x4.87 x10^{24}= 5.1717 x 10^{22} N

Fitting values for all the planets derived correct data in nearly all the identities, indicating validity of Equation 20.2 in expressing sun attractive forces. Neptune actually receives about 891 less energy per kilogram as earth 6.6956x10^{22} N/1.025x10^{26} = 0.0006523 N/kg. WEB pages state that Neptune receives 1/900 energy of the earth and generate 2.7 times the energy received from the sun. Table 7 shows that for all other planets forces of gravitation hold. Generation of self energy by Neptune and to a smaller degree by Jupiter upsets the gravitation induced by the gravitons. Figure 13 is a plot on logarithmic scale of the square of the planet distance to sun versa force of sun gravitation energy per kilogram of the planet mass. Again except for the Neptune and Jupiter the data for other planets follow a straight line. Table 7 shows that the typical force of gravitation applied by sun to Neptune is more than two orders of magnitude smaller than the force acting, as calculated in Table 11. In Table 11 huge corrections had to be applied to fit the data with reference.

Equation 20.2 is the valid force of sun attraction. However the force of repulsion must be different. So far analyses favored the centrifugal force for expressing the universal repulsion, which on atomic scale may represent the unwinding of kinetic energy at least for earth. The force remaining unchanged for all planets is that expressing the product of mass ratio, $3m_{body}/m_{es}$, and the Planck constant h_s. Actually the above value is the vector after space quantization. An electron or proton sends per kelvin in each of six directions energy or

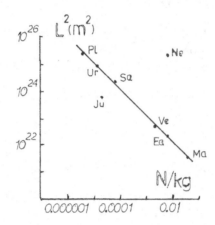

Figure 13. Squared distance to sun versa gravitation force per kilogram planetary mass power of 1×10^{-33} J (or W) reduced by space volume quantization of 1×10^{-30} m³. This is measured with the reference to the sources, both being attractive, but opposing each other if reference is chosen. The discovery of the investigation express energy released per kilogram of body mass in the amount of

$$3h_s \text{xrotational probability} = 1/5xg_{eV} \, x1/(1-1/10) = 3x10^{-33}/10^{-30} = 3 \, x(1/10^2) = 0.03 \text{ eV-kg}^{-1} \quad (20.2a)$$

Replacing 2000 $m_{es} = m_{ps}$ in $3m_{body}/m_{es}$ obtains the force to be $6m/m_{ep}$ $_s x10^{-27}x \, K_{rel}$ meaning that each proton generates in one earth year of time energy of 10^{-27} J in each of six space directions.

Tables for planets may comprise also pure oscillations, which being numerically equal to gravitation forces cannot be distinguished from them.

Non-circular pathways of rotations of planets including earth around the sun are most likely to be the consequence of an external celestial body passing through our planetary system or the effect of an external explosion displacing the planets.

Currents of various planets given in Table 4 normally generate a force **1/6 xu_{os}I**. That basic equation with significant corrections applies to earth, Venus and Mercury.

Venus, Mars and earth are closest to the sun and they are attracted by the sun with the force given by the Equation 20.2. Using data of Table 4 the planets attract the sun with equal force expressed now for:

Venus $F_{Rs} = 1/6 \times \mu_{os} I_{ven} \times cor$ (20.2c)
$= 1/6 \times \mu_{os} m_{ven} v_{ven} \times 1.2 \times 1.04^{3/5}$
$= 1/6 \times 1 \times 10^{-6} \times 4.87 \times 10^{24} \times 34978.5 \times 1.11111^3 \times 1.1547^2 = 5.1928 \times 10^{22}$ N
deriving 0.01066 N/kg mass.

Mars $F_{Rs} = 1/6 \times \mu_{os} I_{mars} \times cor$
$= 1/6 \times 1 \times 10^{-6} \times 1.5483 \times 10^{28} (1.1547^3 \times 1.0328) = 1.6228 \times 10^{21}$ N (20.2d)

deriving 0.00253 N/kg mass.

Earth in Equation 19.33 $F_{Rs} = 3.5445 \times 10^{22}$ N, the rounded up value being 0.006 N/kg.

Mercury is a planet having a very large core with magnetic materials at temperature above Curie value at which increases above μ_{os} value do not occur according to WEB sources. Nevertheless the calculation shows that the force of reflection is increased by a factor of three suggesting a value for relative permeability $u_r = 3$.

Mercury $F_{Rs} = 1/6 \times \mu_r \mu_{os} I_{mer} \times cor$ (20.2e)
$= 1/6 \times 3 \times 1 \times 10^{-6} \times 1.6227 \times 10^{28} \times 1.2^2 \times 1.11111^2 \times 1.04^{3/5}$
$= 1.3044 \times 10^{22}$ N deriving 0.0385 N/kgmass.

The correct classical electromagnetic forces induced by the linear motion of charges contained in the masses of planets moving around the sun could be also observed by applying one half velocity of light to the sun and to earth in their plane of rotation against the veleocity of earth (46, Section 15).

Saturn, Neptune and Jupiter are much further away from the sun that above four planets. They attract (and repel) the sun with forces, which beside relativity and quantum corrections include factors 1/30, 1/25 or 1/10:

Neptune $F_{Rs} = 1/6 \times \mu_{os} I_{nep} \times cor$ (20.2f)
$= 1/6 \times 1 \times 10^{-6} \times 1/30 \times 9.977346 \times 10^{30} = 6.6767 \times 10^{22}$ N
deriving 0.0006545 N/kg mass.

Saturn $F_{Rs} = 1/6 \times \mu_{os} I_{sat} \times cor$ (20.2g)
$= 1/6 \times 1 \times 10^{-6} \times 1/25 \times 4.9887 \times 10^{30} \times 1.11111 = 3.6953 \times 10^{22}$ N
deriving 0.0006505 N/kg mass.

Jupiter $F_{Rs} = 1/6 \times \mu_{os} I_{jup} \times cor = 1/6 \times 1 \times 10^{-6} \times 1/10 \times 2.4834 \times 10^{30} \times 1.04^{1/6}$ (20.2h)
$= 4.1662 \times 10^{22}$ N deriving 0.0002193 N/kg mass

Uranus and Pluto generate forces with factors 1/868.4 and 1/258.24:

Uranus $F_{Rs} = 1/6 \times \mu_{os} I_{uran} \times cor$ (20.2i)
$= 1/6 \times 1 \times 10^{-6} \times 5.9134 \times 10^{30} \times 1.1111 \times 1/781.54 = 1.401 \times 10^{21}$ N
$\equiv 0.00001612$ N/kg mass

Pluto $F_{Rs} = 1/6 \times \mu_{os} I_{pluto} \times cor$
$= 1/6 \times 1 \times 10^{-6} \times 6.2101 \times 10^{25} \times 1/232.46 \times 1.1111 = 4.9472 \times 10^{14}$ N

$$\equiv 0.00000003776 \text{ N/kg mass} \tag{20.2k}$$

nearly with 1/163162 fraction of force generated by earth.

Table 8. Apparent gravitation forces of Jupiter

$$F_G = G \times m_{jup} \times m_{sun}/L_{jup}^2 = 4.1637 \times 10^{23} \text{ N Reference} \tag{20.3}$$

$$F_R = m_{jup} \times v_{jup}^2/L_{jup} = 4.1700 \times 10^{23} \text{ N} \tag{20.4}$$

$$F_{Gs} \equiv 2/3 \times 1/10^3 (h_s \times v_{Hms}/a_{os} \times m_{jup} \times m_{sun})/(X_{jup}^2 \times K_{rel}^2) = 4.159 \times 10^{23} \text{ N} \tag{20.5}$$

$$F_{Gs} \equiv 1/3 \times g_{sJ}^2 \times m_{jup} \times m_{sun}/m_{es}^2 \times 1/(X_{jup}^2 \times K_{rel}^2 \times 1.04) = 4.1726 \times 10^{23} \text{ N} \tag{20.6}$$

$$F_{Rs} \equiv m_{jup}/m_{es} \times g_{eV} \times 10^{-30} \times 1/10^2 K_{rel}^2 = 4.2220 \times 10^{23} \text{ N} \tag{20.7}$$

$$F_{Gs} \equiv \sqrt{2} \times 1(\text{N}) \times L_{jup}^2/K_{rel}^{1/4} = 4.1323 \times 10^{23} \text{ N} \tag{20.8}$$

$$F_{Rs} \equiv 1/10 \times A_{cirjup} \times c_s/K_{rel} = 4.172 \times 10^{23} \text{ N} \tag{20.9}$$

$$F_{Rs} \equiv 2 \times m_{jup}/m_{es} \times 1 \times 10^{-20} \times 1/10^2/K_{rel} = 4.2282 \times 10^{23} \text{ N} \tag{20.10}$$

$$F_{Rs} \equiv 3(m_{jup}/m_{es}) \times h_s/K_{rel}^2 = 4.275 \times 10^{23} \text{ N} \tag{20.11}$$

$$F_{Rs} = 1/2 \times 1/2 \times m_{jup} \times v_{jup}/X_{jup} \times 6000 \times 10/K_{rel} = 4.1446 \times 10^{23} \text{ N} \tag{20.12}$$

$$F_G = 1/2 \times 1/10^2 \times \sigma(5780 - T_{jup})^4 \times A_{cirsun} \times 1.03278 = 4.1526 \times 10^{23} \text{ N} \tag{20.13}$$

$$F_{Gs} = 1/2 \times 1/10^2 \times \sigma_s(6000 - T_{jup})^4 \times A_{cirsun} \times (1/1.11111)^{1/2} = 4.1839 \times 10^{23} \text{ N} \tag{20.14}$$

$$F_{Rs} = 1/2 \times 1/2 \times m_{jup} \times v_{jup}/X_{jup} \times 10^{-30}/(hs/60)]/K_{rel}^{1/2} = 4.0707 \times 10^{23} \text{ N} \tag{20.15}$$

$$F_{Rs} = 2 \times I_{jup}/L_{jup} \times [6000] K_{rel}^{1/2}$$
$$= 2 \times 1.90 \times 10^{27}/7.7833 \times 10^{11} \times 13070 \times 6000 \times 1.07547 = 4.1142 \times 10^{23} \text{ N}$$

$$= 2 \times 1/10 \times I_{jup}/L_{jup} \times 10^{-30} \times K_{rel}^{1/2}/(hs/60) = 4.1143 \times 10^{23} \text{ N} \tag{20.16}$$

$$F_G = F_{Gearth} \times m_{jup}/m_{earth} \times (L_{earth}/L_{jup})^2 = 4.1596 \times 10^{23} \text{ N} \tag{20.17}$$

$$= 3.5438 \times 10^{22} \times 1.90 \times 10^{27}/5.98 \times 10^{24} \times 0.03694 = 4.1593 \times 10^{23} \text{ N}$$

$$F_{Rs} = 1/6 \, \mu_{os} \times I_{jup} \times 1/10 = 4.1390 \times 10^{23} \text{ N} \tag{20.18}$$

Table 9. Apparent gravitation forces of Mars

$$F_G = G \times m_{mar} \times m_{sun}/L_{mar}^2 = 1.6404 \times 10^{21} \text{ N Reference} \tag{20.19}$$

$$F_R = m_{mar} \times v^2/L_{mar} = 1.6361 \times 10^{21} \text{ N} \tag{20.20}$$

$$F_{Gs} \equiv 1 \times 10^{-10} \times 10^3 \times m_{mar} \times m_{sun})/(L_{mar}^2 \times K_{rel}^2)/(1+1/8) = 1.6382 \times 10^{21} \text{N} \tag{20.21}$$

$$F_{Gs} \equiv 2/3 \times 1.111 \ldots \times 1/3 \times g_{sJ}^2 \times m_{mar} \times m_{sun}/m_{es}^2 \times 1/(X_{mar}^2) = 1.6382 \times 10^{21} \text{ N} \tag{20.22}$$

$$F_{Rs} \equiv 1/2 \times 1/3 \, m_{mar}/m_e \times g_{eV} \times 10^{-30} \times 1/K_{rel}^{1/2} = 1.6595 \times 10^{21} \text{ N} \tag{20.23}$$

$$F_{Gs} \equiv 1/5 \times 1/10 \times \sqrt{2} \times 1.111 \ldots \times 1(\text{N}) \times X_{mar}^2 = 1.6328 \times 10^{21} \text{ N} \tag{20.24}$$

$$F_{Rs} \equiv 3/2 \times 1/10 \times A_{cirmar} \times c_s = 1.6414 \times 10^{21} \text{N} \tag{20.25}$$

$$F_{Rs} \equiv 2/3 \times m_{mar}/m_{es} \times 1 \times 10^{-19} \times 1/K_{re}^2 = 1.6164 \times 10^{21} \text{ N} \tag{20.26}$$

$$F_{Rs} \equiv 2/3 \times 3 \times 1.111 \ldots (m_{mar}/m_{es}) \times h_s \times K_{rel} = 1.6475 \times 10^{21} \text{ N} \tag{20.27}$$

$$F_{Rs} = 1/2 \times m_{mar} \times v_{mar}/L_{mar} \times 6000 \times 1/[K_{rel} \times 1/(1-1/10)^{1/2}] = 1.6741 \times 10^{21} \text{N} \tag{20.10}$$

$$F_G = 1/6 \times 1/10^2 \times \sigma(5780 - T_{mar})^4 \times A_{cirsun} \times K_{rel}^{4/3} = 1.6566 \times 10^{21} \text{ N} \tag{20.28}$$

$$F_{Gs} = 1/6 \times 1/10^2 \times \sigma_s(6000 - T_{mar})^4 \times A_{cirsun} \times K_{rel} = 1.6265 \times 10^{21} \text{ N} \tag{20.29}$$

$$F_R = 3/8 \times m_{mar} \times v_{mar}/L_{mar} \times 10^{-30}/(K_{rel}^{2/3} \times hs/60)] = 1.6802 \times 10^{21} \text{ N} \tag{20.30}$$

$$F_{Rs} = 1/2 \times 1/2 \times m_{mar} \times v_{mar}/L_{mar} \times 10^{-30}/(K_{rel}^{2/3} \times h/60)] = 1.6889 \times 10^{21} \text{ N} \tag{20.31}$$

$$F_{Rs} = 1/2 \times I_{mar}/L_{mar} \times [6000] \times 10 \times (1/1.11111)^2 = 1.6574 \times 10^{21} \text{ N} \tag{20.32}$$

$$F_G = F_{Gearth} \times m_{mar}/m_{earth} \times (L_{earth}/L_{mar})^2 = 1.6404 \times 10^{21} \text{ N} \tag{20.33}$$

$F_{Rs}=1/6\mu_{os}\,I_{mar}\text{x}1/K_{rel}{}^3=1/6\text{x}\mu_{on}m_{mar}v_{mar}\text{ x}1/10\text{ x}1/K_{rel}{}^3$ (20.34)

$\quad=1/6 \text{ x } 1\text{x}10^{-6} \text{ x } 1.5483 \text{ x } 10^{28} \text{ x } 1/1.2\text{x } 1/1.33333=1.6128 \text{ x } 10^{21}\text{ N}$ (20.35)

Table 10. Apparent gravitation forces of Mercury

$F_G=G\text{x}m_{mer} \text{ x } m_{sun}/L_{mer}{}^2=1.3064\text{x}10^{22}\text{ N Reference}$ (20.38)

$F_R=m_{mer}\text{x}v^2/L_{mer}=1.3063\text{x}10^{22}\text{ N}$ (20.39)

$F_{Gs}\equiv1/1.11111\text{x}1/10^3(h_s\text{x}v_{Hms}/a_{os}\text{ x}m_{mer}\text{x}m_{sun})/(L_{mer}{}^2\text{ x}K_{rel}{}^2)=1.319\text{x}10^{22}\text{ N}$ (20.40)

$F_{Gs}\equiv2/3\text{x } 1/3\text{x}g_{sJ}{}^2\text{x}m_{mer} \text{ x } m_{sun}/m_{es}{}^2\text{x}1/(X_{mer}{}^2 \text{ x}K_{rel}{}^{2/3})=1.2956\text{x}10^{22}\text{ N}$ (20.41)

$F_{Rs}\equiv1/3m_{mer}/m_e\text{x } g_{eV}\text{x}10^{-30}\text{x}10/[K_{rel}\text{x}(1/1.11111)^{1/2}=1.3046\text{x}10^{22}\text{ N}$ (20.42)

$F_{Gs}\equiv\sqrt2 \text{ x}(1/1.11111)^{1/2}\text{ x}3/K_{rel}\text{ x}1(\text{N})\text{x}X_{mer}{}^2=1.299\text{x}10^{22}\text{ N}$ (20.43)

$F_{Rs}\equiv2\text{x}\sqrt2 \text{ x}A_{cirmer}\text{x}c_s\text{ x}1\text{ eV }\text{x}1/K_{rel}{}^{1/3}=1.3078\text{x}10^{22}\text{ N}$ (20.44)

$F_{Rs}\equiv 3\text{x}m_{mer}/m_{es}\text{x}1\text{x}10^{-30} \text{ x}1\text{x}10^{-2}\text{x } K_{rel}{}^2 = 1.3201\text{x}10^{22}\text{ N}$ (20.45)

$F_{Rs}\equiv10 \text{ x } 3(m_{mar}/m_{es})\text{x}h_s \text{ x } K_{rel}{}^2=1.3198\text{x}10^{22}\text{ N}$ (20.46)

$F_{Rs}=1/6\text{x}10^2\text{x}1/2\text{x}m_{mer}\text{x}v_{mer}/L_{mer}\text{x}6000 \text{ x}10\text{x}1/K_{rel}{}^2=1.3000\text{x}10^{22}\text{ N}$ (20.47)

$F_G=1/6\text{x}1/10^3\text{x}\sigma(5780\text{ -}270)^4\text{x}A_{cirsun}=1.3255\text{x}10^{22}\text{ N}$ (20.48)

$F_{rs}=1/6\text{x}1/10^3\text{x}\sigma_s\,(6000\text{ -}270)^4\text{x}A_{cirsun}\text{x}1/K_{rel}{}^{1/3}=1.3031\text{x}10^{22}\text{ N}$ (20.4 9)

$F_{Rs}=1/2 \text{ x } m_{mer}v_{mer}/L_{mer}\text{x}10^{-30}/(h_s/60)]\text{x}1/K_{rel}{}^{1/3}=1.3000\text{x}10^{22}\text{ N}$ (20.50)

$F_{Rs}=\sqrt2\text{x}1/2 \text{ x}I_{mer}/L_{mer}\text{ x } [6000\text{x}10] \text{ x}K_{rel}=1.3363\text{x}10^{22}\text{ N}$ (20.51)

$\quad= \sqrt2\text{x } 1/2\text{x}I_{mer}/L_{mer}\text{x}10^{-30}/(h_s/60)] \text{ x}K_{rel}=1.3363\text{x}10^{22}\text{ N.}$

$F_G= F_{Gearth} \text{ x m}_{mer}/m_{earth} \text{ x } (L_{earth}/L_{mer})^2=1.3064 \text{ x}10^{22}\text{ N}$ (20.52)

$F_{Rs}=1/6\mu_{os}I_{mer}\text{x}3 \text{ x } 1.1547^3 \text{ x } 1.04^{1.1}=1.3041\text{x}10^{22}\text{ N}$ (20.53)

Table 11. Apparent gravitation forces of Neptune

$F_G=G\text{x}m_{nep} \text{ x } m_{sun}/L_{nep}{}^2=6.6956\text{x}10^{22}\text{ N Reference}$ (20.54)

$F_R=m_{nep}\text{x}v^2/L_{nep}=6.706\text{x}10^{22}\text{ N}$ (20.55)

$F_{Gs}\equiv2/3\text{x}1/10^3(h_s\text{x}v_{Hms}/a_{os}\text{ x}m_{nep}\text{x}m_{sun})/(X_{nep}{}^2)=6.688\text{x}10^{22}\text{ N}$ (20.56)

$F_{Gs}\equiv(1/1.111\ldots)^{1/2}\text{x } 1/3\text{x}g_{sJ}{}^2\text{x}m_{nep} \text{ x } m_{sun}/m_{es}{}^2\text{x}1/(X_{nep}{}^2\text{x}K_{rel}{}^2)=6.6086\text{x}10^{22}\text{ N}$ (20.57)

$F_{Rs}\equiv3m_{nep}/m_e \text{ x}10^{-30}\text{x}g_{eV} \text{ x } 1/10^4\text{x}1.3333 =6.7996\text{x}10^{22}\text{ N}$ (20.58)

$F_{Gs}\equiv(1/1.111\ldots) \text{ x}3 \text{ x}1/10^3\text{x}1/10^2\text{x}X_{nep}{}^2 =6.7409\text{x}10^{22}\text{ N}$ (20.59)

$F_{Rs}\equiv1/10^3\text{x}A_{cinep}\text{ x}c_s\text{ x}K_{rel}=6.674\text{x}10^{22}\text{ N}$ (20.60)

$F_{Gs}\equiv 6\text{x}m_{nep}/m_{es}\text{x}1\text{x}10^{-30} \text{ x}1/10^6\text{x}1.03278 = 6.7810 \text{ x}10^{22}\text{ N}$ (20.61)

$F_{Rs}\equiv2/3\text{x } \text{x}1/10^2 \text{ x } (m_{nep}/m_{es})\text{x}h_s=6.800\text{x}10^{22}\text{ N}$ (20.62)

$F_{Rs}=(1/1.111\ldots)^{1/2}\text{x}m_{nep}\text{x}v_{nep}/X_{nep}\text{x}6000 \text{ x}10=6.7274 \text{ x}10^{22}\text{ N}$ (20.63)

$F_G=2/3\text{x}1/10^3\text{x}\sigma(5780\text{ -}72)^4\text{x}A_{cirsun} \text{ x}K_{rel}{}^{2/3}= 6.7148\text{x}10^{22}\text{ N}$ (20.64)

$F_{Gs}=2/3\text{x}1/10^3\text{x}\sigma_s(6000\text{ -}T_{nep})^4\text{x}A_{cirsun}\text{ x}K_{rel}{}^{1/2}= 6.7308\text{x}10^{22}\text{ N}$ (20.65)

$F_{Rs}=1/2m_{nep}\text{x}v_{nep}/L_{nep} \text{ x}10^{-30} \text{ x}K_{rel}{}^{1/4}/(hs/60)]=6.7331\text{x}10^{22}\text{ N}$ (20.66)

$F_{Rs}=1/2 \text{ x}I_{nep}/L_{nep}\text{x}[6000 \text{ x}10] \text{ x } 1.03278$

$= 1/2 \text{ x}I_{nep}/L_{nep}\text{x}10^{-30}/(hs/60)] \text{ x } 1.03378=6.7083 \text{ x}10^{22}\text{ N}$ (20.67)

$F_G= F_{Gearth} \text{ x } m_{nep}/m_{earth} \text{ x } (L_{earth}/L_{nep})^2 \text{ x}100=6.6956 \text{ x}10^{22}\text{ N}$ (20.68)

$F_{Gs} = 1/6\,\mu_{os}\,I_{nep}x\,1/30 \text{ x } 9.7343 \text{ x}10^{30} \text{ x } 1.111\ldots^2 = 6.7099\,x10^{22}\,\text{N}$ (20.69)

Table 12. Apparent gravitation forces of Pluto

$F_G = Gxm_{plu} \text{ x } m_{sun}/L_{plu}{}^2 \text{ x } 1/(\sqrt{2} \text{ x } 1.04^{3/4}) = 4.9470 x10^{16}\,\text{N Reference}$ (20.70)

$F_R = m_{plu}\,xv^2/L_{plu}\,x1/(1+2/10) = 4.9929 x10^{16}\,\text{N}$ (20.71)

$F_{Gs} \equiv 2/3x(h_s xv_{Hms}/a_{os}xm_{plu}\,xm_{sun})/(X_{plu}{}^2) = 4.9676 x10^{16}\,\text{N}$ (20.72)

$F_{Gs} \equiv 1/3xg_{sJ}{}^2 xm_{plu} \text{ x } m_{sun}/m_{es}{}^2 x1/(X_{plu}{}^2 xK_{rel}{}^2 x1.0357) = 4.9547 x10^{16}\,\text{N}$ (20.73)

$F_{Rs} \equiv 1/4xm_{plu}/m_{es}xg_{eV}x10^{-30}x1/10^3/K_{re}{}^{1/2} = 5.07935 x10^{16}\,\text{N}$ (20.74)

$F_{Gs} \equiv \sqrt{2} \text{ x } (1/10^3)^3 \text{ x}1(N)xX_{plu}{}^2 = 4.9454 x10^{16}\,\text{N}$ (20.75)

$F_{Rs} \equiv 1/2x1/10^3 x\ 1/10^2 x1.111\ldots^{1/2} xA_{cirplu}\,xc_s/K_{rel}{}^2 = 5.0133 x10^{16}$ (20.76)

$F_{Rs} \equiv 3xm_{plu}/m_{es}x10^{-30}x1/10^3 x1/10^3 x1/10x(1/10^3)^3 x1.111\ldots = 5.0421 x10^{16}\,\text{N}$ (20.77)

$F_{Rs} \equiv 3x(m_{plu}/m_{es})xh_s x1/10^3 xK_{rel}\,x1.111\ldots = 5.0421 x10^{16}$ (20.78)

$F_{Rs} = 1.11111^{1/2}x1/2 \text{ } xm_{plu}xv_{plu}/X_{plu}x6000 \text{ x } 1/K_{rel}{}^2 = 4.9819\,x10^{16}\,\text{N}$ (20.79)

$F_G = 1/2x\ (1/10^3)^3 x\sigma(5780\ -T_{plu})^4 xA_{cirsun}\,xK_{rel}{}^{1/2} = 5.0189\,x10^{16}\,\text{N}$ (20.80)

$F_{Gs} = 1/2x\ (1/10^3)^3\ x\sigma_s(6000\ -T_{plu})^4 xA_{cirsun}\ \text{x } K_{rel}{}^{1/3} = 5.0227\,x10^{16}\,\text{N}$ (20.81)

$F_{Rs} = 1/2x\ m_{plu}\,xv_{plu}/L_{plu}\,x10^{-30}/[K_{rel}{}^{3/2} xhs \text{ x } (1/1.111\ldots^{1/2}] = 4.9751 x10^{16}\text{N}$ (20.82)

$F_{Rs} = \sqrt{2}x1/2 \text{ } xI_{plu}/L_{plu}\,\text{x } [6000] = 5.1104\,x10^{16}\,\text{N}$ (20.83)

$\qquad = \sqrt{2}x1/2xI_{plu}/L_{plu}x10^{-30}/(h_s/60)]\,x1/10 = 5.1104\,x10^{16}\,\text{N}$

$F_G = F_{Gearth} \text{ x } m_{plu}/m_{earth} \text{ x } (X_{earth}/X_{plu})^2 = 4.9732 x10^{16}\,\text{N}$ (20.84)

$F_{Rs} = 1/6\,x\mu_{os}I_{plu}x1/258.24 = 1/6\,x1.1111 x10^{-6}x6.210\,x10^{25}\,x1/232.46 = 4.9470\,x10^{16}\,\text{N}$ (20.85)

Table 13. Apparent gravitation forces of Saturn

$F_G = Gxm_{sat} \text{ x } m_{sun}/L_{sat}{}^2 = 3.702 x10^{22}\,\text{N Reference}$ (20.86)

$F_R = 1.24m_{sat}xv^2/L_{sat} = 3.714 x10^{22}\,\text{N}$ (20.87)

$F_{Gs} \equiv 1/1.11111x1/10^3(h_s xv_{Hms}/a_{os}xm_{sat}xm_{sun})/(X_{sat}{}^2 xK_{rel}{}^2) = 3.757 x10^{22}\,\text{N}$ (20.88)

$F_{Rs} \equiv 2/3x1/3xg_{sJ}{}^2 xm_{sat} \text{ x } m_{sun}/m_{es}{}^2 x1/X_{sat}{}^2 xK_{rel}{}^{1/2} = 3.6892 x10^{22}$ (20.89)

$F_{Rs} \equiv 1/1.11111x1/3m_{sat}/m_{es}xg_{eV}x10^{-30}x1/10^2 xK_{rel}{}^2 = 3.7571 x0^{22}\,\text{N}$ (20.90)

$F_{Gs} \equiv \sqrt{2}\ x1/10^2 xL_{sat}{}^2 xK_{rel}{}^2 = 3.8279 x10^{22}\,\text{N}$ (20.91)

$F_{Rs} \equiv 1/10^2 xA_{cirsat}\,xc_s \text{ x } xK_{rel}{}^{1/2} = 3.7146 x10^{22}\,\text{N}$ (20.92)

$F_{Rs} \equiv 1/3xm_{sat}/m_{es}x1x10^{10} xK_{rel}{}^{1/2}\ (1/10^3)^3 x1/10^3 = 3.7109\,x10^{22}\,\text{N}$ (20.93)

$F_{Rs} \equiv 1/10\ (m_{sat}/m_{es})xh_s/K_{rel}{}^{2/3} = 3.7146 x10^{22}\,\text{N}$ (20.94)

$F_{Rs} = 1/3x1/2m_{sat}xv_{sat}/X_{sat}x6000 \text{ } x10 \text{ } xK_{rel}{}^{1/2} = 3.7513\,x10^{22}\,\text{N}$ (20.95)

$F_G = (1/10^3)^3 x1/10\sigma(5780\ -134)^4 xA_{cirsun} = 3.6860\,x10^{22}\,\text{N}$ (20.96)

$F_{Gs} = (1/10^3)^3 x1/10\sigma_s(6000\ -134)^4 xA_{cirsun}\,xK_{rel}{}^{1/2} = 3.7872\ 10^{22}\,\text{N}$ (20.97)

$F_{Rs} = 1/3x1/2m_{sat}xv_{sat}/X_{sat}\,x10^{-30}/(h_s/60)]\,xK_{rel}{}^{1/3} = 3.6743 x10^{22}\,\text{N}$ (20.98)

$F_{Rs} = 1/3x(1/10^3)^3 x1/2I_{sat}/X_{sat}x[6000\ x10]\ xKrel^{1/2} = 3.7622 x10^{22}\,\text{N}$

$\qquad = 1/2\ I_{sat}/L_{sat}x10^{-30}/(hs/60)] = 3.7622 x10^{22}\,\text{N}$ (20.99)

$F_{Rs}=1/6\mu_{os}I_{sat}$ x x1/25 $x4.987$x10^{30} x 1.111111= 3.6953 x10^{22} N   (20.99a)

Table 14. Apparent gravitation forces of Uranus

$F_G=Gxm_{ura}$ x m_{sun}/L_{ura}^2=1.4001x10^{21} N Reference   (20.100)

$F_G=m_{ura}xv^2/L_{ura}$=1.4016x10^{21} N   (20.101)

$F_{Gs}\equiv2/3x1/10^3(h_sxv_{Hms}/a_{os}xm_{ura}xm_{sun})/(X_{ura}^2)$=1.3980x$10^{21}$ N   (20.102)

$F_{Gs}\equiv1/3xg_{sJ}^2xm_{ura}$ x $m_{sun}/m_{es}^2x1/(L_{ura}^2xK_{rel}^2x1.0357)$=1.3948x$10^{212}$ N   (20.103)

$F_{Rs}\equiv1/12m_{ura}/m_exg_{eV}x1x10^{-30}x1/10^2x1/(1-1/10)^{2/3}x K_{rel}^{1/2}$=1.3912 x$10^{21}$ N   (20.104)

$F_{Gs}\equiv\sqrt2$ x $1/10^4 xL_{ura}^2$ x $K_{rel}x 1.04$=1.3997x10^{21} N   (20.105)

$F_{Rs}\equiv1/3x$ $1/10^2xA_{cirura}xc_s/K_{rel}^2 x1/1.04$=1.3894x$10^{21}$ N   (20.106)

$F_{Rs}\equiv 1/3xm_{ura}/m_{es}x1x10^{-20} x10xK_{rel}^2$=1.3452 x$10^{21}$ N   (20.107)

$F_{Rs}\equiv1/6$ x $1/10$ $(m_{ura}/m_{es})xh_s x1/1.0357$=1.3868x$10^{21}$ N   (20.108)

$F_{Rs}=1/6x1/2$ x$m_{ura}xv_{plu}/X_{plu}x6000$ x10 xK_{rel}^2 =1.3731x10^{21} N   (20.109)

$F_G=1/10^2x\sigma_s(5780 -T_{ura})^4xA_{cirsun}$ xK_{rel} = 1.3349 x10^{21} N   (20.110)

$F_{Gs}=1/10^2x\sigma_s(6000 -T_{ura})^4xA_{cirsun}$x $1/1.111 . . .$=1.3695x10^{21} N   (20.111)

$F_{Rs}=1/6m_{ura}xv_{ura}/L_{ura}x10^{-30}/(1/6x$ x$h_s/60)]x 1.111 . . .$ =1.3904 $x10^{21}$ N   (20.112)

$F_{Rs}=1/6x1/2xI_{urd}/L_{ura}x[6000$ x$10] xK_{rel}^2x 1x10^{-10} x1/10^3$= 1.3731x$10^{21}$ (20.113)

$= 1/6x1/2xI_{urd}/L_{ura}x10^{-30}/(h_s/60)] xK_{rel}^2$ (20.114)

$=1/6x1/2xI_{urd}/L_{ura} x10^{-30}x1/10^{-33}/ K_{rel}x1.11111$ x$10^{-10}x1/10^3$=1.4304 x10^{21}

$F_G= F_{Gearth}$ x m$_{ura}/m_{earth}$ x $(X_{earth}/X_{ura})^2$= 1.3996x10^{21} N   (20.115)

$F_{Gs}=1/6$ xμ_{os} $I_{ura}x1/868.4$=1/6 x1x10^{-6} x5.9134 x $10^{30}/781.36$=1.401 $x10^{21}$ N   (20.116)

Table 15. Apparent gravitation forces of Venus

$F_G=Gxm_{ven}$ x $m_{sun}/(1.04^{3/2}L_{ven}^2)$=5.224x$10^{22}$ N Reference   (20.117)

$F_R=m_{ven}xv^2/L_{ven}x1.111 . . .^1$ =5.2242x10^{22} N   (20.118)

$F_{Gs}\equiv2/3x1/10^3(h_sxv_{Hms}/a_{os}xm_{ven}xm_{sun})/(X_{ven}^2xK_{rel}^2)$ x$1.111 . . .^1$=5.2336x10^{22} N   (20.119)

$F_{Gs}\equiv1/3xg_{sJ}^2xm_{ven}$ x $m_{sun}/m_{es}^2x1/L_{ven}^2x K_{rel}^{1/2}$= 5.2067x$10^{22}$ N   (20.120)

$F_{Rs}\equiv 2x1/3$ x$m_{ven}/m_exg_{eV}x1x10^{-30}$ x $K_{rel}^{1/2}$=5.2351 x10^{22} N   (20.121)

$F_{Gs}\equiv\sqrt2$ x$3xL_{ven}^2$ x$1.11111^{1/2}$=5.2669x10^{22} N   (20.122)

$F_{Rs}\equiv3/2A_{cirven}xc_s xK_{rel}$=5.1754x$10^{22}$ N   (20.123)

$F_{Rs}\equiv 1/10m_{ven}/m_{es}x1x10^{-20} xK_{rel}^{3/2}$=5.1659 x$10^{22}$ N   (20.124)

$F_{Rs}\equiv10(m_{ven}/m_{es})xh_s x K_{rel}^{1/2}$=5.2290x$10^{22}$ N   (20.125)

$F_{Rs}=1/2xm_{ven}xv_{ven}/X_{ven}x6000$ x10 x $1.111 . . .^{1/2}$ =5.3432 $x10^{22}$ N   (20.126)

$F_G=1/10^3x\sigma(5780 -735)^4xA_{cirsun}$ x$1/K_{rel}^{1/2}$= 5.2024 $x10^{22}$ N   (20.127)

$F_{Gs}=1/10^3x\sigma_s(6000 -735)^4xA_{cirsun}$ x$1/K_{rel}^{1/2}$=5.2621x10^{22} N   (20.128)

$F_{Rs}=3/2m_{ven}xv_{ven}/L_{ven}x10^{-30}/(h_s/60x1.111 . . .^{/2}x5.2796x10^{22}$ N   (20.129)

$F_{Rs}=1/2xI_{ven}/L_{ven}$ x$[6000$ x$10]$ x$1.111 . . .$

$=1/2xI_{ven}/L_{ven}$ x$10^{-30}/(h_s/60)]$ x$1.111 . . .$ = 5.2395 x10^{22} N   (20.130)

FG= FGearth x mven /mearth x (Learth/Lven)²= 5.5213x10^{22} N

$$F_{Rs} = 1/6 \; x\mu_{os} \, I_{ven} \; x \; 1.1547^2 \; x1.111 \ldots^3 = \qquad (20.131)$$
$$=1/6 \; x \; 10^{-6} \; x1.7035 \; x \; 10^{29} \; x \; 1.1547^2 \; x \; 1.111 \ldots^3 = 5.1928 \; x10^{22} \, N.$$

Thus the charges of rotating masses of the last five planets generate insignificant forces to play any role in creating repulsive forces. Mutually inductance effects nearly disappar and become replaced by self inductance for large distances. The cores of these planets may also carry currents nulling external effects.

The gravitational energy E_{g3} uncovered in Chapter 17, based on the geometric mass of each sun-planet couple, is the same all the sun-planet couples.

The gravitation force of planet attraction by the sun can be essentially attributed to the black body radiation

$$\sigma_s T^4_{sun} = 5 \; x \; 10^{-8} \; x \; 6000^4 = 0.648000 \; x10^8 \, W/m^2 \qquad (20.132)$$

with various multipliers of uncertain origin: $1/3x1/10^3$, $1/2x1/10^2$, $1/6x1/10^2$, $2/3x1/10^3$, $1/2x(1/10^3)^3$, $1/10x(1/10^3)^3$, $1/10^2$. The power density of the source due radiation of non-interacting components is often 1.2^3 larger at $\sigma_s T^4_{sun}$ x $1.2^3 = 1.119744$ x 10^8 W/m^2. That power density is also delivered by the local velocity 1 x 10^8 m/s: 1 J/m^3 x $1x10^8x1.111 \ldots$ x $1.04^{1/5} = 1.119856$ x 10^8 W/m^2. The source of the similariry was not investigated, because scatter is excessive.

Venus generates σ_s x 735^4 x $10^{-3}/(\sqrt{2} \; x1.04^{3/3})=10.019$ W/m^2.

Venus gravitation force is about $\sqrt{2}$ larger that that of earth $5.224x10^{22}$ N/$\sqrt{2}=4.00x10^{22}$ N.

The sun interaction was priorly attributted to temperature of 3000 K, which derives with surface probability 10^{-20} minimum energy pulses 10^{-14} J in four orientations:

$$\sigma_s T^4_{3000} = 5x10^{-8}x3000^4/(1+1/80)=4 \; 000 \; 000 \; W/m^2 \qquad (20.133)$$

The temperatures of planets are badly defined and they were not analysed. Bodies with much lower tempearure cannot generate with black body radiation power densities values, which could balance sun gravitation. However if they rotate they unbalance stability of moments in protons and they discharge 2 W/(kgKm^2) sufficient to balance attraction of the very hot body.

Although energies expended are pronably minute gravitation forces require sources of energy from the sun and from the planet. Planet circulation, as explained in Chapter 7, unwinds angular moments of electrons releasing energy of protons in sufficient amount to equal gravity. These moments are displayed in one plane. It it believed that they generate the force at the atomic level opposing acceleration induced by the centrifugal force and one opposing gravitation.

Tables 8 - 15 clearly indicate in the top two items that for all the planets the force of attraction of the sun is balanced in microscopic terms by the force of acceleration inducecd by centrifugal force of planet rotation.

The concept of mass generating the force of 1 N/kg applies only to earth as displayed in Equations 5.38-9. The force of sun attraction varies with planets according to Equation 20.2. These values are essentially expressed in Equations 20.2c-20.2k confirming the data in Tables and defining also the force in neutons per kilogram of the mass of the planet.

Equation 7.14a derive that unwinding of reduced electron spins of the whole mass of earth generates a force equal to that of sun attraction. Since the earth contains iron the result is taken to mean that because of elevated temperature nearly all earth mass is above Currie temperature converting magnetic permiability of all matter to non magnetic state expressed by all matter at $\mu_{os} = 1$. Furthermore unwinding momenst means that all matter is in solid or semi-glass state sustaining effective all electrons spin for period of one year directed in one orientation towards the sun.

Only two planets, earth and Mercury, display magnetic field. It is suggested that the rotatings centers of mass create a ring of current with verticallly directed magnetic flux at planets centers. Except for earth, Venus and Mercury the materials at the cores of other planets are sufficiently fluid to generate flux cancelling the magnetic field of the source.

Comparing parameters of gravitation forces expressed by different expressions derived several interesting relations implicating temperature to is imbeded in the planet and in the sun mass and in the gravitation constants These are new formulations of gravitation energy and of gravity force unknown previously.

The first two expressions derives gravity forces from mass, temperature and local velocity, while the next two from mass, gravitation constant and local velocity. Corrections used were discussed elsewhere.

$$F_{Gs} \equiv m_{ssun}/6000^4 \times 2/3 \times 1 \times 10^8 /(1.111\ldots \times 1.1547^{1/2} \times 1.2) = 3.5903 \times 10^{22} \text{ N}$$

$$F_{Gs} \equiv m_{searth}/300^4 \times 2/3 \times 1 \times 10^8/(1.11111 \times 1.1547^{3/2}) = 3.5919 \times 10^{22} \text{ N} \qquad (20.134)$$

$$F_{Gs} \equiv m_{ssun} \times 1 \times 10^{-10} \times (2/3)^2 \times 1 \times 10^8 (1.111\ldots \times 1.10^5 \text{ J}) = 3.6000 \times 10^{22} \text{ N}$$

$$F_{Gs} \equiv m_{searth} \times 1 \times 10^{-10} \times 1 \times 10^8/1.111\ldots = 3.6000 \times 10^{22} \text{ N} \qquad (20.135)$$

Mass of sun expressed in terms of temperature and in the light velocity has the following format:

$$m_{ssun} = 2 \times 10^{30} \text{ kg} \equiv 3/2 \times 6000^4 \times 1 \times 10^{15} \times 1.04^{7/10}$$
$$= 3/2 (6000^4 \times 1.111\ldots^{1/2} \times 3 \times 10^8)^2/1.04^{7/10} \qquad (20.136)$$

Above Equation suggests that $m_{ssun} = f(T^8)$ with square of light velocity is treated as rate $m_{ssun} = f(T c_s)^{10}$ generating the following optional sun temperatures from its mass:

(interacting sun mass $1 \times 10^{30})^{1/10}/1.1547^{1/2} \times 1.0328^{1/10} \equiv 1000.62$ K $\qquad (20.137)$

And alternatively $(2 \times 10^{30} \text{ kg})^{1/8}/1.111\ldots^{1/2} \times 1.0328 \equiv 6006.67$ K has also acceptable variance. These identities support the view that the mass of sun is the source on earth of unity pulse per its oscillator.

Gravitation energy $E_{gl} \equiv 2 \times 10^{30}$ kg/6000^4 x 2/3 x 10/ $1.04^{7/10} = 1.00094 \times 10^{16}$ J (20.138)

Gravitation energy E_{gl} is per plane and per K as indicated in Equation 17.25, E_{gl} per three oscillators in three planes times 1.111 . . . becomes $E_{s1kgSI} = 1 \times 10^{17}$ J.

Minimum energy pulse is mentioned on numerous occasions. It is corrected gravity constant per degree. For the sun:

Minimum energy pulse = 1×10^{-10}/6000 x 2/3/1.111 . . . = 1×10^{-14} J (20.139)

For earth: 1×10^{-10}/(3 x 300 x 1.111 . . . x 10) = 1×10^{-14} J

21. *Harmonic of planets and distances to sun*

Kepler discovered the mathematic rules displayed by the particles of the Universe. He believed that the Universe is controlled by harmonic motion. Einstein considered that the Universe is subject of contraction and expansion. S. Hawking suggested initially that Universe is oscillating without singularities like Big Bung and Big Crush. G. LeMaitre proposed that Universe was created by a big explosion, although Hindu mystics wrote about it thousands years ago.

Lars Wahlin in his "Dead Beat Universe" book [48] proposed in detail that the Universe is subject to one oscillation directing all masses with velocity of light to its center. Frank Rubin proposed in the year 2000 on WEB a multi Bang Universe. R.Toleman and Paul Steinharelty were two other researchers dealing with Universe oscillations.

The data presented now confirm strongly the view that the space geometry and equal probability of radiations in all directions has a fundamental role in establishing the planetary system. The importance of sub-harmonics in sun to planets spacings is confirmed by the finding that the corrected (cor.) ratios of (earth to sun) to (planet to sun) distance (or its invert) L expresses an integer or half-integer values:

Table 16. *Sub-harmonics of the radii of planets rotations*

$$L_{earth\ to\ sun}/L_{mars\ to\ sun} \text{ x cor} \qquad = 0.6563 \text{ x } 1.04^{5/12} \approx 2/3$$
$$L_{earth\ to\ sun}/L_{mercury\ to\ sun} \qquad \approx 2.50$$
$$L_{earth\ to\ sun}/L_{venus\ to\ sun} \text{ x cor} \qquad = 1.3826 \text{ x } 1.1547/1.111\ldots^{1/2}=1.5$$
$$L_{jupiter\ to\ sun}/L_{earth\ to\ sun} \text{ x cor} \qquad = 5.2027/1.04= 5.002$$
$$L_{saturn\ to\ sun}/L_{earth\ to\ sun} \text{ x cor} \qquad = 9.538 \text{ x}1.1111^{1/2}=10.0$$
$$L_{uranus\ to\ sun}/L_{earth\ to\ sun} \text{ x cor} \qquad =19.18 \text{ x } 1.04\approx20$$
$$L_{neptune\ to\ sun}/L_{earth\ to\ sun} \text{ x cor} \qquad \approx 33.23/1.111\ldots \text{ x } 1.04^{1/12}\approx 30$$
$$L_{pluto\ to\ sun}/L_{earth\ to\ sun} \text{ x cor} \qquad \approx 32.84/1.111\ldots \text{ x } 1.04^{1/3}\approx 30 \qquad (21.1)$$

Thus neglecting elliptical distortions the ratios of distances of planets to the sun are displayed as aliquots of the earth distance. Two hundred years ago Kepler's view on gravity were universally accepted among scientists. Hence it is likely that the length of one meter established in 1791 was based on a rounded up known then value of $L_{earth\ to\ sun}/1x10^7=$ 15000. For the integer model proposed now this is a perfect choice. Kepler's third orbital rule expressing the cubes of planets distance to sun to be proportional to the square of their period remains valid and mysterious. In the present model, the space gradient of mass in the radiation beam joining the two masses time G expresses the term, 'gravity tension' = $GxMass/L,$ used in some texts.

The space Plank constant $E_{sspaceSI}$ expresses the mass of 1 kG converted to energy and can be expressed in variety of equivalent expressions:

$$E_{sspaceSI} \equiv (\varepsilon_{os}\mu_{os})^{-1} \equiv 10E_{gl} \equiv 1 \text{ x}10^{16} \text{ x}10 \text{ s}\equiv 1 \text{ kg x}c_s^2 \text{ x } 1.111\ldots/10 = 1x10^{17} \text{ J} \qquad (21.2)$$

The other important parameter introduced, the earth mass **1×10^{21} J** pulse is derived in Table 3 from the Planck time $E_{searthSI} \equiv 1/\sqrt{t_p} = 1/\sqrt{10^{-42}} = \mathbf{1 \times 10^{21}}$ **J**. It is the earth mass in one orientation vectorized by three linear probabilities:

$$E_{searthSI} \equiv 1/6 \times 6 \times 10^{24} \text{ kg} \times 1/10 \times 1/10 \times 1/10 \equiv 1 \times 10^{21} \tag{21.3}$$

or for 1 plane $1/3 \times 2/3 \times h_s v_{Lms}/1.1111 = 1/3 \times 2/3 \times 1 \times 10^{-33} \times 5 \times 10^{12}/1.1111 \equiv 1/10^{21}$ J

The signal sent from Bohr atom according to the author expresses the equivalent mass of proton and electron, but because it is so close to electron mass it is mistaken for electron:

$$m_{es} m_{ps}/(m_{es} + m_{ps}) \approx \mathbf{0.999001 \times 10^{-30} \text{ kg}} \tag{21.4}$$

It oscillates with frequency 0.999 $\mathbf{x10^{12}}$ Hz. It cannot be distinguished from their geometric mass $3(m_{es} m_{ps})^{1/2} \times 1.111 \ldots^{1/2} \approx 0.999999999 \times 10^{-30}$ kg. **The mass of 1×10^{-30} kg in 10 orientations of three planes radiating sequentially has probability of $1 \times 10^{-30} \times 1/10^3 = 1 \times 10^{-33}$ of being delivered in any specific orientation. It expresses Planck constant.** There are 32 quantum positions per spherical particle, which with relativity and quantum correction deliver the integer force of gravitation:

$$32 \times 1.111 \ldots \times (1+1/80) \times 1 \times 10^{54} \times 10^{-33} = 3.6 \times 10^{22} \text{ N exactly} \tag{21.5}$$

There are 6 orthogonal orientations with values needing correction for relativity and quantum. Two 1 J/m pulses displaced by 90° at local velocity generate between them $\sqrt{2}$ faster pulse with $\sqrt{2}$ higher amplitude so that from 3 orthogonal directions:

$$(2 \text{ J/m} \times 10^8 \text{ m/s}) \times 3 \times 10^8 \text{ m/s} \times \sqrt{3}/1.04 = 0.99926 \times 10^{17} \text{ W} \rightarrow 1 \times 10^{-19} \approx 0.01 \text{ J} \tag{21.6}$$

One of the fundamental relations is the formation of the unity event on earth at 300 K. It expresses the space gradient of sun power of 1 W/m. It is derived from space probabilities in three planes subjected to inverted space quantization of 10, the linear energy per meter being 10^{-10}. Converted by volume probability to power 3 derives electron mass 1×10^{-30} kg. Electron oscillating in one plane is subjected to two stepped space quantization $1/10 \times 1/10$ deriving muo10_s radiation energy density 10^{-28} kg/m^3 moving through space; energy in 3 planes times $10 \times 1 \times 10^{-10}$, times relativity correction 1.111 . . . and times the velocity of light c_s derives basic unit of power 1 W:

Elementary power $3 \times 10 \times 1 \times 10^{-10}$ J/m $\times 3 \times 10^8$ m/s $\times 1.111 \ldots = 1$ watt $= 1$ J/s $\qquad$ (21.7)

Multiplying by 1 s $\qquad\qquad\qquad$ J/s $\times 1$ s $= 1$ J $\qquad\qquad\qquad\qquad$ (21.8)

In atomic nature space units linear probability converts above unity values to the value of the gravitational constant of the model $G_s = 1.0 \times 10^{-10}$.

In Equation 21.2 $E_{sspace} = 1 \times 10^{17}$ J, which with 1 J basic pulse must involve rate of 1 $\times 10^{-17}$ Hz . It is $(\varepsilon_{os} \mu_{os})$ wtrh paeameters rewferred to 1K, so that 1×10^{-17} Hz refers also to 1 K. At 300 K with 1.111 . . . relativity correction and in 3 planes the product generates:

Minimum SI energy per pulse accumulated at 300 K: 3×10^{17} J-s $\times 300 \times 1.111 \ldots \equiv 1 \times 10^{20}$ J-s(21.9)

In integer space units converted to SI, proton mass time light velocity expresses its energy. The minimum energy per plane with 10 space quantization orientations of Figure 2, the energy per such relativity corrected particle is 1×10^{-10} J:

$$\mathbf{\textit{G}_s = 10 \; \textit{m}_{ps} x(1/3\textit{c}_s^2) = \textit{m}_{ps}/(\varepsilon_o \mu_o) = 1 \times 10^{-27}/(1 \times 10^{-11} \times 1 \times 10^{-6}) = 1.0 \times 10^{-10} \; J} \qquad (21.10)$$

Hence graviton constant is the sum of proton mass conversions from ten local pathways.

It should be apparent by now that the parameter 10^{-10} is the international gravitation constant G_s expressed in terms of a particle, which moving at local velocity of light $1/3 \; c_s$ derives the first Planck radiation law parameter:

$$G_s \times 1/3 \; c_s = 10^{-10} \times 1 \times 10^8 \equiv 0.01 \; J \qquad (21.11)$$

Proton momentum Equation 1.15, $e_s = 10^{-19}$, converted to SI becomes 10^{-9}. That value located in three orthogonal orientation generates the resultant momentum or probability 10^{-27}. Such momentum added vectorially in three orthogonal orientations generates:

$$\sqrt{3} \; (1 \times 10^{-27}) \qquad (21.12)$$

Similarly two $\sqrt{3} \; (1 \times 10^{-27})$ pulses in orthogonal orientations generate 3×10^{-54} J.

Above value will be compared to the smallest energy pulse 2.6812×10^{-54} J derived by Lars Wahlin in his book [48]. Multiplying by relativity and quantum correctors Wahlin's energy derives the same minimum value of energy as derived by the integer cubic model:

$$2.6812 \times 10^{-54} \; J \times 1.11111 \times 1.04^{1/5} = 3.00257 \times 10^{-54} \; J \qquad (21.13)$$

The author shares with Wahlin and Kepler the view that harmonics control the formation of the Universe. The author has no data supporting Universe collapse or expansion, but he congratulates Lars Wahlin on a well numerically presented case for Universe collapse contradicting the popular Universe expansion.

The gravitation force per electron on earth per h_s over relativity and quantum corrections derives 30 eV:

Hartree energy$= 3.5439 \times 10^{22}$ N/$(1 \times 10^{54} \times 1 \times 10^{-33}$ J-s$) \; \mathbf{\times 1/3 \times 1/(1.1547 \times 1.04^{1/6})} = 29.98 \approx 30$ eV
$$\qquad (21.14)$$

The inverted product of space permittivity ε_{os} and linear probability 1×10^{-10}, is $E_{searthSIis}$**:**

$$E_{searthSI} = 1/(G_s \; \varepsilon_{os}) = 1 \times 10^{21} \; \mathbf{J\text{-}a} \qquad (21.15)$$

$E_{searthSI}$ is actually inverted Boltzmann constant $k_s = 1 \times 10^{-23}$ J, with two inverted steps of space quantization 10×10:

$$E_{searthSI} = 1/(k_s \times 10 \times 10) \equiv 1 \times 10^{21} \; J \qquad (21.16)$$

ε_{os} and μ_{os} are energies per pulse energy densities in SI units. The values expressed in natural atomic units per local cycle of infrared frequency 1×10^{12} are

$$\mu_{os}' = 1 \times 10^{-6} \times 1 \times 10^{-30}/1 \times 10^{12} = 1 \times 10^{-48} \text{ J/cycle and} \tag{21.17}$$

$$\varepsilon_{os}' = 1 \times 10^{-11} \times 1 \times 10^{-30}/1 \times 10^{12} = 1 \times 10^{-53} \text{ J/cycle} \tag{21.18}$$

With linear probability of 1/10, the product $\mu_{os}' \times \varepsilon_{os}'$ is as in Equation 25.6:

$$\text{The probability} = \mu_{os}' \times \varepsilon_{os}' = 10 \times 10^{-100} \tag{21.19}$$

Values of the energy 1×10^{100} J derived in Equations 25.1 and 25.2 were first considered unbelievable or mysterious and they were inserted in the Appendix. Consistence of their derivations indicates that the pulse produced by the energy of 1×10^{100} J times probability 1×10^{-100} derives energy 1 J or 1 eV in atomic terms as the model requires.

At 6000 K, with σ_{sa} = unity at atomic level, the power generated by the sun is:

Sun power $= 1 \times 6000^4/1.1111^{1/2} \times 1.04^{1/5} = 1.0015 \times 10^{15}$ W (or energy in J or force in N).

The above value time local average velocity $1/3 \ c_s$ in one direction requires 1/3 multiplier and it derives:

Sun gravitation force $= 1.0015 \times 10^{15} \times 1/3 \times 10^8 \times 1.111 \ldots^{1/2} \times 1.04^{1/5} = 3.5466 \times 10^{22}$ N (21.20)

The variance from the reference 3.5438×10^{22} N is 3×10^{19} N time 1×10^{-19} generating energy of 3 eV, which a quantum pulse.

The relation between the integer model Boltzmann constant and the integer Planck constant within 0.1% accuracy is given by:

$$\sqrt{3} \ k_s T_{300}/1.04 = h_s v_{Lms}, \text{ deriving } v_{Lms} = 4.9963 \times 10^{12} \text{ Hz} \tag{21.21}$$

The value very close to $5 \times (1-1/1000) \times 10^{12}$ Hz oscillations of the proton-electron pair.
    With progression of oscillations established in one direction they continue to take place in a similar manner as in a gyroscope. Probability of sequential oscillations in the same five pathways involves display of sequential probability of energy in that local pathway. The local probability should be expressed per second by $c_s \times 3 \times G_s = 3 \times 10^8 \times 3 \times 10^{-10} = 0.01 \text{ s}^{-1}$. The overall probability of five sequential occurrences bring the power of five, identified now with the model gravitation constant for masses at the same temperature

$$\text{Gravitation constant } G_s \equiv (0.01)^5 = 1 \times 10^{-10} \text{ N-m}^2/\text{kg}^2 \tag{21.22}$$

For the sun-to-earth system G_s becomes effectively $2/3 \times 10^{-10}$ N-m^2/kg^2. Thus power of five, represents probability of sequential display of a pulse moving forward. Hence Planck radiation law parameter 0.01 J derives gravitation constant G_s. The integer space model and the classical SI derive the same energy values, but the integer model delivers more details about the physics and about gravitation.

An initially incomprehensive relation, the **inverted time of earth annual rotation in seconds to power of three, corrected for quantum and relativity, expresses the value of three Boltzmann constants:**

$$[1/(31.5576 \times 10^6 \text{ s})]^3 \times 1/(1.111 \ldots^{1/3} \times 1.15470^{1/6}) \equiv 2.99950 \times 10^{-23} \text{ J} \approx 3k_s \quad (21.23)$$

Another finding is that earth period cubed corrected for relativity and quantum derives Special Relativity gravitation force:

$$(31.5576 \times 10^6 \text{ s})^3 \times 1/(1.111 \ldots^{1/2} \times 1.04^{1/6}) \equiv 3.00104 \times 10^{22} \text{ N}$$

Suggesting that $3/(10k_s)$ expresses Special Relativity gravitation force.

The age of the Universe may be derived because $(\varepsilon_{os}\mu_{os})^{-1} \equiv 10$ J $\mathbf{x}(1/3 \; c_s)^2 \equiv 10$ J $\mathbf{x}(10000$ K $\mathbf{x}$ 10000 K $\mathbf{x}$ 10000 K $\mathbf{x}$ 10000 K $) \equiv 1 \times 10^{17}$ J-s is the radiation energy power in 4 directions. Three orientations converted to SI units with 1 year unit time:

$$t_o = (3 \times 10^{17} \text{ J-s}) \times 1/(1 \text{ J} \times 31.5576 \times 10^6 \text{ s/year}) \times 1.15470^{5/2} \times 1.04^{1/4} = 13.7546 \times 10^9 \text{ years. } (21.24)$$

Above value compares excellently with NASA's seven year 2010 data release of $(13.75 \pm 0.011) \times 10^9$ years $\qquad\qquad (21.25)$.

Correcting for relativity and quantum $13.75 \times 10^9/(1.1547^2 \times 1.04^{39/50}) = 1.00018 \times 10^{10}$ years, which fits well with the integer model proposed. **Uncorrected age of the Universe in the integer cubic model is 1×10^{10} years.** That figure is equal to velocity of light and relativity corrected energy 30 J the resultant parameter seemed not to be connected with age of the Universe:

$$1.0 \times 10^{10} = 3 \times 10^8 \times 1.111 \ldots \times 30 \text{ J (W-m)} \qquad (21.26)$$

The value quoted by Wikipedia is $(13.75 +- 0.11) \times 10^{10}$ years follows Lambda—CDM concordance model. The meaning of the age of the Universe becomes clearer from the following two **preferred now derivations attributing the Universe age** to the period of time to transmit 1 J/m^3 or 1 J/pulse or atom vectorized or 1 J incoherent signals charging or discharging parameters generating ε_{os} and μ_{os}:

Age of the Universe$=\varepsilon_{os}$**x1 x10^{21} x1.1547^{2}x1.04$^{18/25}$ $\equiv$ 13.715x10^9 years** $\qquad$ **(21.27)**
Age of the Universe$=\mu_{os}$ **x 1x10^{17}x1/10 x1.111 . . .^{3} $\equiv$ 13.717 x10^9 years** $\qquad$ **(21.28)**

Atomic units here are years because full earth rotation is the time needed to unwind the reduced electron spin causing destruction of the proton balance converting proton into 2001 independent electrons.

Discharging the angular momentum of the electron spin of h_s may be the origin of most forces in the Universe. For example the minimum energy$=10^{-30} \times 10^{-17}/10^{-33}=10^{-14}$ J.

Further analyses indicates that **energy of Planck constant h_s is related to the release of energy Bohr atom structure meaning dissociation and its destructure.** Radiation pulse 2 J $\approx$ 20.07 J x 1/10 =13.6 eV x $\sqrt{3}$ /(1.1547x 1.0328$^{1/2}$) and 2 J at atomic level per kelvin is for two oscillators of Bohr atom:

$$2 \text{ J}/(300 \text{ } x \text{ } 3 \text{ } x \text{ } 1.11 \ldots) \text{ } x \text{ } 10^{-30} = 2 \text{ } x \text{ } 10^{-33} = 2 \text{ } h_s \qquad (21.29)$$

The sun mass 10^{30} kg generates the force of 1x10^{31} N decreased to 1/6 per orthogonal orientation and by 1/10x1/10x1/10 of space quantization formulating the vector and further corrected by quantum and relativity 1.04/1.1547^2. Dividing above force by distance to the edge of the Universe expressed by its age in years, seconds per year and velocity of light derives the numerical value of model Planck constant 1x10^{-33} J-s, which is the electron spin for 6 independent orientations, Equation 22.54:

$$1 \text{x} 10^{31} \text{ N} \text{x} 1/6 \text{x} 1/10 \text{x} 1/10 \text{x} 1/10 \text{x} 1.04/1.1547^2 \text{x} 1/(13.7165 \text{x} 10^{10} \text{x} 31.557 \text{x} 10^6 \text{x} 3 \text{x} 10^8)$$
$$= 1.00118 \text{x} 10^{-33} \text{ N/m} \equiv 1 \text{ x } 10^{-33} \text{ J-s} \quad (21.30)$$

The variance is believed to be h_{selec}/1.1547=1.02 x 10^{-36} J the electron Planck constant.

Planck constant is hence derived as the gradient of force, otherwise as second space gradient of energy *W: dF/dx =dW^2/dx^2*. Integrating with distance *dF/dx* =2x10^{11} N/m derived in Equation 16.41 with the sun gravitation force becoming the corrected square of the distance to earth. In this evaluation the effects of other celestial bodies have been neglected because **gravitation is believed to be carried in the beam interaction between two bodies, the center star and a rotating planet. Without rotation the two bodies attract each other and collapse.**

The product of the velocity of light and of the age of the Universe in seconds derives the expanse of the Universe at: 3x10^8 x 13.756 x10^9 x 31.447x10^6 = 1.30229x10^{26} m. The product of the space expanse in 6 orientations and of the vectorized space pulse per plane 10^{21} J and of relativity and quantum corrections:

$$1.30229 \text{ x} 10^{26} \text{ m} /1 \text{ x} 10^{21} \text{ J x } 1.1547 \text{ x} 1.04^{3/50} = 0.904378 \text{ x } 10^6 \text{ m/J} \qquad (21.31)$$

Above velocity, corrected by 1.111 … μ_{os}:

$$0.904378 \text{ x} 10^6 \text{ m/s x } 1.111 \ldots \text{x} 10^{-6} \equiv 1.00486 \text{ J or eV} \qquad (21.32)$$

At atomic level the deviation is 0.00486 J x 1 x 10^{-19} $\approx$ 50 x 10^{-23} J $\equiv$ 50 k_s expressing most probably uncertainty related with sun or space pulse energy of 50 eV. The sun data may unduly suggest that values of 6000 K energy are particularly important in the Universe.

Linear gradient of energy or mass expressed by the gravity constant 10^{-10} time the linear probability 10^{10} defines the unit of energy, ratio of mass to velocity defines the velocity of light. So **the source of gravitation in physics is from the rate of expiration of Bohr structures the energy being absorbed by the receiving mass or dispersed in**

infinity. Balance of forces in the sun-planet couples is generally obtained by the equality of the force of sun attraction and the planet repulsion induced by its rotation.

The linear probability 1×10^{-10}, otherwise the international gravitation constant G_s times space pulse energy 3×10^{21} J, derives diameter of the circular earth rotation:

$$1 \times 10^{-10} \times 3 \times 10^{21} \equiv 3 \times 10^{11} = 2 \times 1.5 \times 10^{11} \, m \qquad (21.33)$$

These results suggest that the oscillation of the mass of the Universe, which expresses parameters of hydrogen, may have a fundamental role in establishing the planetary system and in the value of the elementary physical constants.

Newton' s three laws of motion and the conservation of momentum in the Universe, confirmed by the above data show that the Universe has been formed and run by mathematical probability and geometry rules generating statistics expressed broadly by the old term Newton's Clockwork Universe meaning that the accuracy of prediction is high, but not perfect.

We live in an environment of infrared frequencies generated by the average 300 K on earth, where radiations are essentially those of proton and (proton+electron) at mean values of (1 or 5) $\times 10^{12}$ Hz generating energies 0.05 eV or locally 0.01 eV or J; the sun delivers frequencies (1 or 5) $\times 10^{15}$ Hz and energies (10-50) J or eV. An average spectrometer records the middle wavelength at 400 nm displaying energy 400 $\times 10^{-9}$ x $3 \times 10^8/1.2 = $ **2 x 50 J or eV received** may express two H_2 molecules each atom having oscillators discharging energy 13.65 eV of three directions deriving

$$13.65 \, eV \times 3 \times 2 \times 1.1547 \times 1.111\ldots^{1/2} \times 1.04^{1/2} = 100.01 \, eV \qquad (21.34)$$

The model does not deal with a wide range of frequencies and energies of the electromagnetic waves. The same energy reference event of 1 J or eV of the sun pulse on earth at 300 K is used in the classical SI system as is used in the integer cubic model with the same value of the force between two long conductors of 2 $\times 10^{-7}$ N/m per ampere. Expressed per K per plane it is the Planck radiation term 0.01 W. The energy 1 J or power of 1 W on earth and sun emission of 1 unit/kg with volume probability converter derives the sun mass of 10^{30} kG interacting in one orientation. Energy of 1 J converted by volume converter and three step quantization expresses the Planck radiation constant $h_s \equiv 1$ J x 10^{-30} x 1/10 x 1/10 x 1/10 $= 1 \times 10^{33}$ J-s.

Minimum energy of pulse at atomic level converted to SI 1 x 10^{-44} J $\rightarrow$ x $10^{30} \equiv 1$ x 10^{-14} J; energy 1 x 10^{-44} J comprises 10 orientations of plane each of 1 x 10^{-45} J, which are sourced from three planes of $(1 \times 10^{-45})^{1/3} = 1 \times 10^{-15}$ J of Equation 11.10-12 and 2.25-6.

Coulomb Law applied to derive special relativity force of gravitation

Force in terms of Coulomb's Law is the product of two charges over the square of their separation and $4\pi\varepsilon_o$. Earth balances gravitation is assumed to occur with the same number of electron charges from the sun and from the earth. That number for 6 orientations 6 $\times 10^{54}$ has been calculated in Equations 13.2 and 17.34. The gravitation force is the product of two pulses, the exact special relativity value needed correction for relativity and quantum (1+1/80) with π replaced by three. The converter to SI units 10^{-75} is the product of mass conversion of two pulses 1×10^{-30} and further conversion of the force, which is $(1 \times 10^{-30})^{1/2}$ because squaring force derives energy.

Using adapted Coulomb Law Special Relativity gravitation force derives for two equal forcers:

$$F_{Gs}=2\text{x}(6\text{x}10^{54})\text{X}(6\text{x}10^{54})/[4\text{x}3\varepsilon_{os}\text{x}(1.5\text{x}10^{11}\text{m})^2]\text{x}1.111 \ldots \text{x}(1+1/80)=3.00 \ldots \text{x}10^{22}\,\text{N}$$
(21.35)

Special Relativity gravitation expresses also earth gravitation force as three pulses generated in orthogonal orientations each of which is the product of two pulse in wave function format expressed by product energy of 1 J from earth, proton frequency 1×10^{12} Hz and linear probability 1×10 generating:

$$F_{Gs} \equiv 3 \times (1\,\text{J} \times 1 \times 10^{12} \times 1/10)\text{X}(1\,\text{J} \times 1 \times 10^{12} \times 1/10) = 3 \times 10^{22}\,\text{N}$$
(21.36)

Ultimate find

The proton mass at the atomic level of the classical SI value 1.6726×10^{-27} kG corrected for absence of electon oscillations, for relativity and for quantum derives 1.6726×10^{-27} kG/$(1.1111111 \times 1.04^{9/100})\text{x}2/3=1.0000298\text{x}10^{-27}$ kg. This is the SI integer model value with uncertainty $30h_s$ equal h_s per each of 30 quantum orientations. The product of the integer model radiation mass 1×10^{-27} kg and squared velocity of light corrected by relativity derives exactly the gravity constant

$$1\text{x}10^{-27}\,\text{kg} \times (3\text{x}10^8)^2\text{x}1.111 \ldots = 1 \text{ x}10^{-27}\text{x}10^{17}=1 \times 10^{-10}\,\text{J}$$
(21.37)

Reminder $1/(\varepsilon_{os}\mu_{os}) \equiv 10^{17}$. Hence the international gravity constant is either **1/2 mass of proton or ten masses of muo10_s in integer SI units converted to energy by the squared velocity of light.** This statement is also not quite correct because conversion should be expressed at the local velocity of light. The $G_s =1\text{x}10^{-10}$ J constant has to be again multiplied by 2/3 to express sun to earth gravitation because this factor in classical terms applies to both interacting masses. **Energy $1\text{x}10^{-27}$ J or W-s spread in 4 space orientations** times the life of the Universe in seconds and by relativity and quantum corrections **derives the international integer model gravitation constant** power inclusive of independent electron energy:

(21.38)
$$\textbf{1 x } 10^{-27}\textbf{x13.716 x } 10^9 \textbf{ x 31.557 x } 10^6\textbf{/[4 x 1.1547}^{1/2} \textbf{ x (1+1/136)]=(1.0-0.00035) x } 10^{-10}$$

Thus age of the Universe is derived from the stability of the proton structure, which according to the integer model, is a congregation of electrons, as shown in Fig.6. The variance $4\text{x}10^{-14}$ J is the minimum energy in four directions.

Minimum Entropy Production Principle generate powers of Universe, Universe is in a very long lasting equilibrium and not in chaos

Generation of forces requires energy sources and **the source of energy remained unclear up to the end of the project.** The integer cubic space model postulates presence in a Universe being unaccounted for **Energy and Power.** The sun is the source of energy most likely caused by the thermonuclear processes within its body. This is displayed by the sun critical temperature of 6000 K, with pulses combining to 10000 K.

Classical SI suggests that mass generates waves and particles in different forms, without detailing the source of energy. **Most energy exchanges involve reductions of energy in one body being equal the increment of energy of the other body.** Equations presented in the book indicate accurately that **the gravitation force from earth expresses energy released due to conversion of mass of protons into equivalent electrons, which oscillate at doublet infrared frequency of protons and (proton+ electron) pair and display expected radiation kinetic energy. The energy generated**

$$G_s = 1\text{x}10^{-27} \text{ kg} \times (3 \times 10)^2 \times 1.111\ldots = 1 \times 10^{-10} \text{ J} \rightarrow \times 1 \times 10^{-23} = h_s = 1\text{x}10^{-33} \text{ J} \qquad (21.39)$$

Since all SI terms express the probability of occurrence of the chosen reference term the meaning of some equivalencies becomes apparent. Thus $\varepsilon_{os} = 1\text{x}10^{-11} = (h_s)^{1/3}$ is the probability of three orthogonal orientations converted to one direction.

Energy of 1 J per kilogram at 300 K, expressing h_s at the atomic level per orientation, is generated by most of the masses of the Universe. Energies and moments of the planets are cancelled by the energy of the sun, with part of the vectorized energy of the sun being dispersed into infinity, as predicted by the Second Law of Thermodynamics. The age of the Universe is derived from energy available in the protons of the Universe appearing in Universe creation and dispersed in the above manner.

The unaccounted energy and power in the Universe is derived from the Minimum Entropy Production Principle discussed by ET Jaynes [65] and I Prigogili [66]. These authors explain that thermodynamic equilibrium expresses expenditure of energy of 'least dissipation' related with minimal increase of entropy. The system is in equilibrium and still expends energy. The value expressed per 1 K and per atom is minute 1 h_s. It becomes very significant at higher temperatures and masses and becomes the source of gravitation and exhaust the proton specifying the age of Universe existence.

The Universe is in equilibrium and in perfect balance as specified by the laws of physics. The concepts of minimum entropy production have been developed nearly 100 years ago. **The important conclusion is that the Universe is not in chaos, it is in a very long term perfect balance**. In relation to the number of stars in the Universe their explosions are extremely rare.

With 6×10^{54} electrons on earth (Equation 17.34) Sun pulse of 6000 K expressing its rate delivers minimum entropy vectorized energy h_s generating the exact gravitation force:

$$6000 \text{ K} \times 6 \times 10^{54} \times h_s \times 1\text{x}10^{-3} = 3.6 \times 10^{22} \text{ N} \qquad (21.40)$$

SUN GRAVITATION EXPRESSES MINIMUM ENTROPY PRODUCTION PRINCIPLE

22. *Fitting integer cubic space physical constants into the classical rules of physics*

Several assumptions of the model could be treated as additions to the physics not requiring major changes to the classical SI system. The good agreement of the model with the experimental data presented in the three books indicates a very high probability of the basic concepts being correct. Classical physics attributes energies in hydrogen atom to electrons, while the model appropriates equal energies to the electron, the proton and the (electron+proton) pair, creating an inconsistency requiring clarification. Some textbooks admit that the reduced electron is responsible for energies, but do not explain the reason for the infrared frequency and the proposed energy balance is hazy. Hartree derives total hydrogen energy to be 27.24 eV value with some scientists identifying it with twice the ionization 13.62 eV. The integer model attributes three 10 eV values in three directions making a total 30 eV. However the 10.2 eV represent also the first stage excitation energy for n1→n2, both identifications being correct, but representing different events. To validate the integer model the proposed integer physical constants of the cubic geometry will be applied to several classical SI equations of physics textbooks. Equations quoted were collected from different sources, representing the efforts by their authors to find new derivations of physical expressions. To obtain the required values corrections will be applied justifying if possible any disparity observed.

Energy of Bohr atom and the atomic force

To obtain 10 eV of the integer model the basic expression for energy quoted by Beiser [34, Equation 4.14] and others is:

$$E_n = -e/(8\pi\varepsilon_o \times 1/2a_o) = -13.60569224 \text{ eV}.$$

In integral model units E_{n10} has to be multiplied by $5 \times 1.04^{1/7}$

$$E_{n10} = -5 \times 10^{-19}/ (8\pi \times 1 \times 10^{-11} \times 1/2 \times 10^{-10}) \times 1.04^{1/7} = -5 \times 2.0006 \text{ eV} \approx -10 \text{ eV} \qquad (22.1)$$

Thus in the integral model SI energy in each of the five pathways is:

$$E_{n2} = -e_s/(8\pi\varepsilon_{os} \times 1/2 \times a_{os}) \approx 2 \text{ eV} \qquad (22.2)$$

The atomic force in Bohr atom $F_{atomic} \times 1/2a_o$ should also derive atomic energy. For best accuracy the following relation is chosen confirming the above calculation, where 10 eV is in three orthogonal directions:

Atomic energy $= 1 \times 10^{-7} \times 1/2 \times 10^{-10}/\sqrt{3} \times 1.04 = 3.002$ eV, which for three planes time 1.111 . . . derives 10.007 eV (22.3)

Energy 13.6 eV in two orientations time relativity correction derives 30 eV:

$$13.6 \times 2 \times 1.111\ldots/1.04^{1/5} = 30.0 \text{ eV} \tag{22.4}$$

Classical atomic force in SI units and in terms of the integer model SI units is:

$$\tag{22.5}$$
$$F_{atmclas} = e^2/[4\pi\varepsilon_o \times (1/2\ r_{at})^2] = 1\times10^{-38}/[4\pi \times 1 \times 10^{-11} \times 1/4 \times 10^{-20}] = 1/\pi \times 10^{-7} = 0.318310 \times 10^{-7} \text{N}$$

Clearly this force is 1/3 of $F_{atomics}$. $F_{atomics}$ with quantum corrector $1.04^{5/4}$:
$$F_{atomics} = 0.318310 \times 10^{-7} \times 3 \times 1.04^{5/4} = 1.00095 \times 10^{-7} \text{ N}$$

$$\tag{22.6}$$

$$F_{atomic} = \mu_{os}/2\pi \times 1/(1.1547^3 \times 1.0325) = 0.99909 \times 10^{-7} \text{ per conductor}$$

$$\tag{22.6a}$$

Using $F_{atmclas}$ and replacing π by 3:

Energy $= 1/3 \times 10^{-7} \times 1/2 \times 10^{-10}/(1 \times 10^{-19} \times 1.111\ldots^2 \times 1.04^{1/5}) = 13.61$ eV $\tag{22.7}$

Now 13.60 eV $\times 3 \times 1.111\ldots^2/1.04^{1/5} = 49.98$ eV ≈ 50 eV $\tag{22.7}$
and energy is:

$$E_{ns} = h_s v_{Hms}/e_s = 1.10^{-33} \times 5 \times 10^{15}/1 \times 10^{-19} = 50 \text{ eV} \tag{22.8}$$
While

10 eV$\times 1.333\ldots \times 1.039721^{1/2} = 13.5955$ eV, which is within 0.07% of the classically value derived above. However Ingram [30] quotes value 13.597 eV.

The expression for Bohr atom energy using integer cubic space constants derives

$$E_{ns} = m_{es} e_{as}^4/(8\varepsilon_{os}^2 h_s^2) = 1 \times 10^{-30} \times (10^{-19})^4/(8 \times 10^{-22} \times 10^{-66}) = 0.125 \times 10^{-11} \text{ Jat}$$

Converting to SI units, to 300 K with relativity correction derives:

$$E_{ns} = 0.125 \times 10^{-11} \times 1 \times 10^{30} \times 1 \times 10^{-19} \times 300 \text{ K}/1.333\ldots = 28.125 \text{ J} = 30 \text{ J} - 1.75 \text{ J} \quad (22.9)$$

The variance expresses 116 DOF of $1/2 \times 10^{-23} \times 300/1 \times 10^{-19} = 1.74$ J. Radiation is a two stepped process involving two atoms involving 4 events of 29 DOF =116 DOF from spherical H atoms of 32 DOF. The integer space model identifies the mass 1×10^{-30} kg with 3 geometric masses of (proton+electron) pairs which vibrate with infrared frequency 1×10^{12} Hz. The Hartley energy corrected for relativity per kelvin displays energy over integer value 3O eV (27.2138 $\times$ 1.1547/1.0328 - 30)/300 K =32.6 k_s expressing k_s per quantum position of H atom. Components of quantum energies $2/3 \times 1/2 \times 10^{-27} \times 300/1\times10^{-19} = 0.01$ J and (30-30/1.1111)= 0.01 J represent both probability and the pulse formation originating from several components and involving velocity changes from $\frac{1}{4} c_s$ to $\frac{1}{2} c_s$ to c_s. These involves immeasurable component energy steps of relativity and uncertainty and the measurable total.

Stephan-Boltzmann constant and radiation energy in space

Incorporating the integer cubic constants into the classical expression given by Beiser [34, p. 308] and others:

$$\text{Energy density } \sigma \times T^4 = 8\pi^5 \, k^4/(15 \, c^3 h^3) \times T^4 \tag{22.10}$$

with c_s, being in the model equal 1/3 of velocity of light c_s, 10^8 m/s, and dividing by the quantum corrector 1.04^2 and times linear converter to SI 10^{10} and 2/3 electron absence factor derives with integer constants:

$$\text{Energy density} = 8\pi^5 \times (10^{-23})^4/(15 \times 10^{24} \times 10^{-99}) \times 10^{10} \times 2/3 \times T^4/1.04^2 = 1.006 \text{ J/m}^2 \times T^4 \approx 1 \text{ J/m}^2 \times T^4 \tag{22.11}$$

It derives power density = Energy surface density/(velocity 1×10^8 m/s) = 5×10^{-8} W/m^2 $\times$ T^4. Above value is the Stephan-Boltzmann constant of $\sigma_s = 5 \times 10^{-8}$ W/m^2 proposed in Table 1.

Stephan-Boltzmann radiation from the surface is directed in all star-way directions in one orientation, only components at right angle to the surface being able to generate the signal along the straight source-receiver axes. These radiations $\sigma_s \times T^4$ form a semi sphere and the signal moving along the source-receiver axis is represented by the volume of the semi sphere over the area of its base. At 1 K, particles depart at velocity c_s and move generating the radiation power, being treated as converted from atomic to integer SI units:

$$2/3 \, \pi \, (1 \text{ m}^3)/[\pi(1 \text{ m}^3)] \times \sigma_s \times T_{1K}^4 \times c_s = 2/3 \times 5 \times 10^{-8} \times 1^4 \times 3 \times 10^8 = 10 \text{ W/m} \tag{22.12}$$

At 300 K: $5 \times 10^{-8} \times 300^4/(1.33 \ldots \times 1.04^{1/3}) = 299.8$ W/m^2, which per kelvin degree is ~1W/m^2

Stephan-Boltzmann radiation expresses wave function format of two pulses:

$$(\sigma^{1/2} \times T^2) \text{ X } (\sigma^{1/2} \times T^2) \text{ each generated by } (\sigma^{1/4} \times T) \text{ X } (\sigma^{1/4} \times T) \tag{22.12a}$$

where $\sigma_s^{1/4} = (5 \times 10^{-8})^{1/4} = 0.014953487 \rightarrow \times 2/3 = 0.009969 \approx 0.01$ W (or eV/per atomic event) Planck radiation reduced due to absence of electrons in radiations on earth per K. Also $\sigma s^{1/4}/1.111 \ldots = (2 + 1/80) \times 10^{-4}$ J.

The variance is 3.1×10^{-24} J$\rightarrow \times 1/1000 \times 1.111 \ldots \approx 3.333 \ldots \times 10 \times 10^{-27}$ J $\approx 1/3 \, k_{selectr}$ expresses the newly proposed true electron 1/3 Boltzmann constant.

Space impedance, wave impedance, characteristic impedance and surge impedance

Physics textbooks are shy in deriving this value. Classical SI units derive
$$Z_c = \sqrt{(\mu_o/\varepsilon_o)} = 376.730 \ldots \text{ ohms}$$
In integer cubic SI units
$$Z_c s = \sqrt{(\mu_{so}/\varepsilon_{so})} \times 1.1547 \times 1.0328 = 377.125 \text{ ohms} \tag{22.13}$$

Bohr magneton

The classical Bohr magneton energy $eh/(2m_e \times 2 \times \pi) = 0.927415 \times 10^{-23}$ JT^{-1} corrected for relativity and quantum identifies in integer constants the energy of Bolzmann constant:

Bolzmann constant $\equiv 0.927415 \times 10^{-23}$ JT^{-1} x $1.1547^{1/2}$ x $1.04^{2/25} \equiv 0.99973 \times 10^{-23}$ J $= k_s$ (22.14)

    The variance 28×10^{-28} J expresses 28 DOF of orientations in the spherical structure energy of one mu10ns per DOF.

Bohr magneton equation expressed in terms of integer cubic SI constants with relativity and quantum corrections $e_s h_s/(2m_{es} \times 2 \times \pi)$ and π replaced by 3 is:

(22.15)

$u_{Bs}' = e_s h_s/(2m_{es} \times 4 \times \pi) \equiv 1 \times 10^{-19} \times 1 \times 10^{-33}/(2 \times 10^{-30} \times 4 \times 3)$ x1.33 . . . /1.11 . . . =0.5000 . . . x10^{-23}J=1/2k_s

expresses energy of one degree of freedom.

Rydberg constant

The value of the Rydberg constant in integer units converted to SI and per one direction =
$m_e e_s^4/8\varepsilon_o^2 c_s h_s^3 = 10^{-30} \times (10^{-19})^4/[8(10^{-11})^2 \times 3 \times 10^8 (10^{-33})^3] \times 10^{-30}/4 = 10.01666 \times 10^{-7}$ m^{-1} (22.16)

Table 1 show the Rydberg constant is essentially integer model value times 1.111 . . . over $1.04^{1/3}$. The value calculated using integer cubic constants, with c_s replaced by 1/3 of the light velocity, derives ~1/8 x10^{-7} m^{-1}, which could be connected with each of 8 corners of the cube their sum making the 1x10^{-7} m^{-1} value, which is some 10% smaller than the SI value.

    It is not easy to balance of forces within the Bohr atom, because although the forces between the proton and electron are equal, the third oscillator (electron+proton) is neutral and it carries energy. The equality in atomic terms is between the momenta of the three oscillators and it is not instantaneous because it takes some 1000 interactions to equalize momenta between the electron and proton. This balance is lost at the instance of radiation.

Particle energy between two rigid walls L m apart

Using SI units derives for n = 1 the minimum energy: (22.17)

Energy=$n^2 h^2/(8m_e L^2) \times 1/e = (6.626 \times 10^{-34})^2/(8 \times 9.1094 \times 10^{-31} \times 1.1201 \times 10^{-20})/1.602 \times 10^{-19} = 33.57$ eV.

    For a deep potential well R. Einsberg [49, Equation 8-97) using Schrodinger Equation derives the minimum energy level, with L equal the Bohr diameter, the value of a_o. The equation quoted by Einsberg using the integer cubic model parameters derives:

Energy= $h_s^2/(8m_{es} L^2) \times 1/e_s = 1 \times 10^{-66}/(8 \times 1 \times 10^{-30} \times 10^{-20}) \times 1/1 \times 10^{-19} = 125$ eV $\equiv 1/8 \times 10^{-16}$ J $\equiv$ $1.04^3/c_s^2$, the value $1.04^3/c_s^2$ varying from 1/8 x10^{-16} J only by uncertainty h_s (22.18)

It is suggested that 125 eV expresses n=2 deriving 4 x33.57 eV=134.29, which is 125eV relativity corrected 125 x 1.1547$^{1/2}$ = 134.

Energy 1/8 x 10^{-16} J time linear probability derives in space cubic unit's radiation pulse of

$$1/8 \times 10^{-16} J \times 10^{-10}/1.111 \ldots^2 = 1.003 \times 10^{-27} J \qquad (22.19)$$

Cubic Bohr atom wavelength

Classical textbook SI value of $\lambda = h/e \times \sqrt{[(4\pi \times \varepsilon_o \times 1/2 \times a_o)/m_e]}$ derives 3.3 x 10^{-10} m, that is the circumference of the electron orbit deriving classical atom diameter at 1.050422 x10^{-10} m, which relatively corrected by 1.111 . . .$^{1/2}$ derives ~1 x10^{-10} m. Inserting space cubic constants with 3 for π into the classical expression:

$$\lambda = 1 \times 10^{-33}/1 \times 10^{-19} \times \sqrt{[(4 \times 3 \times 1 \times 10^{-11} \times 0.5 \times 10^{-10})/1 \times 10^{-30}]} = 7.7460 \times 10^{-10} m \qquad (22.20)$$

Inspection of the cube indicates that the correct $\lambda = 1 \times 10^{-10}$ m should fit between the side of the box in six directions or along twelve edges. Since all those dimensions are uncertainties it is really unimportant how they fit, but the signal is 1 D so they should be vectors. Thus in cubic space units:

$$\lambda_s = 7.7460 \times 10^{-10} m/(6 \times 1.11 \ldots^2 \times 1.04^{8/7}) = 0.9999 \times 10^{-10} \approx 1 \times 10^{-10} m \qquad (22.21)$$

The wave length is related to electron velocity $\lambda = h/m\square$, which in integer cubic units is:

$$v_{el} = 1 \times 10^{-33}/(1 \times 10^{-30} \times 1 \times 10^{-10}) = 1/2 \times 10^7 m/s \qquad (22.22)$$

The energy is:

$$1/2 \times 1 \times 10^{-30} \times 1/4 \times 10^{-14}/1 \times 10^{-19} = 125 \text{ eV of incoherent energy as above,}$$

$$\text{which time } 1/10 \times 1.1547^{1/2} \times 1.04^{1/3} = 13.609 \text{ eV} \qquad (22.23)$$

The 125 eV energy is also derived using reformulated Einstein's Equation for deep potential well with $n = 1$ expressed in integer space units:

$$E_n = h^2/(8m_e a_o^2)/e_s = 1 \times 10^{-66}/8 \times 10^{-30} \times 10^{-20})/1 \times 10^{-19} = 125 \text{ eV} \qquad (22.24)$$

$$\text{The energy 125 eV} \times 8 = 1000 \equiv 1/6 \times 6000 \text{ K} \qquad (22.25)$$

The vector of energy is directed along the longest axis of the cube formed by discharges of a cube of 100 Bohr atom group.

The longest wavelength λ_p of the Pashen series according to the classical estimate using Rydberg constant and $n = 3$ and 4 is:

$$\lambda_p = 1.097 \times 10^7 (1/9 - 1/16) = 18,752 \text{ A} \qquad (22.27)$$

The estimate using the model integer constants at 1 K is as follows:

$$\lambda_p = h_s/m_{ps} \, v_{Lms} = 2 \, x10^{-19} \, \text{m} \tag{22.28}$$

That value at 300 K for three planes times quantum correction $10^{10} \, x300x3x1.04 = 18{,}752$ A is identical with the previous value. For three planes it expresses energy of three oscillators, corrected by relativity and quantum $1.111\ldots^2 x1.04^{1/20}$ and it derives the sun constant:

$$18{,}752 \, \text{A} \, x3x3x10^8/1.111\ldots x \, 1.04^{1/20} \equiv 1370 \, \text{W/m}^2 \tag{22.29}$$

Another expression for the invert of electron wavelength $1/\lambda_{es}$ with cubic space values inserted converted to 300 K is:

$$1/\lambda_{es} = m_{es} \, e_s^{\,4}/8 \, \varepsilon_o^{\,2} \, c_s \, h_s^{\,3} = 1x10^{-30} \, x10^{-76}/(8 \, x10^{-22}x10^{-99}) = 0.41666\ldots 10^6 \, \text{m}^{-1} \tag{22.30}$$

$\lambda_{es} \, x \, 300 = 720 \, x10^{-6}$ m and the electron frequency energy is closely relayed to the infrared frequency:

$$E_n = \lambda_{es300} \, x \, 5 \, x \, 10^{15} = 5 \, x \, 10^{12}/(1.1547^2 \, x \, 1.04) \, \text{J} \tag{22.31}$$

Ingram [30, Equation 2.27] quotes from Rayleigh-Jeans Law the energy contained in the volume $a_o^{\,3}$ of black body as $E_n = 8\pi kT\lambda^{-4}$. Using space cubic constants, replacing π by 3 applying relativity corrector and space quantization $1/10^2 \, x \, 1/10^2$ obtains:

$$E_n = 8x3 \, x2x10^{-23}x300 \, x \, (1/18752 \, x10^{-10})^4 \, x \, 1/10^2 \, x \, 1/10^2 \, 1/1.1547 = 0.993 \approx 1.00 \, \text{J} \tag{22.32}$$

The temperature of 10000 K expresses 6 corrected sun constants:

$$10000 \, \text{J} \equiv \sqrt{2} \, x \, 6 \, x \, 1370/1.1547 \, \text{W/m}^2 \tag{22.33}$$

The wavelength 18,752 A times electron frequencies, velocity of light, referred to 300K, converted to electron volts over $\sqrt{3}$ delivers energy of 5.000 eV:

$$18{,}752 \, x \, 10^{-10}x5 \, x \, 10^{15} \, x \, 3 \, x \, 10^8 \, x \, 1.04^{\,2/3}x \, 10^{-19} \, x \, 300/\sqrt{3} \equiv 5.00092 \, \text{eV} \tag{22.34}$$

The relation $10 \, \text{eV}x\sqrt{3}x10x10x10 \equiv 18752$ A is taken to indicate, that once the potential barrier is broken, nearly all incoherent radiation is converted into a vector directed in one direction.

Diffraction of particle beam by a crystal

According to de Brogie rule a beam of electrons having energy 54 eV, on striking a single nickel crystal, diffracts generating its maximum intensity at an angle 50° displaying a wavelength of 0.165 nm.

The measured experimental values of 0.167 nm matches the prediction of Braggs Law

$n\lambda= 2d\ x\ \sin(90-\theta/2)$ for $n=1$, $\theta=50°$ and the spacing of nickel crystal of 0.091 nm

On striking a single nickel crystal the beam of electrons having energy 54 eV diffracts according to De Brogie formula, $\lambda=h/m_e v_e$. The formula used with the integer space units and converted to SI derives:

$$\lambda_s=h_s/m_e v_{es}, =1\times10^{-33}/(1\times10^{-30}\ x1\times10^8)= 0.1\times10^{-10}\ m \qquad (22.35)$$

The energy expresses the minimum value $E_n =m_e v_{es}{}^2=1\times10^{-30}\times1\times10^{16} =10^{-14}$ J as indicated earlier. Since wave length increases with square root of the energy:

$$\lambda_{s54}=0.1\times 10^{-10}\ m\ x\ \sqrt{54}= 0.7385\times 10^{-10}\ m \qquad (22.36)$$

Using multiplier $\sqrt{2}$ derives the Bohr atom diameter $\sqrt{2}\times0.7385\times10^{-10}/1.04^{1.1}=1.0003$ $\times10^{-10}$ m, while relativity corrections $1.1547^2 \times 1.1111^2$ derives:

$$0.7385\times 10^{-10}\ m\ x\ 1.\ 1547^2 x\ 1.1111^2=1.646\times 10^{-10}\ m \qquad (22.37)$$

That value is predicted by de Brogie relation and it is confirmed by the measurements of Devidson and Germer [see WEB].

The frequencies of atomic oscillators expressed by space cubic constants

Proton infrared frequency is defined by:

$$v_{Lms}=1/2\pi\ x\ \sqrt{\text{(Boltzmann constant/mass of object)}}$$

Replacing π by three, applying the mass to proton moving at velocity $1/3\ c_s$ with its relativity correction 1.1111 . . . and applying the probability 10^{-20} derives:

$$v_{Lms}=1/(2x3)x\sqrt{[2x10^{-23}/(2\ x10^{-27}x1.111\ldots10^{-20}]}=5x10^{12}\ Hz \qquad (22.38)$$

The need to use the probability 10^{-20} suggests that the forward movement with oscillations have to be treated as having two degrees of freedom during formation.

The squared product of: energy 30 J, relativity correction 1.111 . . .$^{3/2}$, velocity c_s over space frequency $3x10^{17}$ Hz (expressing probability $1x10^{-9}$) and c_s derives the minimum energy $1x10^{-14}$ J with uncertainty $1/3\ k_s$ J:

$$h_s/e_s=[30\ Jx1.111\ldots^{3/2}\ x3x10^{-17}x3x10^8/3x10^{-7}]^2= 0.999\ 999\ 999\ 7\ x10^{-14}\ J \qquad (22.39)$$

In ten pathways $1x10^{-14}$ J becomes the input of $1x10^{-13}$ J, which converted to electron volts derives:

$$\text{Sun pulse energy} = 1x10^{-13}\ J/1x10^{-19} = 1\ MeV \qquad (22.40)$$

$$\text{Energy 1 MeV}x1/10x1/10/\sqrt{3}x1.04 \equiv 6002.8\ K \qquad (22.41)$$

The energy required to bring the particle from 0 to L_{earth} is approximately given by Hooke's law:

$$E_n = 1/2 x k_{sp} x L_{earth}^2, \text{ where } k_{sp} \text{ is the spring constant} \qquad (22.42)$$

Bohr atom is the Boltzmann constant, being equal to unity at the atomic level. For three planes the resultant is exactly the reference:

$$\text{Force of gravitation} = 3/2 (1.5x10^{11})^2 x1.11111^{1/2}/1.04^{1/10} = 3.5436x10^{22} \text{ N} \qquad (22.43)$$

and the force density of 1 N/m² is obtained at the surface of sun from the radiation from earth:

$$5.9742x10^{24}kg/[\pi(1.4961x10^{11})^2]x1/10x1/10x1.1111^2x1.04^{3/2} = 1.0012 \text{ N/m}^2 \qquad (22.44)$$

In quantum terms the energy at the first level above absolute zero, with n = 0, is

$$E_n = (n+1/2)h_s v_{Lms}/e_s = 1/2 \ x \ 10^{-33} \ x \ 5 \ x \ 10^{12}/1 \ x \ 10^{-19} = 2.5 \ x \ 10^{-21}/1 \ x \ 10^{-19} = 0.025 \text{ eV}$$
$$(22.45)$$

And for electron frequency:

$$E_n = (n+1/2)h_s v_{Lms}/e_s = 1/2 \ x10^{-33} \ x \ 5x10^{15}/1x10^{-19} = 2.5x10^{-18}/1 \ x \ 10^{-19} = 25 \text{ eV} \qquad (22.46)$$

For two dimensions these expressions become 0.05 eV and 50 eV:

$$E_n = h_s v_{Lms}/e_s = 1 \ x \ 10^{-33} \ x \ 5 \ x \ 10^{12}/1 \ x \ 10^{-19} = 0.05 \text{ eV} \qquad (22.47)$$

$$E_n = h_s v_{Hms}/e_s = 1 \ x \ 10^{-33} \ x \ 5 \ x \ 10^{15}/1 \ x \ 10^{-19} = 50 \text{ eV} \qquad (22.48)$$

Otherwise atomic force $1 \ x \ 10^{-7}$ over distance $1/2 \ x \ 10^{-10}$ derives:

$$E_n = 1 \ x \ 10^{-7} x \ 1/2 \ x \ 10^{-10}/1 \ x \ 10^{-19} = 50 \text{ eV} \qquad (22.49)$$

Relation between the mass of sun and the mass of earth

The mass of sun, which in kinetic atomic units is also its energy, is equal to the quantized $1/10 \ x \ 1/10$ momentum of 1/3 of the mass of earth time the local velocity of one third of the velocity of light; the first one third is likely to express the interaction occurring only in one plane:

$2 \ x \ 10^{30} \text{ kg(mass of sun)} \equiv 1/3 \ x \ 6 \ x \ 10^{24} \text{ kg(mass of earth)} \ x \ (1/3 \ x3x10^8) \ x \ 1/10 \ x \ 1/10$

This derives the sun gravitation force in one plane $\sqrt{3} \ x2x10^{30}/1x10^8x1.04^{3/5} = 3.5466x10^{22} \text{ N}$

The nine local velocity terms form three groups of three, each in three radiation orthogonal orientations. Velocity expresses the rate being the number of 1 J pulses. The pulse wavelength is 1 m with 1 J/m and earth must generate this number in the square of its linear velocity:

$$3 c_s = 3 \times 3 \times 10^8 \equiv 30000^2 \tag{22.50}$$

Schrödinger Equation

The first energy level, $n = 1$, is for Bohr atom in steady state using space cubic constants:

$E_n = 4m_{es} e_s^4/(32 \varepsilon_o^2 h^2) \times 1/e_s \times 1/(1.1547 \times 1/1.04^{1/3})$
$= 1/8 \times 1 \times 10^{-30} \times (1 \times 10^{-19})^4/[(1 \times 10^{-11})^2 \times (1 \times 10^{-33})^2] \times 1/1 \times 10^{-19} \times 1/(1.11 \ldots \times 1/1.04^{1/3}) = 0.9994$ eV

The variance is a very acceptable 6 k_s (22.51)

If the relativity and quantum corrector used are $1/(1.1547 \times 1/1.04^2)$: $E_n = 1.00086$ eV. Since $1/2 k_s T/e_s = 1/2 \times 2 \times 1 \times 10^{23} \times 300/1 \times 10^{-19} = 0.03$ eV the above results are accurate within Heisenberg uncertainty.

Replacing electron mass by radiation proton mass $1/2 m_{ps}$ derives value $1/1000$

$$E_n = 0.001 \text{ eV} \tag{22.52}$$

Planck radiation law

Planck had great difficulty himself deriving and describing his law. One form of expression, which fits perfectly with integer model constants and has been quoted by Serway and ales [32 p. 63] is the energy per unit volume per unit frequency within the cavity:

$$E(f, T) = 8\pi f^2 kT/c_s^3$$

Replacing for the cube π by 3, dividing by relativity square and using infrared frequency derives for 300 K a perfectly accurate value of $e_s \equiv 1 \times 10^{-19}$ J $\equiv 1$ eV:

$E(f, T) = 8 \times 3 \times (5 \times 10^{12})^2 \times 2 \times 10^{-23} \times 300/(2.7 \times 10^{25} \times 1.33333) = 1.00000 \ldots \times 10^{-19}$ J $= 1$ J

Energy converted to radiation vector at atomic level 0.001 J and to SI 0.01 J   (22.53)

Relating author's infrared Bohr atom with classic atom based on electron energies

Author's conduction and relaxation experiments [2, 3, 4], applying the Bohr atom model to divalent impurity in a NaCl crystal and to other alkali halides derived valid atom dimensions, valid dissociation energies, valid frequencies, and valid specific heat values. They were expressed in Bose-Einstein Statistics implying sole infrared frequencies to the atom being in total conflict with standard textbook attributing with some reservations atomic energy to electron energies. This conflict has been now resolved. Present research was initiated by such conflicts.

When 0.9994 eV of Equation 2.51 calculation is repeated using proton mass $m_{ps} = 2 \times 10^{-27}$ kg its value has to be multiplied by 5 to express space quantization for a moving particle. The result gives:

$$E_n = 4\, m_{ps} \times 5 e_s^4/(32\, \varepsilon_o^2 h_s^2) \times 1/e_s \times 1/(1.1547 \times 1/1.04^2) = 0.0100086 \text{ eV} \tag{22.54}$$

How can these two values be related? It is suggested that the concept of equation expressing equivalent, or reduced, mass m1m2/(m1+m2) applies equally well to equivalent energy.

The two energies 1 eV and 0.001 eV are represented then by 1x0.001 (1+0.001) = 0.000999 eV. Consequently the electron and proton produce very nearly the same effect as the electron itself. The present model uses both frequencies for the proton and for the electron. Using frequency of slightly corrected value is sufficient to describe all the phenomena for hot hydrogen, but fails at room temperature. This is the reason for classical physics not being incorrect by neglecting usually proton oscillations in describing Bohr atom energies; the procedure is a good approximation. Values added by the model dealing with both oscillations do not conflict with present physics, but add new information.

Energy of 10 J coming from the sun in classical SI units is 1 J at the atomic model level at 300 K. At 1 K they become 0.01 J of Planck radiations per orientation in classical SI and 0.001 J at the atomic model level. They are the same inferred infrared energies.

Classical SI electron spin is in the integer model equivalent electron spin$_s$

The classical SI angular electron spin momentum measured by Goldsmith and Uhlenbeck in 1925 measurents:

$$\text{Electron spin} = \sqrt{3}/2 \times h/2\pi \approx 0.1378\, h \tag{22.55}$$

In model SI cubic units π is replaced by 3, h by h_s deriving with relativity correction:

$$\text{Electron spin}_s \text{ energy} = \sqrt{3}/2 \times 1/6\, h_s \tag{22.56}$$

The value $\sqrt{3}/2$ is the invert of relativity correction 1.1547 for the increased mass (reduced or equivalent) and momentum of the electron+ proton couple decreasing the displayed value of h_s because static equivalent electron in the radiation pulse is moving at integer model velocity 1/2 c_s:

$$\sqrt{3}/2 = 0.866025403 \rightarrow \times 1.154700538 = 0.999\,999\,999\,5 \tag{22.57}$$

Thus for 6 independent orientations **the total value of spin$_s$ is h_s, the value generating the correct value of the force of gravitation in Equations 7.50-51.**

Thus 1925 spin experiments can be considered to be a demonstration of the spin display of $h_s/6$ per orientation and of the presence of the local electron linear

velocity 1/2 c_s, **the basic postulate of the model. This is the spin of the equivalent (reduced) electron meaning the electron-proton couple oscillating at proton frequency. It suggests that local velocities can be treated both as independent components of velocity of light in orthogonal orientations and contained in the atomic structure.**

In the model integer units the angular electron momentum is given by electron mass of 1×10^{-30} kg, its spin velocity 1×10^{-7} m/s and radius of rotation $1/2 \times 10^{-10}$ m generating $1/2 \times 10^{-33}$ kg-m^2/s $\equiv 1/2\ h_s$. Self spin adds another $1/2\ h_s$ making **a total h_s.**

Larmor formula

In classical SI units Larmor formula expresses total power radiated by a point charge:

$$P = e^2 a^2/(6\pi\varepsilon_o c^3);\ \text{in integer cubic model units } P_s = (10^{-19})^2(1/137.067)^2/(6\pi \times 10^{-11} \times 10^{24}) = (\sqrt{3}e_s)^3.$$
(22.58)

Thus from Equation 1.15, P_s takes now a simpler form and a more specific meaning:

$$P_s = (\sqrt{3}e_s)^3 = (\sqrt{3}m_{ps1D} \times 1/3 \times c_s^2)^3 = (\sqrt{3} \times 10^{-19})^3 = 5.19615 \times 10^{-57}\ \text{W; corrected for relativity}$$

$$P_{sr} = (\sqrt{3}e_s)^3 \times 1.154700538 = 5.999\ 999\ 998 \times 10^{-57}\ \text{W} \approx 6/(\sqrt{3}\ e_s)^3$$
(22.59)

while it's classical value is:

$$P = (9.10938 \times 10^{-19})^2(1/137.036)^2/(6\pi \times 8.85418 \times 10^{-12} \times 27 \times 10^{24}) = 9.80607 \times 10^{-57}\ \text{W}.$$

It is suggested that to bring the integer value to the classical SI value a correction is needed for relativity and quantum expressed by multiplication by $1.154700^4 \times 1.111 \ldots^{1/2} \times 1.04^{1/6}$ deriving 9.80114×10^{-57} W. The product 9.80×10^{-57} W $\times 10 \times 1.04^{1/2} \times (1 \times 10^{15})^2 \approx 1 \times 10^{-27}$ J is the radiation energy, while $6/(\sqrt{3})^3 = 1.154700538$ is the relativity correction.

Thus **the integer cubic model displays P_s essentially as the cube of e_s or of proton momentum moving at local velocity 1/3 c_s, $(m_{ps1D} \times 1/3 \times c_s^2)^3$, confirming the** presence of 1/3 c_s and $P_s \approx e_s^3$; it shows that P_s is the product of probabilities of creation of momenta $m_{ps1D} \times 1/3 \times c_s^2$ **in three orthogonal directions; the latter expression should replace the classical mass x c_s^2, which hides the true events.**

Larmor precession frequency, (Serway 32, p. 302), expressed in integer cubic units is:
(22.60)
$$f_s = e_s B/2m_{es} \times 1/(2 \times 3) \times 1.1547 \times 1.04 = 1.00007 \times 10^{10}\ \text{Hz is the same value as in Equation 3.28.}$$

Compton Effect

Compton wavelength in classical SI units derived in 1921 is given as
$$\lambda_c = h/m_e c = 0.243 \times 10^{-11}\ m$$
while in integer cubic model SI units the local value λ_{cs} is ε_{os}:

$\lambda_{cs}=h_s/(m_{es}\text{x}1/3c_s) =1\text{x}10^{-33}/(1\text{x}10^{-30}\text{x}1\text{x}10^8)=1\text{x}10^{-11}$ *m* $\equiv \varepsilon_{os}=$ *energy of permittivity of space*
(22.61)

It is suggested that $4 \lambda_c$ **x1.04**$^{18/25}$=0.99984x10^{-11} *m* $\approx \lambda_{cs}$.

Expressing Compton Effect in terms of electron model SI Planck constant h_{se}:

$\lambda_{cs}=h_{es}/(m_{es}\text{x}1/3c_s) =1\text{x}10^{-39}/(1\text{x}10^{-30}\text{x}1\text{x}10^8)=1\text{x}10^{-17}$ *m* $\equiv$ **1/space frequency** v_{os} (22.62)

Bohr magnetron μ_{Bs}

Let μ_{Bs} refer to one orientation, and not to five as given in Table 1, and let the classical SI value, corrected for relativity and quantum with 2π be replaced by 6 as in most model SI expressions. Bohr magneton μ_{Bs} derives then equality with Boltzmann constant. Boltzmann constant k_s is μ_{Bs} in one dimension:

$k_s\equiv\mu_{Bs}=e_s h_s/(2\text{x}10^{-30}\text{x}6)=10^{-19}\text{x}10^{-33}/(2\text{x}10^{-30})$ **x1.111 . . . x1.04**$^{1/12}\equiv1.00074\text{x}10^{-23}$ J (22.63)

Current of one ampere according to the integer cubic model

The definition of one ampere has been changed a few times. Webster's New World New York Encyclopedia, 1990, defines one ampere as representing the flow of about 6.28x 10^{18} electrons per second. Encyclopedia Wikipedia, reports that the value proposed for adoption at the International Conference for Weight and Measures in 2013 will be the flow of 6.245093 x10^{18} of electron per seconds.

At temperature 300 K the unity values of Equation 1.2 do not contain electron radiations and have values 2/3. Unity value express mass conversion to energy of the equivalent electron mass 10^{-30} multiplied by the square of local velocity $(10^8)^2$, times equivalent infrared radiations of proton or (electron plus proton pair), times 1D size atomic converter to SI units 10^{10}, times frequency 10^{12} Hz to express per second value, over 10^8 atomic to SI converter deriving power of $10^{-30}\text{x}10^{16}\text{x}10^{12}\text{x}10^{10}/10^8 =1.000$ W. The product of that value over the electron charge 1 x10^{-19} and times 2/3 over the product of relativity, quantum corrections and electron charge derives $2/3\text{x}1/(1.111 \ldots^{1/2}\text{x} 1.04^{1/80})/e_s =$ 6.2414x10^{18} electrons/s the value agreeing within 6 parts in 10^4 with the proposed value at the above Conference. Dividing by the space frequency and by 1.2 derives ~51 eV, which is the sun pulse.

The force between long parallel conductors of length L [m] carrying current I [A] $F = \mu_o IL/2\pi$ [N] is a derivative from Ampere's Law. At $I= 1$ A and $L= 1$ m it generates the earlier Paris Reference force 2 x 10^{-7} N. The integer cubic model expressed in SI units generates the force $F_s = \mu_{os}I/2\pi$ x 1.111 . . . x 1.04$^{9/20} = 1.99986$ x 10^{-7} N.

The variance between the value proposed for the year 2013 and the model value $(6.2451-6.2414)\text{x}10^{18}/(3$ time electrical local frequency 1x10^{15} Hz x decrement 1.2 x1.1111$^{1/4}$) = 1.001 electrons/s expressing the limit of model accuracy in terms of quantum correction 1.04$^{1/40}$ = 1.00098, which is $5k_s$.

Gauss Law

Wikipedia states that Gauss Law is well represented by specifying the capacitance of parallel plate capacitor C = ε_o x Plate area (m^2)/Plate separation (m) F/m. For deriving essentially the same values of capacitance ε_o of the classical SI system has to be multiplied in three planes by relativity corrector $(1.111 \ldots^{1/2})^3$ and divided by quantum corrector $1.04^{37/4}$:

$$(22.64)$$

$$\varepsilon_{os} = \varepsilon(1.111\ldots^{1/2})^3/1.04^{37/40} = 0.885415 \times 10^{-11} \times (1.111\ldots^{1/2})^3/1.04^{37/40} = 1.00006 \times 10^{-11} \text{ F/m}$$

Temperature and velocity of light expressed as pulse rate

The reference of 1 W/m^2 is expressed as unity probability, a value that cannot be exceeded. The reference unit is expressed now by the unity pulse rate generating local velocity of light. Two 6000 K sun pulses required in wave function format structure to generate velocity of light are ten expressed in four orientations of 10^8:

$$(6000 \text{ K}) X (6000 \text{ K}) \times 1.111 \ldots \equiv 4 \times 10^7 = 1/10 \times 4 \times 10^8 \text{ m/s} \qquad (22.65)$$

The temperature of 6000 K time relativity correction and the decrement of transmission 1.2 in the source-receiver direction derives input of 10 000 K:

$$6000 \text{ K} \times 1.333 \ldots \times 1.2 = 10\ 000 \text{ K} \equiv (1/3\ c_s)^{1/2} \text{ m}^{1/2}/\text{s}^{1/2} \qquad (22.66)$$

$$\text{And } (10\ 000 \text{ K})^{1/2} = 100 \equiv 10 \times 10 \qquad (22.67)$$

expresses two planes having each ten optional orientations as derived by Stern and Gerlach in 1921 and shown in Fig.2. The input pulse rate of the source requires simultaneous three pulses for output of 10000 K is:

$$[(1/3\ c_s)^{1/3}]^3 \equiv 1 \times 10^{12} \text{ Hz} \qquad (22.68)$$

expressing the proton local frequency, that being very close the frequency of (electron + proton pair). Electron oscillation generate a local pulse rate of 1×10^{15}. Thus the sun temperature, proton frequency and local velocity of light form set of values connected logically to other parameters.

Meaning of the basic radiation formulas

In SI integer units the energy $q_s = h_s\ v_{Lms} = 10^{-33} \times 1 \times 10^{12} = 1 \times 10^{-11} \equiv \varepsilon_{os}$. At the atomic level the value decreases in 1D by 1×10^{-10}; after converting to electron volts it derives

$$1 \times 10^{-11} \times 1 \times 10^{-10}/1 \times 10^{-19} = 0.01 \text{ eV} \qquad (22.69)$$

expressing the first term of the Planck Radiation Law. Thus the Law refers to infrared frequency of 10^{12} Hz, inherently the proton frequency.

The mean wavelength of a spectrometer scale of 400 nm expresses 1/10 of the integer SI units the output energy of $10q = 100W$ (or eV) decremented by 20% of its no interacting input (2 out of 12 units) :

$$q = \lambda_s c_s /1.2 = 400 \times 10^{-9} \times 3 \times 10^8/1.2 \equiv 10 \text{ W} \equiv 10 \text{ J} \qquad (22.70)$$

Wavelength of 400 nm displays $\sqrt{2} \times 10/1.04=13.6$ eV or J pulses in one plane of Figure 2 expressing dissociation energy of the Bohr atom structure of 8.1 eV subjected to relativity correction $1.111 \ldots^2$ studied in Chapter 25 in relaxation of NaCl. Two pulses at SI level of energy of 100 W generate energy expressed by the $(100)^2 \equiv 10000$ K with probability $1/10 \times 1/10$ generate a vector pulse moving at the velocity of light with the value:

$$100 \text{ W} \times 1/10 \times 1/10 = 1 \text{ J} \qquad (22.71)$$

At the atomic integer level the 1D pulse is 1×10^{-10} smaller expressing the value of the gravitation constant $G_s = 1 \times 10^{-10}$ of Table 1 between two bodies both displaying infrared and electron frequency oscillations. Physically the temperature of the sun of 6000 K is spread in one plane of three planes of space and decremented by 20% of no reactive components so that 5000 K is spread in ten 500 K directions as thermal pulses converted to vectors of 50 J by linear probability 1/10 and waves augmented to 100 W as pulses with reflection. Thus the sun has in three planes 36 beams of 100 J which individually interact one with each of the planet generating at the speed of 3×10^8 m/s power of 3×10^{10} W. The values of 1 J, 10 J and 100 J are at the atomic level and the rotational conversion to SI units requires 10^{20} multiplier. This bring the value of the total sun power to 3×10^{30} W of which 2×10^{30} W interacts with earth as explained in Section 5.

Energy of 20 J or eV is quoted in some textbooks for hydrogen atom energy. Three such pulses would need an input $20^3 \equiv 8000$ K = 6000 K $\times$ 1.2 $\times$ 1.111 . . ., expressing true values because relativity and the missing radiation factor 1.2 are very real. **This consideration disclosed that 1.2 x 1.111 . . .= 1.1547^2 and the possibility of misjudgments, which have not been corrected.**

De Brogie Relation in integer SI model units

$$p_s = h_s k_s \equiv h_s k_s/(2 \times 3)= 1 \times 10^{-33}/(2 \times 3) \times 1 \times 10^{-23}=1.2 \times 1.111 \ldots^3 \times (1+1/80) \times h_s/10 \times k_s \qquad (22.72)$$

Crossed Planck constant h_s is expressed in SI model units by $h_s/10$ with corrections; $(1+1/80)$ is another form of quantum correction.

$$\text{Please note that } (h_s)^{1/2} \equiv 3 \times 10^{-17} \text{ Hz}/1.111 \ldots^{1/2} \qquad (22.73)$$

Paris International Force reference

Application of the force 2×10^{-7} N/A-m to the circumference of earth carrying earth current 1.8×10^{29} A in the integer model units derives energy 1×10^{29} J. The energy per meter in one of six directions appears to be $E_{sspaceSI}$:

Energy in earth circumference=Atomic force x distance
$$= 2 \times 10^{-7} \text{N/(A-m)} \times 2 \times \pi \times 1.5 \times 10^{11} \text{ m}/[1.04^{47/40} \times (10^{10})^{1/2}]$$
$$\equiv 2.00006 \text{ J/m or eV/m with variance } k_s \qquad (22.74)$$

Using, special relativity gravitation energy density for two oscillators remains 2 J/m²:

Energy $= 3 \times 10^{22}$ N/$L_{earth}^2 = 3 \times 10^{22}$ N/$(1.5 \times 10^{11}\text{m})^2 \times 1.154700/1.04^{2/3} = (2 - 0.0002)$ J/m²
with variance 1/5 k_s $\qquad (22.75)$

The amazing mass

In the c.g.s. system of units 1 Ampere is replaced by 10 A units called abampere. The factor of ten or 1/10 appears several times in the model: there are ten quantum orientations per plane; the force of 10 N/kg decreases to 1 N per orientation; the loss of sun mass in SI atomic units of 10^5 kg $\equiv 10^5$ W $\equiv 10^5$ J converts by $10^{20} \times 1 \times 10^{-19} = 10$ to 10^6 W, J kg pulse. This is conversion of electron volts expressed per atomic event to joules referring to one kilogram. Thus expressing 1 J at atomic level per radiation event represents 10 J at SI level per 1 kg, or SI Planck radiation term 0.01 J is expressed at atomic terms by 0.001 J. The model applied to the spherical structure has 32 quantum momenta orientations, ten per plane with quantum vectors moving at velocity c_s, derived from 1/3 c_s spinning each with h. For the molecule H2 the total energy:

$$64 \times 10^{-33} \times 1.154700^3 \times (1+1/67) = 1.00005 \times 10^{-31} \text{ J} \qquad (22.76)$$

Converted to 300 K by $(300)^2 \times 1.111 \ldots = 10^{17}$ K² derives minimum energy $1 \times 10^{-14} \times 10$ J in SI units expressing ten 1×10^{-14} J values in one plane (22.77)

Stability requires that energy density of a rotating body of fluid remains the same throughput the volume. Hence the centre of mass of the rotating sun is located at 2/3 of its radius $R_{sun} = 2/3 \times 6.9099 \times 10^8$ m from sun centre. The period of rotation 25.38 days expressed in seconds over circumference of the sun mass derives mass rotational velocity $v_{sun} = 25.38 \times 24 \times 3600 \text{ s}/(2\pi \times 2/3 \times 6.9099 \times 10^8\text{m}) = 7.57742 \times 10^{-4}$m/s.

Hence sun current $= 1 \times 10^{30}$kg $\times 1 \times 10^{54}$ charges/kg $\times 7.57742 \times 10^{-4}$ m/s $= 7.57742 \times 10^{80}$ A.

Applying probability factor to current derives temperature at atomic level, which has to be multiplied by 10 and relativity corrections to obtain 6000 K with variance of ~1.5 kT per interacting orientation:

$$(7.57742 \times 10^{80} \text{ A})^{1/32} \times 1.154700^{3/2} \times 1.111 \ldots^{3/2} \times 1.0328^{1/2} \times 10 \times 1.2 = 6017.3 \text{ K} \qquad (22.78)$$

The same procedure for earth requires multiplication for 3 planes besides other corrections:

$$(1 \times 10^{29} \text{ A})^{1/32} \times 1.2^{1/2} \times 1.111 \ldots \times 10 \times 3 \times 1.04^{1/2} = 300.0773 \text{ K} \qquad (22.79)$$

The sun current should be also created with reference to earth, which sees sun of mass 1×10^{30} kg with 10^{54} charges per kilogram moving at velocity 30000 m/s generating current 3×10^{88} A. Since SI values were used in the calculation the current has to be divided by linear factor 1×10^{10}. This leaves a factor $7.57742 \times 10^{80}/3 \times 10^{80} = 2.5258$ to be explained. This is very close to $(\sqrt{2})^2 \times 1.1547^2/11111^{1/2}$ indicating that balance is needed between momenta of radiations; they are also subject to a formative atomic process and relativity corrections.

Electron spin in the model unit has been assessed in Equation 22.57 at $h_{st} = 1 \times 10^{-33}$ for six orientations with h_s being identified to be the energy converter at the atomic level and also temperature converter in Chapter 2. Chapter 10 represents radiation as sequential chain of linear moments moving at velocity of light converting to clockwise rotation reversing linear direction and rotation and becoming anticlockwise. Such a chain is generated by the hydrogen molecule with atomic energy being considered, the structure representing average positions of quantum vectors. There are 32 quantum momenta with 10 per plane changing their orientation and converted to rotary clockwise and anti clockwise motions. The basic signal at lower temperature does not display independent electron oscillations. Spin angular momentum according to textbooks is S=2, which requires total oscillating energy $3/2 \, h_s$ per quantum orientation position. With 32 positions the energy is:

$$64 \times 10^{-33} \times 1.1547 \times 1.111 \ldots^{3/2} = (1.0000 - 0.00056) \times 10^{-31} \text{ J-s }^{33} \qquad (22.80)$$

Multiplying by 10^{30} converter to SI derives 1 J per event at atomic level and 10 J per kg. The pulse energy 1×10^{-27} J is obtained on converting 1×10^{-30} to 300 K temperature $10 \times (300)^2 \times 1.111 \ldots = 1 \times 10^6$ W deriving the sun and the earth pulse.

Applying the probability factor 1/32 to radii of earth, hydrogen molecule and the sun derives the energy radiated to be 2 eV, 2 eV and 2 eV; independent electron oscillations are not considered:

Two pulses of earth combined:

$$2 \times [(\text{Circular earth area}/\pi)^{1/2} \text{ m}]^{1/32} \times 1/(1.2 \times 1.1111^3) \times 1.04^{11/50} = 2 \times (1 - 0.00014) \text{ J with variance } \sim 3 \, k_s \qquad (22.81)$$

The sun pulse: $(6.907883 \times 10^8 \text{ m})^{1/32} \times 1.11111^{1/2} \times 1.04^{0.113} = 2.000014$ J; variance is $1/6 \, k_s$
$$(22.82)$$

Four pulses from hydrogen molecule, two from each atom generate also 2 J:
$$(22.83)$$

Pulse from hydrogen molecule: $4 \times (1/2 \times 10^{-10})^{1/32} \times 1.11111^{1/2}/1.04^{0114} = 2.00028$ J; variance is $\sim 3 \, k_s$.

Please note that the probability and the input range of values generate asymptotic output. This probably means that the output pulse value is indignant of the distance to the receiver the varying part being the repetition rate. That is the reason for being able to identify elements radiating form the stars (22.84)

Energy density from sun radiation remains a puzzle $(6000 \text{ K})^3/L_{earth}^2 = 2.16 \times 10^{11}/(1.5 \times 10^{11})^2$
$$\equiv \varepsilon_{os} \times (1-0.00003)/1.04^{1.04} \text{ J/m}^3.$$

With each of 64 quantum orientations in three planes and in four directions the input energy per 1 J or eV reference is

$$4 \times 64^3/1.04^{1.2} = 1.00037 \times 10^6 \text{ J or eV} \tag{22.85}$$

Further analysis considering most likely variances $3k_s$ $3.5k_s$ and 4 k_s derived $3.5k_s$ in the following
$$(1/4 \times 1.00035 \times 1.04^{1.2})^{1/3} = 63.9985 \approx 64 \tag{22.86}$$

There are 16 quantum orientations from the surface of the semi sphere or from a side of a cube of 64 quantum orientations of H2 molecule (4x4x4), which may be formed from two spheres representing H atoms each having 32 orientations. This may caused by the high pressure around the sun and **the cubic shape may represent a physical reality and not only independent orientations created by setting the source-receiver axis**. The masses and configurations deriving correct input energies to provide output of 1 J/m or 1 eV/atomic event are as follows (22.87)

For earth $[(6 \times 10^{24} \text{ kg})^{1/16}]^2/(1.111 \ldots^2 \times 1.04^{17/54}) = (1000+0.086)$ J, eV expressing $4k_s$ variance

For sun $(1 \times 10^{30} \text{ kg})^{1/16} \times 6000 \text{ K}/300 \text{ K} \times 2/3 = 1000$ J, (22.88)

For radiation particle involving proton $1/2 \times (1 \times 10^{-27})^{1/16} \times 1/(1.1547)/1.04^{0.66} = 0.0100052$ J
For radiation involving electron particle $(1 \times 10^{-30})^{1/16} \times 1/(1.2 \times 1.111 \ldots) = 0.0100014$ J
(22.89)

The variance of the last evaluation of 3 $k_s/100$ is more reliable

With five orientations per plane per direction the input energies are:

For sun $(1 \times 10^{30} \text{ kg})^{1/5} \equiv 1.0 \times 10^6$ J, eV (22.90)

For earth $[(6 \times 10^{24} \text{ kg})^{1/5} \times 1.11111/1.04^{1/12} \times 10 \equiv (1-0.000063) \times 10^6$ J, (22.91)
For $(1 \times 10^{-27} \text{ kg})^{1/5} \times 1.04^{1/8} \equiv 1 \times 10^{-6}$ Note: $1 \times 10^{-6} \times 10^{12}$ Hz $= 10^6$ J (22.92)

SI input energy of 1 MJ or 1 MeV is 10 times smaller at 10^5 J for SI atomic terms just as is 10 J pulse versa pulses of 1 J at the output. Two pulses in phase and the same orientation of 1000 J require the total input 1×10^6 J:
$(1 \times 10^{30} \text{ J})^{1/5}]^{1/2} =$

$= (1 \times 10^{30})^{1/10}$ and $(1 \times 10^{30})^{1/10} \times (1 \times 10^{30})^{1/10} = (1 \times 10^{30})^{1/5} = 1000 \times 1000 = 1.0 \times 10^6$ J.
Thus
$(1 \times 10^{30}$ kg$)^{1/10} = 1000$ J ; $3(6 \times 10^{24})^{1/10} \times 1.111 \ldots = 1001.59897$ J with var$=1.5k_sT_{300}$;

$$1/2 \times (1 \times 10^{30} \text{ kg})^{1/9}/(1.1547 \times \sqrt{3} \times 1.04^{3/50}) \equiv (1000-0.28) \text{ J} \qquad (22.93)$$

And
$$(1 \times 10^{-27} \text{ kg})^{1/9} \equiv 0.001 \text{ J}$$

$2(6 \times 10^{24}$ kg$)^{1/9} /(1.1547 \times 1.04^{1/2}) = 1000.46$ J ; while this results appears accurate enough it does not repeat the earlier proposed process expressing addition of two 300 K pulses with three in orthogonal orientations generating the resultant 1000 J. The expression producing better accuracy is:

$$(6 \times 10^{24})^{1/9} \times 1.111 \ldots^{1/2} \times \sqrt{3}/[1.0328 \times (1+1/800)] \equiv 1000.0087 \text{ J} \qquad (22.94)$$

Number of orientations in one of six space directions is the sum of five in one plane and the four in the other lane generating the pulse probability of 1/9 and an accurate resultant energy values.

$$\text{And } 1/2 \times (1 \times 10^{-27} \text{ kg})^{1/10} \approx 0.001-0.0000024 = 0.001 - k_s/40 \qquad (22.95)$$

Sun constant and SI energy reference 10 W expressing 1W at atomic level and black body radiation

Sun constant derives within calculator accuracy the 10 W reference, or vice-versa:

$$(1371.742112 \text{ W/m}^2)^{1/3}/1.111 \ldots = (10 - 0.0000\ 0002\ 5) \text{ W/m}^2$$

$$(10 \text{ W/m}^2)^3 \times 1.111 \ldots^3 = 1371.743112 \text{ W/m}^2 \qquad (22.96)$$

Sun constant varies with sun to earth distance during earth rotation and should be compared with other values derived in books 1 and 2. It derives well the sun temperature (rate) interacting with earth:

$$3 \times (1371.74 \text{ W/m}^2) \times 2/3 \times 1.111 \ldots/1.04^{2/5} = 3000.86 \text{ K} \qquad (22.97)$$
However $\qquad (10 \text{ W/m}^2)^3 \equiv 1000 \text{ W} \equiv 1000 \text{ K}$
And $\qquad 1000 \text{ W} \times 1.1547^{3/2} \times 1.111 \ldots^3 = 1378.6 \text{ W/m}^2$

Two expressions appear to be used to express black body radiations $8\pi v^2/c^3$ and $2\pi k_s T v^2/c^2$

Using integer model units: $v_s^2 = 1 \times 10^{24}$ Hz, $c_s^3 = 2.7 \times 10^{25}$ and $v_s^2/(c_s^3 \times 1.111 \ldots^2) = 0.03$ J at 1 K inclusive of energy of independent electron oscillations reducing to 0.02 J or 0.01 J per oscillator expressed by Planck radiation Law.
Energy 0.01 J times local velocity of light 1×10^8 m/s becomes the energy of the sun and of earth pulse energy 1 MJ. In the integer model $8\pi = 8 \times 3 = 24$.

In the integer model units the expression

$$2\pi k_s T v^2/c^2 \equiv 6 \times 10^{-23} \times 300 \text{ K} \times 1 \times 10^{24}/9 \times 10^{16} = 20 \times 10^{-14} \text{ J} \qquad (22.98)$$

Total number of 10 J at 300 K requires further squared temperature increment (total T^3) deriving energy, which multiplied by the minimum energy pulse, derives 3 J:

$$10 \text{ J} \times T_{300}{}^2 \, c_s \times 1.111 \ldots = 3 \times 10^{14} \rightarrow \times 1 \times 10^{-14} = 3 \text{ J} \qquad (22.99)$$

Some physics textbooks quote 20 eV instead of 13.62 eV when discussing Bohr atom energy. This energy is generated by $\sqrt{3} \times 10$ eV $\times 1.154700 = 19.99999$ eV.

Unity reference probability and its input and output energies

Vector addition shown in Figure 5 increases the relativity corrected velocity by a factor of two. It is suggested that the receiver records the rate of the arriving pulses and the sum of the momenta of the incoming pulses. The pulses sourced from the centers of three planes directed towards the receiver have to occur at the same time. With ten optional orientations per plane probability of the pulse repeatedly discharging in the same orientation is $P = 1/10 \times 1/10 \times 1/10 \ldots = 1/10^{10}$. The probability for such pulses to be discharged together generating one output along the long axis of the cube is: $(1/10^{10})^3 = 1 \times 10^{-30}$. If the pulse is started from the interacting mass of the sun $(1 \times 10^{30}$ kg$) \times 10^{-30} \equiv 1$ J at the atomic level and 10 J at SI level. In the classical SI units the total input is highly exaggerated because the reference is 1 m which is 10^{10} larger on linear scale than the atomic dimension and 1×10^{30} larger on energy scale. These figures are further increased by each formative process. It is suggested that each of three planes has three quantum orientations locations plus one in the centre of the plane making a total of 27 positions. It is believed that this value generates the radiation mass of proton of 10^{-27} kg because each quantum position generates 1 kg and probability of them being generated together is

$1/27 \times 1/27 \times 1/27 \times \ldots = (1/27)^{1/27} = 1/729$. Three such pulses form the three planes generate a pulse along the long axis of the cube and derive probability corrected for relativity and quantum:

$$p = 1/729^3 \times 1.2 \times 1.111 \ldots \times 1.03279559 \equiv 1/3 \ 0009 \ 4909 \approx 1/\text{velocity of light} \qquad (22.100)$$

The number of quantum locations on three sides of a 3x3x3 cube is only 19 expressing the charge of the electron 1×10^{-19} C.

Probability applied to mass derives pulse energy

Using a variety of probabilities applied to masses derives a wide range of pulse energies, from which the structure of the event can be often deduced. The probabilities are applied to the masses of sun, of earth, of electron and of proton. Some results are spurious, but left for reference; probability of 1/8 expresses radiation being emitted from eight corners of the cube, but actually it is the sum of energies of pulses generated by centers of three planes directed towards the receiver. For the sun the probability of 1/8 derives the sun temperature. For the earth the result derives the energy of sun constant. For the electron, the proton or two protons of H the output is 0.001 J:

$$(1 \times 10^{30} \text{ kg})^{1/8} \times 1.111\ldots^{1/2} \times (1+1/80) \equiv 6001.693 \text{ K} \tag{22.101}$$

$$(6 \times 10^{24} \text{ kg})^{1/8} \times 1.111\ldots \times 1/(1+1/80) \equiv 1372.88 \text{ J/m}^2 \tag{22.102}$$

$$6(1 \times 10^{-30} \text{ kg})^{1/8}/(1.111\ldots^{1/2} \times 1.04^{3/10}) \equiv 0.00100036 \text{ J} \tag{22.103}$$

$$2(2 \times 10^{-27} \text{ kg})^{1/8}/(1.1547^{1/2} \times 1.04^{3/10}) \equiv 0.00100004856 \text{ J} \tag{22.104}$$

$$2(4 \times 10^{-27} \text{ kg})^{1/8}/1.04^{3/16} \equiv 0.001000513 \text{ J} \tag{22.105}$$

The pulses in 8 different orientations are each produced from three pulses generated in orthogonal orientations directed towards one of the corners of the cube, making the total probability 1/24, which generates the following energies from the sources:

$$(1 \times 10^{30} \text{ kg})^{1/24} \times 1.1547/1.04^{27/40} \equiv (20 - 0.0028) \text{ J or eV} \tag{22.106}$$
$$(6 \times 10^{24} \text{ kg})^{1/24} \times 1.1547/1.04^{27/40} \equiv 10.0274 \text{ J} \tag{22.107}$$
$$1/5 \times (1 \times 10^{-30} \text{ kg})^{1/24}/ [1.111\ldots \times (1+1/80)] \equiv (0.01 - 0.000003) \text{ J} \tag{22.108}$$
$$(2 \times 10^{-27} \text{ kg})^{1/24} \times 1.1547^2/(1.04^{18/25}) \equiv 0.100058 \text{ J} \tag{22.109}$$
$$4(4 \times 10^{-27})/1.111\ldots^{1/2} \times 1.04^{1/9}) \equiv 0.30018 \text{ J} \tag{22.110}$$

The last value represents input per plane 0.3 J in SI units, 10 times 0.03 J.

There is a wide range of probability values that can be applied to the interacting mass of sun and they all generate energies derived in the book in different ways. The author may be accused of fixing these numbers, but this is the geometry of the space, which fixes these values:

$$(1 \times 10^{30} \text{ kg})^{1/4} \equiv (1000 - 0.00007) \text{ K} \tag{22.111}$$
$$1 \times 10^{30} \text{ kg})^{1/12}/1.111\ldots^{1/2} \equiv 300.00000 \text{ K} \tag{22.112}$$

Probability 1/12 represents one of four sun orientations in the sun to earth rotation planes contributing to three orthogonal directions generating the pulses of 300 J ≡ 300K

$$(1 \times 10^{30} \text{ kg})^{1/11} \equiv (1000 - 0.00007) \text{ J or eV} \tag{22.113}$$
$$(1 \times 10^{30} \text{ kg})^{1/13}/1.04^{2/5} \equiv (200 - 0.07) \text{ J or eV} \tag{22.114}$$
$$(1 \times 10^{30} \text{ kg})^{1/14} \times 1.111\ldots^{1/2} \equiv (40 - 0.0215) \text{ J or eV} \tag{22.115}$$
$$(1 \times 10^{30} \text{ kg})^{1/30} \equiv (10 - 0.00000023) \text{ J or eV} \tag{22.116}$$
$$(1 \times 10^{30} \text{ kg})^{1/32} \times 1.1547 \equiv (10 - 0.00071) \text{ J or eV} \tag{22.117}$$
$$(1 \times 10^{30} \text{ kg})^{1/27} \times 1.111\ldots^{1/2} \equiv 13.614 \text{ J or eV} \tag{22.118}$$
$$(1 \times 10^{30} \text{ kg})^{1/5} \equiv 100000 \text{ J or eV} \tag{22.119}$$
$$6 \times 2 \times (6 \times 10^{24} \text{ kg})^{1/5}/1.04^{1/2} \equiv 100171 \text{ J or eV} \tag{22.120}$$

Above two 10^5 eV energies are 1 MeV pulses of Equations 2.11, 2.21, 2.22 and others are expressed in SI model units at the atomic level and they are 1/10 smaller.

$$3(4 \times 10^{-27} \text{ kg})^{1/54} \times 1.04^{0.69} \equiv 1.00054 \text{ J or eV} \qquad (22.121)$$
$$2 (6 \times 10^{24} \text{ kg})^{1/15} \times 1.111 \ldots \times 1.04^{3/10} \equiv 100.0847 \text{ J or eV} \qquad (22.122)$$

After longer than usual search for quantum corrector probability, 1/21 applied to the sun mass, derived unexpectedly a 30 eV value. It fits well the triatomic hydrogen atom H_3, which normally is unstable or in an excited stage. Since $2k_s T_{300}/e_s = 2 \times 10^{-23} \times 300/1 \times 10^{-19} = 0.06$ eV, the variance below is 0.05 eV, $5/6 \ kT \approx kT/(1.1547 \times 1.1111^{1/2})$:

$$(1 \times 10^{30} \text{ kg})^{1/21} \times 1.04^{0.104} \equiv 30.05135 \text{ J or eV} \qquad (22.123)$$

The above expression represents a pulse formed from three pulses, each of the three positioned in orthogonal orientations being also the product of three pulses in orthogonal orientations, the total probability for generation such formation being $1/(3 \times 3 \times 3) = 1/21$.

For masses interacting in one direction probability might be 1/16, 1/15 or 1/14. The late reassessment below displayed all important energies 100 J, 0.01 J, 0.02 J, 6 J and 0.001 J:

$$(1 \times 10^{30} \text{ kg})^{1/16} \equiv 74.9894 \rightarrow \times 1.1547^2 = 99.98577 \text{ J or eV}$$
$$(1 \times 10^{30} \text{ kg})^{1/15} \equiv 100 \text{ J}$$
$$(1 \times 10^{30} \text{ kg})^{1/14} \equiv 5.39145 \rightarrow \times 1.11111 \approx 6 \text{ J}$$

The resultant of probability $\mu_{os}\varepsilon_{os} \equiv 1/E_{s1kgSI} = 1/10^{17}$ J applied to $\frac{1}{2} \ m_{sun}$

$(\frac{1}{2}m_{sun})^{\mu\downarrow osx \ \varepsilon\downarrow os} = (1 \times 10^{30} \text{ kg})^{1/17} \times 1.04^{4/5} \equiv$ graviton power or energy 1/60 W or J=0.01665947 eV/s with variance $5 \times 10^{-3} \ k_s$ which takes a very acceptable value 1.5 J or eV/s on multiplication by 300 K;

$$(1 \times 10^{-27} \text{ kg})^{1/16} \equiv 0.02053525 \text{ J}$$
$$(1 \times 10^{-27} \text{ kg})^{1/15} \equiv 0.0158489 \text{ may on multiplication by } 1.0328 \text{ be } 0.0166 \text{ eV. The}$$
energy step in NCl vacancy experiments of 0.0174 eV is probably $0.0166 \times 1.0328^{3/2}$

$$(1 \times 10^{-27} \text{ kg})^{1/14} \equiv 0.0117876 \rightarrow \times 1/(1.1547 \times 1.04) = 0.0100102 \text{ J}$$

Electron mass or electron mass equivalent generates 0.01 J with probability 1/32:

$$(1 \times 10^{-30} \text{ kg})^{1/32}/1.1547^2 \times 1/10 \equiv 0.0100071 \text{ J}$$
$$(1 \times 10^{-30} \text{ kg})^{1/16}/1.1547^2 \equiv 0.01000142 \text{ J}$$
$$(1 \times 10^{-30} \text{ kg})^{1/15} \equiv 0.01 \text{ J}$$
$$(1 \times 10^{-30} \text{ kg})^{1/14} \times 1/6/(1.1547 \times 1.04) \equiv 0.001 \text{ J} - 1 \times 10^{-6} \text{ J} \qquad (22.123a)$$

The kinetics having no units of time and distance justify applying probability to mass for obtaining energy. In kinetics the ratio, distance over time (velocity), must be unity and so its square converts mass to energy. So in kinetic atomic interactions mass can be treated as energy. Such rule does not apply to SI interactions.

Higg's boson$_s$

WEB pages describe Higg's boson to be a very common energy particle of space needed to complete the Standard Model. It is very difficult to be measured. On September 2012 CERN announced that several measurements established its definite value at (1.23-1.24) x 10^9 kg/c^2. The current integer model provided such energy for one dimension in Equation 2.14b (now 23) in the spring 2012 first printout of the book withheld from sale for corrections and additions. Above Equations express corrected for relativity the mass of proton converted to energy and actually derive the gravitation constant G_s = 1 x 10^{-10} N-m^2/kg^2 in presence of independent electron oscillations.

The exact value proposed now for Higg's boson$_s$ in integer model units is:

$m_{ps}c_s^2$ x1.111 . . .^{3}/e_s =1x10^{-27}x (3x10^8)^{2}x1.111 . . .^{3}/1x10^{-19} = 1.234567901x10^9 eV per event
or in SI units kg/cs^2 (22.124).

Triple relativity correction 1.111 . . . indicate that at the atomic level the pulse has originated in three independent orthogonal orientations directed towards the source along the long axis of the cubic representation of the atom. Third root of (1.234567901x10^9)$^{1/3}$ = 1.072765175x10^3. This value is the product of the 1000 J pulse of Equations 10.18-19 (and many others) and relativity and quantum corrections 1.1547$^{1/2}$ x 1.04$^{0.045}$.

In terms of integer model SI units the Higg's boson$_s$ is

$$10\varepsilon_{os}/e_s \text{ x1.111 . . .}^2 = 1.234567901\text{x}10^9 \text{ eV} \qquad (22.125)$$

It sill remains unclear whether beside expressing pulse energy, $10\varepsilon_{os}/e_s$ x1.111 . . .^{2} also expresses the energy capacity of the space or the actual energy of space activated by the disturbance.

The other two equally important parameters at atomic level are: (22.126)

Radiation energy = 1/10 x Planck radiation = 0.01 J = $\mu_{os}^{1/3}$ = $(\varepsilon_{os}/10)^{1/6}$

The elementary energy at atomic level = **0.01 J/10 = 0.001 J has no identifiable electrical or magnetic features; it could be a rotating gradient of the field.** These are the energies of elementary progressive oscillations, packets of vibrating fields in two dimensions and two orientations characterized by four interlinked loops in two planes each interlinked loop represented the Maxwell curl ▼ of the function expressing both magnetic and electric flux ▼ X F = 1/c_s x $\partial F/\partial t$. Parameter F does not display distinct electrical or magnetic features. With 32 quantum orientations of Chapter 10 and 4 different oscillations the probability is 1/(4 x 32) reducing radiation energy of hydrogen atom to

$$(1 \text{ x}10^{-27})^{1/128} \text{ x } 1.2^3/[1.111 . . .^{1/2} \text{ x } (1+1/86)] = 0.9970 \text{ eV} \qquad (22.127)$$

generating huge uncertainty of 70 k_s attributed to 140ty energy quanta elements 1/2 k_s comprising 128-ty probabilities and 12 directions with variance k_s per direction.

The uncertainty at level derived further in the text was quantum less at

$$60 \, k_s = 70 k_s / (1 + 1/6) \qquad (22.128)$$

Rayleigh and Jeans Law expresses infrared radiation without the independent contribution of electron oscillations at 1×10^{15} Hz. The Law $8\pi v^2/c^3 \; x \; kT$, with $v_s = 10^{12}$ Hz and $c_s = 3 \times 10^8$ m/s, expressed in integer model SI units:
(22.129)

$8\pi v_s^2/c_s^3 x k_s T \quad x \quad$ correction$=8 \quad x 3 x 1 x 10^{24}/2.7 x 10^{25} \quad x 1 x 10^{-23} x T \quad x 1.111 \quad . \quad . \quad . \quad x$
$(1+1/80)=k_s T = 1 x 10^{-23} T$

$k_s T$ at SI level it converts to $1x \; 10^{-23} x \; 10^{-30} \; x \; 1/10 = 1 x 10^{-44}$ J/K $= B_{pulse1}$ at atomic level. at 300K in six orientations to $\frac{1}{2} x \; 6 \; x \; 300 x \; 1.111 \ldots = 1000$ J,

for the earth pulse at atomic level, 10000 J at SI level, the product of two generating local $1/3 c_s = 1 x 10^8$ m/s.

New Planck constant $h_{selec} = 1 x 10^{-36}$ J-s and new Boltzmann constant $k_{selec} = 1 x 10^{-26}$ J-s for electrons; high frequency $v_o = 10^{17}$ Hz
Each of 1 J of 10 pulses in the 17 degrees of freedom in space has very high event probability of $1/10$ making a total $1/10 \; x$ (seventeen times)$= 1/10^{17}$. Plank constant in Equation 2.2 of

$$\text{space } E_{sspaceSI} \equiv \lambda_s x \, v_o = \lambda_s \, x 10^{17} \equiv 1 \, x \, 10^{17} \text{ J deriving } \lambda_s = 1 \text{ m} \qquad (22.130)$$

applying linear size conversion 10^{-10} derives diameter of hydrogen atom $2a_{os} = 1 x 10^{-10}$ m.

Frequency v_o is essentially $v_o \equiv 1/(\mu_{os}\varepsilon_{os}) \equiv 10^{17}$ Hz
    The reference 1 J follows the sequence:

$$\text{Resultant energy } v_o \, E_{sspaceSI} \, h_s = 10^{17} \text{ Hz } 10^{17} \text{ Jx1 } x 10^{-33} = 10 \text{ J} \qquad (22.131)$$

Frequency $v_o = c_s^2 \; x \; 1.111 \ldots = 1 \; x \; 10^{17}$ Hz² suggests that it is formed by two pulses of light velocity, or rather $(\sqrt{3} \; x \; \frac{1}{2} \text{ cs } x 1.1547)^2 \; x \; 1.111 \ldots = 1 \; x \; 10^{17}$ Hz².
Boltzmann constant is derived as magnetic permeability of space per cycle of v_o frequency:

$$k_s = \mu_{os}/v_o = 10^{-6}/10^{17} = 1 \; x \; 10^{-23} \text{ J} \qquad (22.132)$$

Boltzmann constant is ten magnetic permeability's of space to power 4:

$$k_s = 10\mu_{os}^4 = 10 \; x \; 10^{-24} = 10^{-23} \qquad (22.133)$$

This can be expressed per space direction: $k_s^{1/4} = \sqrt{3} \; \mu_{os} \; x 1.04^{0.67} = 0.99975 \; x 10^{-23}$ J more interesting presentation because it derives $\frac{1}{4} \, k_s$ uncertainty.

    Electron Boltzmann constant has been discovered from analyzing the uncertainty in the identity equating magnetic permeability μ_{os} with four relativity corrected square roots of 6_s:

$$4 \, x \, 1.111 \ldots x \, (6_s)^{1/2} = (0.001 - 0.00009938) \text{ J or eV per charge event} \quad (22.134)$$

The variance expresses $62x10^{-26}$ J value identified with 2 x 30 DOF deriving the probability $1/30$ discussed in several equations. Uncertainty 1 $x10^{-26}$ J has identified electron Boltzmann constant at $k_{selectron}$=**1x10^{-26} J-s**. The corresponding Planck constant for electrons is 1000 smaller $h_{selectron}$=**1x10^{-36} J-s.**

Electron charge e_s in terms of the integer cubic SI model Planck constant over minimum energy

$$e_s = h_s/(\text{minimum energy}) = 1x10^{-33}/1x10^{-14} = 1x10^{-19} \text{C} \quad (22.135)$$

Further attributes of physical constants

Gravitation Energy without electrom controbution is $1/60 = 0.0166 \ldots$ J or eV per atomic event/K. At 300 K $0.0166 \ldots$ J x 2/3 = 0.01 J becoming 3 J. The input of 1 J is divided in six space directions and 5 quantum orientations with reflections deriving $1/6$ x $1/5$ x $1/2 = 1/60 = 0.0166 \ldots$ J per orientation event. Probability of event occurance with four options of each of three independant planes is $1/4^3 = 1/64$. This is not 6 independant directions with 5 orientations generating probability $1/30$. Energy input corrected by relativity and quantum needed to obtain $1/60$ J output per orientation is
$0.0166 \ldots^{1/64}$ x $1.1547^{1/2}/(1+8/1000) = 0.999978 = (1 - 1/3x10^{-23})$ J and expreasses var $=1/5 \, k_s$
Thus 0.01 J represents enegy per orientation of 1 J at 300 K. Energy $0.01/(300$ x $1.111 \ldots) = 50 \, x10^{-6}$ J times $1x10^6$ pulse derives 50 J or eV creating no conflict.

New concepts need clarification. The new parameters were energies derived from integral cubic physical quanta. These are energies for space $E_{sspaceSI} = 1$ x 10^{17} J-s, for earth $E_{searthSI} = 1x10^{21}$ J-s and for sun $E_{ssunSI} = 1x10^{21}$ J-s The last two value are gravitation forces per orientation in Specialized Relativity terms at SI atomic level. The value multiplied by ten and three becomes 3 x 10^{22} N. It is still suggested that there are too many physical constants and some may be redundant. Reassessment displays that $E_{sspaceSI} = 1$ x 10^{17} J-s is the classical SI value for the gravitation energy of all planet-sun couples at SI atomic integer units of Table 4 1 x 10^{16} J. This leads to a significant reduction of the number of physical parameters expressing energies.
Two earth pulses of 300 K to power 4 times permiabiliy 1 x 10^{-11} with cor. derives 1 J:

$$1 \text{ kg x } 600^4 \varepsilon_{os}/(1.1547^{3/2} \text{ x } 1x.04) = 1.0043 \approx 1 \text{ J or eV} \quad (22.135a)$$

Above variance 0.0043 is taken to be 45 k_s expressing 1.5 k_s for each of 30 orientations. Each 1.5 k_s represents ½ k_s for each 200 K per plane. The earth temperature is:

$$\sqrt{2} \text{ x } 200/1.0328 = \textbf{273.86 K}, \text{ which times } 1.1111 \ldots \text{ obtains } 304.28 \text{ K} \quad (22.135b)$$

Variance per kelvin $4.28/300 = 0.0147$, which compares with $1/2 \, x10^{-23}/1$ x $10^{-19} = 0.0147$, so using temperature 300 K is fully justified in clasic expressions. High temperature accu racy is obtained in $T300 = (1x10^8)^{1/3}/\sqrt{3}$ x $1.111 \ldots x1.0328^{1/40} = \textbf{299.999954 K} \quad (22.135a)$

Earth and sun share equality earth gravitation force: $2 \times 600^8 \times 1.111 \ldots = 3.5409 \times 10^{22}$ N. and 300 K $\times 10^8 \times \mu_{os} = 0.03$ J.

$$\text{Muo}10_s \text{ force or mass} = (\mu_{os} F_{satomic})^2 \times 1/10 \times 1/10 = 1 \times 10^{-28} \text{ kg} \qquad (22.136)$$

The relation $\mu_{os} \times c_s \equiv 300$ K defines the meaning of most important physical constant; magnetic permeability per degree kelvin is the inverse of the velocity of light expressing **seconds per meter** (22.137)

Linear earth velocity 30000m/s $\equiv 300$ K $\times 10 \times 10$; $G_s \equiv \varepsilon_{os} \times 1/10$; $(1000)^4 \equiv 1 \times 10^{12}$ Hz (22.138)

Three Planck radiation pulses in orthogonal orientation of 0.01 J generate as a product of probabilities energy 0.01^3. Two such pulses in wave function formation generate velocity o light and energy $0.01^6 = 1/10^{12}$, which times frequency 1×10^{12} Hz derive 1 J or W.

The Special Relativity gravitation force for sun-earth couple uses only integer cubic values with geometric mass 1×10^{16} and distance 1.5×10^{11} m and its accuracy is limited by true uncertainty. The relation was not tested for other couples having different μ_{os}:

$$F_{Gs} = (2/3)^2 \times 1/\mu_{os} \times m_{geom.mass} /\text{Distance} \times 1 \times 10^{12} \text{ Hz}(1+1/80) = 2.999 \, 999025 \times 10^{22} \text{ N} \qquad (22.139)$$

Sun constant per cycle of the local velocity of light, corrected for relativity and quantum expresses energy 1×10^{-15} W of Equations 2.36-7:

$$1372.5 \text{ W/m}^2/1 \times 10^8 \text{ m/s} \times 1.1547005^2 \times 1.04^{3/4} = (1 - 0.00046) \times 10^{-15} \text{ J/ m}^3 \qquad (22.140)$$

Interacting mass over squared distance to earth, corrected by two relativities and converted to 1 K quantum derives the sun pulse exactly at 10^5 W, J, or N and at 300 K that value increases to 3×10^7. Any inaccuracy is most probably produced by the calculator:

$$1 \times 10^{30} \text{ kg}/(1.5 \times 10^{11} \text{ m})^2/(1.111 \ldots \times 1.154700538^2 \times 300) = 100 \, 000.0001 \text{ W, N/m}^2 \qquad (22.141)$$

The force in SI units of 1N/kg converted to atomic units by multiplier 1×10^{-30} and expressed per kelvin ($6000/6 = 1000$ K) derives the Planck constant in cubic integer units:
$$1\text{N/kg} \times 1 \times 10^{-30}/1000\text{K} \equiv 1 \times 10^{-33} = h_s \qquad (22.142)$$

The force 0.006 N/kg for 6 space orientations/K and 0.006 N/kg $/6 \times 1 \times 10^{-30} = 1 \times 10^{-33} = h_s$ The energy at 300 K per cycle of 1×10^{12} Hz is 1×10^{-45} J which divided by 1×10^{-19} derives 1×10^{-26} J per atom expressing Planck constant for electron k_{selec} delivering in each of six space directions with relativity and quantum corrections

$$1.66 \ldots \times 10^{-27} \text{ J/} (1.1547^{3.5} \times 1.04^{1/5}) \equiv (1-0.0005) \times 10^{-27} \text{ kg equal to the mass of}$$
10.005 muo10$_s$ with variance 5×10^{-23} J $= 5 k_s$ that is ½ k_s per muo10$_s$.

Lorentz force

The integer model gives additional interpretations to Lorentz force taking $v_e = 1 \times 10^7$m/s and $v_e = E = B = 1$

$$F_{Lorentz} = e_s E + e_s/c_s \times v_e B \equiv 1 \times 10^{-19} + 1 \times 10^{-19}/3 \times 10^8 = 1 \times 10^{-19} + 1 \times 10^{-28} \qquad (22.143)$$

Multiplying 1×10^{-28} by probability 10^9 derives 1×10^{-19} and $F_{Lorentz} = 2 \times 10^{-19} = 2 e_s = 2$ eV, J, W
Equating two Lorentz force components using $v_e = 1 \times 10^7$m/s derives:

$$E = 1 \times 10^7 (\text{m/s})/3 \times 10^8 \times B = 1/30 \times B \qquad (22.144)$$

This is interpreted that the ratio of 30 expresses 30 units of 1 eV, J, W adding to 30 eV, J, of total atomic atom energy expressed by B.

The ratio of $\mu_{os}/\varepsilon_{os} = 10^5$ is the energy pulse. Divided by number $(30 \times 1.111 \ldots^2)$ derives:

$$30\,000 \text{ m/s the rotation velocity of earth around the sun} \qquad (22.145)$$

This expresses 6000 K each creating 5 J quantized in three orientations so that:

$$6000 \text{ K} \times 5 \text{ J} \times 1/10 \times 1/10 \times 1/10 \equiv 30000 \text{ m/s} \times 1/10 \times 1/10 \times 1/10 \equiv 30 \text{ J, eV, W} \quad (22.146)$$

And $5 \text{ J} \equiv 3 \times \sqrt{3}/1.04 = 4.9963$ eV implying uncertainty of 27 $k_s T$; with 34 beams this is 0.8 $k_s T$ per orientation, equal 1.2 $\times$ 2/3 $k_s T$.

Vectorization of time in all directions derives velocity of light. Unit of time in space of earth year has been justified earlier. Presently the life of the Universe is treated as time in all directions and likely consequences of Time = Velocity are proposed. The time expressing the life of the Universe vectorized in one plane and corrected for relativity derives the local velocity of light 10^8 m/s satisfies the above condition. The value of the variance 30000 = 300K $\times$ 10 $\times$ 10 remains puzzling:

13.7546×10^9 years$/1.1547005^2 \times 1/10 \times 1/10/1.04^{4/5} = (1 - 0.0003) \times 10^8$ m/s $\qquad (22.147)$

Gravitation forces are quoted with the reference to sun or to earth. Actually each body pulls the other body with one half of the value calculated. The formula deriving that value is:

$$F_{Gs} = 1 \times 10^{-10} \times 6 \times 10^{24} \times 1 \times 10^{30}/[\sqrt{3}(1.5 \times 10^{11})^2 \times 1.0328] \times 1.04^{1/6} = 1.500918 \times 10^{22} \text{ N} \quad (22.148)$$

These results display high accuracy values obtained using integer model SI values.

Classical velocity of light versa integer cubic value

Classical velocity of light $c = 1/\sqrt{(\mu_o \varepsilon_o)} = 2.99792 \times 10^8$ m compares well with the value obtained using the integer cubic physical constants decreased by the relativistic correction:

$$c_s = 1/\sqrt{(\mu_{os} \varepsilon_{os})} \times 1/1.111 \ldots^{1/2} = 1/(10^{-6} \times 10^{-11}) \times cor = \sqrt{10^{17}} \times cor = 3.0000\ 0000 \times 10^8 \text{ m/s}$$

$$(22.149)$$

The variance 5700 is identified with sun temperature of 5700 K. Alternatively one could divide the integer model value by the quantum corrector $1.042^{91/77}$ obtaining $c_s = 2.99842 \times 10^8$ m/s. The choice of the appropriate procedure is arbitrary.

White noise

The space radiation of wavelength 0.2 m discussed in Section 10 was attributed to the model frequency of proton 5×10^{12} Hz. The white noise is quoted on WEB pages with frequency 4080 MHz. It has been checked to originate from space and not from stars or galaxies. Applying to that frequency two parameters of the model derives 4080 MHz x 1.2 x $1.0^{1/2} \approx 5 \times 10^{15}$ Hz, which is the electron frequency. It is suggested that it actually represents 4.9979×10^{15} Hz of electron frequency of the equivalent mass expressing five quanta of energy

Our personal experience of infrared reality

Gravitation experienced by everybody is according to the integer model the vectorized thermal flux generated at infrared frequency of 5×10^{12} Hz essentially by the proton, which is accompanied by the electron. Electron's effects appear numerically minute, they may be missed at times, but they are very powerful.

Temperature on earth at around 300 K is generated by the oscillation at 5×10^{12} Hz of the (proton+electron) pair. Human body temperature over quantum corrector of $(273.15 + 36.6)/1.04^{4/5} = 300.18$ K $\equiv 300.18$ W expressing an ideal uncertainty variance of $6k_s T_{300}$ of $k_s T_{300} = 0.03$ J or eV per orientation of space. Thus electrons play little role in the human experience, unless special event take place on the sun or the body is exposed to sun radiation for an over extended period.

Are components of light velocities measurables?

Early measurements by the author in the late 60-ties, found that experimental data display energy step predicted by 1D quantum Bose-Einstein distribution. The difference between the classical value and quantum value was attributed to the relativity velocity corrections about seven years ago. Content of the relativity correction in Goldsmith and Uhlenbeck spin definition was uncovered in 2012. There are very many identities deriving numerically correct data using relativity corrections and component velocities including Equations 19.7, 19.28, 19.45, 19.56 displaying component of light velocities.

Requirements for the gravitational balance between the sun and a planet

Final review revealed three independent requirements to balance the forces between the sun and the planet with some aspects remaining obscure. The absence of units of time and of length in kinetics establish that in atomic interactions the following unit equivalence for elastic kinetic interactions that take place both within the atoms and in the space **Mass = Force** and **Time = Velocity.** The Newton gravitation force of 3.5438 x10^{22} N is $(1 + 1/63)$ smaller than the force expressed by the model using integer cubic SI units of

$$4/3 \ x1x10^{-10}x6 \ x10^{24}x1x10^{30}/(1.5 \ x10^{11})(1+1/80) = 3.600000000 \ x10^{22} \ N \qquad (22.151)$$

It is believed that Newton gravity and energy of 1 J in the above relation is the energy converted from mass expressing 1 W/kg due an inherent instability of protons and hence of Bohr structures both in the sun and in the planets. This is the principal main source of mutual gravitation forces with presence of other small factors at times.

The second balancing requirement is the sustenance of the long energy containing beams of the sun so that they are not shortened dissipating energy. This is secured by the planet rotation inducing presence of the centrifugal force opposing sun linear gravitation as expressed by Equations 6.6-7.

The third balancing requirement is sustenance of stable field of forces in around the planet. Two sources of force were proposed for this purpose. Numerous data for four planets suggested that electron spin of the masses of the planets remained in one orientation for a period in excess of one year and unwinding the spin displayed the force explained in Chapter 7 and 13. For distance further away from the sun either the mutual inductance coefficient become to small fraction or the gaseous bodies do not sustain directions of their momenta balancing similar vectors from the sun. It is not possible to give reasons for fractions of forces attributed to vectorized thermal flux not the unwinding of the electron spin.

The four balancing requirement deals with summing rotating moments to zero. WEB pages give a wide range of opinions.

Acceleration on earth is affected by earth self rotation

In integer units earth equatorial circumference is 40 x10^6 m generating earth radius R=40 x10^6 /2π= 6.36620 x10^6 m; with 24 h =86400 seconds linear surface velocity v_s=462.963 m/s. Centrifugal force per kilogram converted to 300 K = 1 kg x v_s^2 x300 K/R=10.10029 N and acts radially inwards, when earth is taken as reference. The total inwards force of gravitation and acceleration on earth surface 10 m/s from Equation 2.17 plus above 10.1003 m/s making a total of 20.10029 m/s or N.

Sun raises on earth surface also the gravitation force from the whole of its mass with probability 1/30 generating per kg

$$(2 \ x10^{30} \ kg)^{1/30} \equiv 20.23374 \ N \qquad (22.152)$$

This leaves on earth the residual force of 20.23374 -10.10029 =10.13334 N.

Above value divided by quantum corrector (1+1/75) derives 10.000066 N.

This force is supplied by the gravitation of earth expressed in integer SI units:
$$6 \times 10^{24} \times 2/3 \times 10^{-10}/R^2 = 9.8696 \text{ N} \qquad (22.153)$$
which compare with Wikipedia data of 9.833 m/s?
The mathematical models quote 9.78 m/s.

Comparison of classical and integer physical constants not listed in Table 1

Classical formula in integer units had to adjust by changes caused by velocity of light having $1/3c_s$ component in 1D.

Name and formula	Integer value	~ classic value	Remarks
Conductance quantum $G_o = 2e_s^2/h_s$	2×10^{-5}	7.748×10^{-5} S	
Magnetic flux quantum $O = h_s/2e_s$	5×10^{-15}	2.067×10^{-15} Wb	
Nuclear magneton $\mu N = e_s h_s/2m_{ps}$	2.5×10^{-27}	5.050×10^{-27} J T	h_s replaced by $h_s/10$
Electron radius $r_e = e_s^2/4\pi\varepsilon_{os}m_{es}c_s^2$	1×10^{-15}	2.818×10^{-15} m	π rep by 3, $\rightarrow \times 1.04^2$
Quantum circulation $h_s/2m_{es}$	2×10^{-4}	3.637×10^{-4} m^2 s^{-1}	
Thompson X - section $(8\pi/3)$	8×10^{-30}	6.652×10^{-29} m^2	π rep by 3
First radiation constant $c1 = 2\pi h_s c_s^2$	6×10^{-17}	3.742×10^{-16} W m^2	π rep by 3, c_s rep by $1/3c_s$
Second radiation constant $c2 = h_s c_s/k_s$	0.01	0.01439 m-K	c_s rep by $1/3c_s$
Wien displacement constant $b = h_s c_s/k_s$	0.03	0.002897 m-K	(22.154)

Please note that most of the integer relativistic physical constants derived in the above list are used in various expressions in the book deriving valid relations and energies.

The world famous author of student books on physics Arthur Beiser quoted in his 1973 book [34, p116]: "The law of physics that are valid in the macroscopic world do not hold true in the microscopic world of the atom". The introduction of the integer list of physic constants and presented new interpretations of findings show that the laws of physics remain valid.

Hartree's Energy

Hartree's Energy can be represented by several values starting with 30 eV/event or J and terms derived using relativity corrections and wave function format formation:

Energy 30/1.111 . . .= 27 J exactly; $30/(1.1547 \times 1.0328)^2 = 21.0939$ eV; 21 eV expresses 7 eV in three orthogonal directions, 7 x 1.111 . . .≈ 0.78 eV is the mean value of 40 researchers activation energy for mobility in NaCl crystals derived by Whitman and

Calderwood [12] identified by the author as dissociation energy of hydrogen structure. Energy 7 x 1.1547=8.08 eV is the above dissociation energy 8 eV given by Dekker [53, Table 5.2]. Energy 0.093 eV $\approx$ 63 quantum positions of 1.5 k_s each less residue 0.3 $\equiv$ 300 $k_{selectron}$ at 1 $k_{selectron}$ per kelvin.

Energy $30/1.0328^3 = 27.2316$ eV less the classical 27.2138 eV leaves residue 0.0178 eV $\approx$ 18 k_s which represents space quantization of 36 positions at 1/2 k_s. The residue of 0.2 $k_s \equiv 200\,k_{selectron}$ is expressing 2/3 $k_{selectron}$ per kelvin.

Actually 27 eV is most representative of Hartree's Energy. Corrected by two relativities: Energy= 27 x 1.1547 x 1.0328 = 32.1995 eV expressing 32 of 1 eV quantum positions on the hydrogen sphere plus 0.19995 expressing 200 k_s, which divided by 300 K derives 2/3 k_s per kelvin. The negative residue 500/300 x10^{-25} eV expresses per kelvin $500/300=5/3 k_{selectron}$; that is five quantum 1/3 $k_{selectron}$ values.

Wave packet

Interference of two frequencies generates wave packet moving with its own phase and group velocititiesa. This is de Broglie wave of a moving particle. In the model the wave packet is formed by the frequencies of the proton and of the electron corrected by relativity, by quantum, by n-interacting energy components 1.23:

Wave packet frequency = $\sqrt{(5\times10^{12}\times5\times10^{15})}/1.23$ x1.111.../(1+1/60) = 1.000012x10^{14} Hz

Above value is consistent with electron rotation velocity $v_{es}= 1\times10^7$ m/s and its cube expressing muo10$_s$ energy: $(1/1\times10^{14})^2$ - 1x10^{-28} J

23. Conclusions

The book's title, Digital Physics and Gravitation *Probabilistic Relativistic Kinetic*, describes new bases of electromagnetic reality discovered by the author after eleven years of trying to improve understanding of gravitation and physics. Physics was found to display a very strong Mandelbrot effect expressed by fractal geometry.

The physics are based on unity values used as references, the four unity values being sufficient to describe the SI system. The integer cubic space SI system proposed is based on the unity sun radiation arriving on earth at temperature 300 K, which in the classical SI system expresses exactly the energy of the 1st term of Planck Radiation per one kelvin of 0.01 W after relativity correction is applied to temperature.

The unity value is derived directly by vectorization of the sun pulse per direction 1000 W/m²:

$$1000 \text{ W/m}^2 \text{ x} 1/10 \text{ x} 1/10 \text{ x} 1/10 = 1 \text{ W/m}^2.$$

The sun constant corrected by two relativity factors obtains also 1 W/m² incremented by $6k_s$ expressing k_s per space direction and $1.5k_s$ from two pulses needed for forming light velocity:

$$1372.5 \text{ W/m}^2 \text{ x} 1/10 \text{ x} 1/10 \text{ x} 1/10/(1.1547^2 \text{ x} 1.0328^{2/3}) = 1.007454 \text{ W/m}^2 \qquad (23.\text{a})$$

Opposite momentum and power 1 W/m² has to have generated also by the mass of earth radiating 1N/kg per square meter at 300 K per oscillator with variance $3k_s +2/3k_s$: $\qquad$ (23.b)

$$6 \text{x} 10^{24} \text{ kgx} 1 \text{ N}/(4\pi \text{ x} 1637806 \text{ m}^2) = 1.1737 \text{x} 10^{10} 1 \rightarrow 1 \text{x} 10^{-10}/(1.1547 \text{x} 1.0328^{1/2}) = 1 - 0.00376 \text{ J}.$$

Above result is accompanied by another find, an accurate Equation (23.c) defining sun constant power with variance $1/3k_s + \sim 1/16k_s$. Here the power of sun constant is attributed to radiation from earth because $1.1737 \times 1.1547 \times 1.0328 = 1.3773$ of Equation (23.b):

$$1372.5 \text{ W/m}^2 \text{x} 1/10 \text{ x} 1/10 \text{ x} 1/10/[1.1547^2 \text{x} 1.0328^{1/2} \text{x} (1+1/80)] = 1.000393632 \text{ W/m}^2 \qquad (23.\text{c})$$

Equations 2.92, 15.28 and others used quantum corrector $(1+1/80)$, $1/60$, $1/40$, $1/20$ and derived accurate data. Atom of oxygen has atomic weight 16 relativity will increase its mass to $16 \text{x} 1.111 \ldots \text{ x } 10^{-27}$. The gravity force of sun acts in one plane and similarly the earth rotates in one plane. Consequently the interacting moments are oscillating in one plane directed toward the sun. Sixteen protons will generate to the right and to the left to satisfy conservation of momentum rule. To increase velocity from $\frac{1}{4} c_s$ to c_s four steps are needed when pulse resultant is generated by two pulses in orthogonal orientations. The probability of the first step is changed from earlier $\frac{1}{4}$ to $\frac{1}{2}$ because with forces in one plane the proton oscillates only forward and back. For the next three steps the probability is $\frac{1}{4}$ per step, because the proton can radiate in any of four directions. The total probability is $1/2$ x $1/4$ x $1/4$ x $1/4 = 1/128$. Several other probabilities have been evaluated as mathematically or physically feasible, but only the one with the highest probability will effectively appear in reality. Hence the pulse generated by the mass of earth corrected for relativity and quantum is:

$(6 \times 10^{24})^{1/128} = 1.561634/(1.1547^3 \times 1.0328^{1/3}) = 1.00346$ J or eV $\quad\quad$ (23.d)

displaying an acceptable variance of 3 k_s + 1/5 k_s.

For two oscillators at 300 K the pulse from earth is 2 J (Equation 21.29). These pulses add in three orthogonal orientations √3 x 2/1.0328 generating 5 J. The first two of 5 J pulse derive 10 J, two sets of 10 J derive 20 J, two sets of 20 derive 40 and two sets of 40 derive 80. The term 80 in the quantum corrector of Equation 23.c expresses 80 J which times $1.1547^2/1.0328^2 = 99.9990775$ J with the variable from 100 J of 0.000925 = 0.0000115 per proton ≈ $12k_{select}$ expressing energy from 32 quantum positions interacting with the other body 3x12 = 36. The 100 J is at 300 K allowing for absence of electrons, over relativity and converting from atomic to SI by converter of 10: 100/300 x 2/3 1/1.1111 x 10 = 2 J value appearing in several Equations. Signal per proton 100/8 =12.5 J; 12.5 x x $1.1547^{1/2}$ x $1.03281^{1/2} = 13.64$ J. Mass of 16 protons atoms of oxygen converted by relativity to 18 units 16 x 1, 1111 x $1.0328^{2/5} = 18.000$ units. Atomic mass of 18 oxygen protons expressing 18 x 1.007825 = 18.14085 u less mass of six created helium atoms 6 x 3.016829 =18.09662 generate residue mass of 0.04423u or kg in SI units. That mass converted to velocity of light, relativity delivers 5 J discussed above:

Energy released= $0.04423ux(3 \times 10^8)^2 \times 1.1547^2/1.0328^2 = 49751 \times 10^{15} \approx 5 \times 10^{15} \rightarrow x1 \times 10^{-1} = 5$ J.

A new digital relativistic description of physics is needed based on event occurrence probability derived from the equality of probabilities in the six independent orientations of space. Generation of velocity of light requires two three pulses (not waves) or four sets of two pulses to be discharged in the wave format function. Velocity of light is the resultant velocity of three orthogonal components (proton+electron) and proton moving with velocity 1/4 c_s generating √3 x1/4 c_s x1.1547005 = 0.4999998c_s m/s and three such resultants converted again generating velocity of light √3x 0.4999998 c_s x 1.1547005 = 0.99999957c_s m/s. The last expresses a very acceptable uncertainty of ½ m = ½ λ. The velocity of equal components in opposite directions at right angle to the propagation orientation null each other, while those in oppositee directions likely reflect each other; three 1/3 c_s **components along orthogonal direction of a cube** form one dimentional velocity of light c_s along long axis of the cube. **The first hint of the presence of another velocity of light was the ratio of energy terms in the author's experiments and in the classical SI system *1/3 kT* and *½ kT*. The presence of relativity correction for ½ c_s was noted last year in the experiments of year 1925 (Equations 22.55 and 22.57) and in the theoretical term √3/2.**

Five quantum orientations of Stern and Gerlach were applied to the volume of an atom. With 12 positions per plane there may be 36 positions per a spherical volume, provided discharges occur sequentially in all directions. Other typical values are 28, 30, 32, and 34 depending on the circumstances. These quantum numbers determine energies exchanged between the source and the receiver and 32 k_s is most common.

Boltzmann constant for electron k_s at 300 K represents three equivalent (reduced) electron masses of (proton+electron) pairs displaying essentially the mass of an electron, but oscillating at infrared frequency 1 x 10^{12} Hz. The created probability is 1/11 x 1/11 x 1/11= 1/33. If the output is $h_s \equiv 1 \times 10^{-33}$ J the input is 1×10^{-33} J $\times 10^{11}$ $\times 10^{11}$ $\times 10^{11} = 1$ J. If the input is 10000 W the output is $(10000 \text{ W})^{1/33}/1.15470053^2 \times 1.04^{1/5} = 1 - 0.0008$ J, with variance expressed by an acceptable energy 8 k_s. For linear interactions there are 10 k_s per plane. Boltzmann constant $k_s = 10^{-23}$ J-s for the integer cubic model is the probability uncertainty per unit volume, with 10^{-22} J-s and 10^{-11} J-s becoming uncertainties per unit surface and per unit length. Linear probability 10^{-10} converts SI k_s to atomic level of Planck constant $h_s = 10^{-33}$ J-s value. Earth pulse is shown below Equation 23.4 to be 10^5 J. The product $10^5 k_s$ generate in 1D muo$10_s = 10^{-28}$ J and the product $(10^5)^2 k_s = 1 \times 10^{-33}$ J.

The Principle of minimum entropy production, was developed nearly 100 years ago. The loss of energy expressed by the minimum entropy production is attributed by the author in the integer model to the instability of protons caused by their rotation around a source of strong field keeping the atomic moments directed at the source. This is not expressed in an equation. The loss of energy by protons is considered in the integer model to express a balanced thermo dynamical condition. In terms of integer SI units with 6×10^{54} electrons per kg of mass the Principle of minimum entropy production the loss of energy per kg is $6 \times 10^{54} h_s$. Entropy is disordered energy unavailable for useful work. The author postulates that probability converts its fraction 1/1000 into vectors. For sun according to Equation 21.40 the resulting gravitation force 6000 K$\times 6 \times 10^{54} h_s \times 10^{-3} = 3.6000 \times 10^{22}$ N. Earth interacts with sun in one plane and energy conversion to vectors requires only two factors 1/10. Hence with other correctors earth temperature requires factors $\sqrt{3} \times 1/10 \times 1/10 \times 1.1547$ to generate 3.6×10^{22} N:

$$\sqrt{3} \times 300 \text{ K} \times 6 \times 10^{54} h_s \times 1/10 \times 1/10 \times 1,1547 = 3.6000 \times 10^{22} \text{ N} \qquad (23.\text{e})$$

Thus even low earth temperature appears able to generate 3.6000×10^{22} N opposing sun attractions in the process of minimum entropy production. Although Equation 23.e quotes correct energy it does not state that energy is released. This occurs during earth rotation and it is not expressed in the formula. In macroscopic terms acceleration of earth has been demonstrated accurately in Equations 6.9 and 6.10 both in classical SI units and in integer SI units to generate repulsive force to sun attraction.

Since in common view entropy expresses irreversible waste energy, its use in gravitation can be severely objected. On the other hand James Kay's interpretation of the Principle (see WEB), by 1977 Nobelist I. Prigogine, demonstrates simply that thermodynamic system can be in steady state, while not being in equilibrium. The integer model uses numbers and probabilities; numbers fit the data showing that small fraction of entropy energy can be converted by probability into vectors.

The excellent accuracy obtained in Equations 17.50 and 17.51 needs further explanation. In the absence of unit of length and of time there is no acceleration. Stationary observer on earth feels no acceleration and no velocity. The variable is replaced by the rate of pulse production at atomic level and the probability of the formation of pulse at the velocity of light at the atomic level The alternative, but accurate derivations of gravitation have to be looked from different perspective of time and of place of formation within the atom or between the bodies of the Universe, be static or relativity corrected, with acceleration or in absence of acceleration, generating purely by thermal effect, involving instability of proton, using principle of production of minimum entropy, using mass and radiation provided by mass with resultant probability being needed to generate velocity of light from space geometry.

Each of these methods derives correct and accurate gravitation values from, which some new rules of physics appeared evident.

The important conclusion is that the Universe is not in chaos, it is in a very long term steady state. The conversion of proton mass into radiation defines the Universe age, it is the source of thermal gravitation and it activates vegetable and animal growth in the local areas of high field intensity and might have effects on their demise. This source of energy is the cause of attraction between elementary masses on earth and the large bodies in the Universe.

The book starts with the Table 1, where the proposed integer values of the physical constants are listed. They have been initially guessed to express integer values and latter derived from probability considerations and directly from integer relativity derived Planck units. They also fit the experimental values in the same way as the classical SI units. The model Planck's units derived integer values for parameters, and the classical values were shown in most cases to be related to them by relativity and quantum corrections, as indicated earlier. The overwhelming importance of quantum probability as the enumerator of the elementary formative radiation process in physics has been realized only in the final book review.

Since earth is cool radiations from bodies on earth express infrared frequencies 1×10^{12} Hz or 5×10^{12} Hz expressing oscillations of equivalent electron which is the (proton+electron) pair or proton. Electron oscillations are man made or derived from lightning. Radiations from space are not considered now. Both Planck and Boltzmann constants express infrared parameters derived many times in the integer model. Classical SI system does not express its view, researchers being satisfied by deriving correct energies. Hence two new parameters need to be added to the list. **Planck constant for electrons $h_{selec} = 1 \times 10^{-36}$ J-s and Bolzmann constant for electrons $k_{select} = 1 \times 10^{-26}$ J-s have been derived in Equation 25.57 and 25.59a. Both new constants have 1000 smaller values than k_s and h_s.**

Probability establishes the relations listed in conclusion 64 between the model SI units and the atomic events. In comparison to their energy sources directed radiation **particles have reduced mass and momentum** by a factor $1/1.2$, $1/1.2^2$, $1/1.2^3$, or $0.5/1.2^3$ because radiations components discharging at right angle to the propagation orientation are not transmitted. Particles of **space radiation in integer cubic units are smaller by above factors than the classical SI particles and they are increased by relativity contraction because they move at the average velocity $1/3$ c_s**. That figure replaces c_s in several classical expressions. Since the observer and particles are usually at relative velocity $1/3$ c_s they require relativity corrector 1.1111 . . ., which is present in most relevant expressions in Table 1, Table 2 and in Equations or Identities. In this book the integer values were applied also to the mass of sun, to the mass of earth, to the earth velocity and to their space separations. The integer cubic SI units are fitted to the classical SI system by applying to it relativity and quantum corrections. Single elementary atomic events in local pathways are expressed by probability of unity and non-events by zero in the same way as in computer language. The computers are copying the most secret language of the Universe events as one and its non-events as zero; the living nervous system uses a similar system. **The space geometry displays events of unity probability** with a wide range of energies moving locally with components of velocity of light smaller than velocity of light. Component velocities in orthogonal orientations travel within atoms or molecules, during the formation of the wave packet and all along the radiation beam. The classical view that energy cannot be created or disappear within the Universe is not questioned. Term nulling incorporates the concept of redirection. Velocity components at right angle to propagation axis are nulled. Coordinating these values with the geometry of space and the classical SI system generates probabilities of pulses traveling in the correct direction. Energy pulses are converted into vectors of the integer space model SI units. The application of the integer model SI units to the most common laws of physics derives the correct numerical results.

All integer value proposed are average values of individual pulses displaying Bose-Einstein or Normal distributions not discussed in the book. In that sense the reality of distributions is replaced in the integer model by a non-existent exactness. Physical constants are per kelvin so uncertainty of measurement at temperature T becomes at least T times greater. When terms are expressed per second the uncertainty increases with the square of frequency, e.g. $G_s \equiv h_s(1/5v_{Lms})^2 = 1 \times 10^{-33} \times (1 \times 10^{12})^2 = 1 \times 10^{-10}$ J-m^2/s. The number of degrees of freedom, DOF, establishes the measurement accuracy of events.

Integer values were found to apply to momenta and to several energy parameters calculated in the book, i.e. 1 eV, 1000 K, 3 eV, 10 eV, 30 eV, 0.01 eV, 0.001 eV and many others. Four limbs, ten fingers of both hands and toes indicate the ancient maxima fields inducing growth in the directions of the limbs and digits in the early formative stages of the fetus. These numbers four, five and ten control also event probability. These features were most likely created when sizable vertebrates moved some 400 million years from sea water on land and were subjected to unmodified forms of electrical fields from the sun.

All values in this Chapter are the integer cubic model values unless otherwise described. Some arguments are repeated in the book to stress the correctness of the concepts postulated. The book presents research results needing corrections and contains old research material needed for building the model.

Generation of the resultant radiation pulse involves several steps and each step involves the following characteristic features:

1. Formation of wave function formation expressing vector addition of elementary velocities.
2. Arithmetical addition of energy components in each step.
3. Formation of the products of probabilities of all events taking part in the stepped process.

After the very end of research it was realized that interactions within the atom and between radiation particles are subject to the rules of **elastic kinetics having no units of time or of length**. It explained lack of dimensional agreement in the relativity corrected integer relations proposed in the book expressed in SI units. With elimination of units of length and of time the atomic level equivalences have been now extended and demonstrated to apply to atomic interactions for absolute values of

$$\textbf{Force = Mass = Power = Momentum = Impulse = Energy} \qquad (23.1)$$

It means that the numerical magnitude of power, momentum or mass specifies fully above other parameters. The mass parameter appears in space at different rates specified by the second parameter expressed by equivalences of

Time = Velocity = Frequency = Rate = Stress = (Temperature)² (23.2)

The product of two above parameters establishes input magnitude of the signal. To generate velocity of light vector addition in three or two planes has to be repeated two to four times, because the velocity of proton is a quarter of the velocity of light. Appearance of radiation in specific direction is expressed by the input value and number of other directions creating probability of the event. Repetition of the event creates product of probabilities. The product of probabilities of the parameter in each orientation generates the output radiation value. Several steps involves creation of the products of low probabilities reducing the average output value to a chance of a lottery win.

Above equivalences with two sets of SI values are the reason for being able to express apparent gravitation in ninety six expressions using physical constants in Table 5. For example radiation energy of 2 J was derived in Equations 10.45, 12.2, 16.2, 16.6, and four others and it is derived as energy pulse from the sun $\mu_{os}/\varepsilon_{os} \times 3 E_{gl}/X_L = 10^{-5} \times 3 \times 10^{16}$ J/1.5 $\times 10^{11} = 2$ J. The basic theory can be presented in a much shorter form with the present research book presenting the development of new ideas step by step.

The above equivalences are the cause of defects and the lack of a clear gravitation theory in the classical SI system and in a smaller extend in the proposed integer cubic mode SI system, which introduced relativity and quantum corrections. Expressing observations by probability are correct with details of construction of events being lost, as events take place in space.

Consequently above equivallencies bring havoc to SI dimensions for problems dealing with atomic interactions. The clarification of the lack of dimensional agreement in expressions justifies the proposed model. The proposed independance of events in orthogonal orientations is very reasonable as well, as is the presence of 1/3 c_s velocity components, of structure of hydrogen. They are all justified by numerical data. Nevertheless Universe was still missing its source of power. Book prematurely printed in the spring of 2012 had errors and dimensional problems, which were not cleared. These books were not released for sale.

At the end of Chapter 22 four independent conditions have been proposed for sustaining sun to earth gravitation balance.

A. In the vicinity of earth vectorized thermal flux of earth must be equal and opposite to that of sun that is spread radially from the sun in all directions with intensity decreasing with the square of the distance. The vectorized forces of thermal flux from earth are directed radially outwards from earths in all directions. Only the direct linear sun beam reaches the earth. This is shown in Figure 8 of Chapter 7.

Using earth power expressed by the product $6_s T_{earth}{}^4$ volumetric surface area of earth seemed initially to derive much smaller powers that sun gravitation. Table 5 does not disclose any gravitation forces using earth temperature. However temperatures, being independent rates in the three orthogonal orientations, add and three 300 K pulses in orthogonal orientations corrected by 1.1547 relativity generate 1000 K or 1000 J pulse of the sun of Equations 2.17-21 and others. Two such pulses would create rate 2000 and increase the initial propton frequwency of ¼ c_s to c_s. Actually the relativity component $1.1547^{1/2}$ should be applied twice. Then the surface temperature of the volumetric earth surface delivers sufficient power to oppose sun attraction:

The force of earth attracting sun = $6_s T_{earth}{}^4$ times earth surface, relativity and quantum correction:

$$= 5 \times 10^{-8} \times (2000\ K)^4 \times 4\Pi \times (6.3662 \times 10^6\ m)^2 \times 10 \times 1.0328^{15/112}/1.1547 = 3.5438 \times 10^{22}\ N. \tag{23.2a}$$

This is within the accuracy limits of the reference 6.1. The number 112 represents 28 quantum levels times 4.

It took some time to derive similar relation attributing the gravitation to the force generated purely by the temperature of the mass of earth. Mass of earth has also to be referred to temperature to power 1/2, the two pulses from three orthogonal orientations are vectorized by 10^{-6}, and relativity correction applied:

$$6 \times 10^{24}\ kg \times (300\ K)^{1/2} \times 10^{-6} \times 1.1547 = 3.600000 \times 10^{22}\ N \tag{23.2b}$$

The above equation is accepted on its accuracy. While Equation 23.2a requires temperature to have normally power of four, power of ½ is acceptable with force being a linear parameter.

It is believed that because of equivalence of meaning of different parameters both above equations are valid. Thus the vectorized thermal flux of earth creates from the whole earth surface the gravitation force opposing sun gravitation. The energy for that force is provided by the converted mass of destabilized protons.

The value $m_{earth} T^{1/2} = 6 \times 10^{24}\ kg \times (300\ K)^{1/2} = (1 - 0.0074) \times 10^{26}$ expresses number of events. Since the event is an integer total force its element must be integer $muo10_s = 1 \times 10^{-28}$ J generating the 1st term of Planck radiation 0.01 J or eV/per atomic event. Converted to atomic level by surface converter 1×10^{-20} and 1/10 derives the Boltzmann constant $k_s = 1 \times 10^{-23}$ J.

Thus

0.01 J $\times 1 \times 10^{-20} \times 1/10 \equiv k_s = 1 \times 10^{-23}$ J, suggests that 0.01 has two degrees of freedom   (23.2c)

B. Independent force of the sun beam is carrying 1/30 of the total sun radiation with power, momentum, energy and mass sustaining the same value all along one plane of the sun-earth axis. The gravitation is balanced by the outwards centrifugal force of earth rotation around the sun. This is shown in Figure 7 of Chapter 4. The sum beams are subjected to magnetic condensation around earth.

C. Thirdly all linear and rotational momenta should add to zero to sustain stability of the planetary system. This totally mechanical explanation is outside the author's interests, but there are books, which attributes gravitation solely to such equilibrium.

D. The number of atoms taking part in the radiation must be sufficient to generate from their reduced mass energy of 5 J for five orientations of 1 J:

$$m_{ps}c_s^2 \text{ x cor} = 1 \text{ x } 10^{-27} \text{ x } (3 \text{ x } 10^8)^2 \text{ x } 1.1111\ldots = 1 \text{ x } 10^{-10} \text{ J} \rightarrow \text{x}1 \text{ x } 10^{10} = 1 \text{ J SI} \qquad (23.2d)$$

The minimum pulse converts 18 atoms of hydrogen into helium having atomic weight 4.

The above defines the gravitation. There is no room in the integer probabilistic model for cosmological constant, or for other Universes. The age of the Universe is derived in Equations 21.24, 21.27-8 and 21.38 from the amount of energy discharged from mass. This process can proceed in different ways. Thermal radiations from the sun contains some twenty photons, which have to be vectorized to generate velocity of light through atomic mass to energy conversion. It is suggested that **vectorized thermal flux takes energy out of the sun mass and radiates it radially to infinity with its fraction being being nulled by the gravitation vectors from the planets. The term "nulling", used by the author, implies that moments are probably mutually reflected.** The end of the Universe is identified with the exhaustion of energy in the mass of the Universe consumed by the gravitation vector forces from all bodies. In equivalent terms momenta of protons are dissipated. The energy discharged by earth and by sun are the minimum rates of production of entropy discussed earlier.

It is suggested that **on collapsing the proton discharges its energy expressed by its mass and velocity of light squared and by relativity correction**:

$$m_{ps} \text{ x } c_s^2 \text{ x } 1.111\ldots = 1\text{x}10^{-27}\text{x}9\text{x}10^{16} \text{ x } 1.111\ldots = 10\varepsilon_{os} = 1 \text{ x } 10^{-10} \text{ J} \qquad (23.3)$$

m_{ps} x c_s^2 expresses the wave function format of two orthogonal pulses $(3\text{x}10^{-14} \text{ x } \sqrt{3}\text{x}10^4)^2$ in atomic units and Equation 23.3 is really in atomic terms. To express it in integer SI units requires multiplication by $1\text{x}10^{10}$ making it 1 J per oscillator. It follows that atomic masses refer to 300 K. It seems that 10000 K, expressing $\sqrt{3}\text{x}6000/1.04$, being in three orthogonal orientations generates rate $\sqrt{3}\text{x}10^4$.

At temperature of one kelvin the mass of one kilogram collapses one proton per second and generates the above power or energy of $1 \text{ x } 10^{-10}$ W, N or J. Physics books do not state that SI masses refer to 1 K. Nevertheless this appears to be true for earth mass! It is surprising that earth mass has to be referred to 300 K! The Special Relativity gravitation supplied by earth has been identified with generation of energy by the collapse of protons structure with probability $1 \text{ x } 10^{-10}$ W/kg per quantum direction expressing at SI level probability 1 W/kg. For sources in three orthogonal directions the probability of outcome becomes $1 \text{ x } 10^{-30}$ W/kg and the probability at 300 K as derived in Equation 25.56 becomes Planck constant h_s.

Earth attracts sun with Special Relativity force of:

$$6 \text{ x } 10^{24} \text{ kg x } 300^3 \text{ x } 1 \text{ x } 10^{-10} \text{ W}/1.1547^{1/2} = 1.50758 \text{ x } 10^{22} \text{ J or eV} \qquad (23.4)$$

The multiplier 300^2 x $1.111 = 100\ 000 \equiv 10^5$ J has been identified earlier as the earth pulse. The temperature is an indicator of rate of pulse generation with energy or power coming from the destruction of proton structure. It follows that the popular particle and radiation wave equivalence of the old form should be abandoned.

With 12 positions per plane there may be 36 positions per spherical volume, if discharges occur sequentially in all directions, with typical values being 28, 30, 32, 34 depending on the circumstances. These quantum numbers determine energies exchanged between the source and the receiver and above 32 k_s is a good example.

Minimum entropy production principle generates its powers from most masses of the Universe by the gravitation forces and their vectors generated through probability. This process occurs at low probability establishing events of the Universe by delivering thermal energy for many processes. Probably most of this energy disappears at infinity around a semi-spherical Universe with a fraction running the events in the Universe. It is claimed that the collapse of proton structure is a constant source of energy for $1.1547^2 \times 1.0328 \times 10^{10}$ years equal to 13.75 billions years calculated by CERN.

Masses of the planets very far from the sun may not sustain their molecules in line with the electrical field of the sun for the whole year. Consequently there is significantly less energy loss than 1 J/kg/year as discussed in Chapter 20. This does not affect the life of the Universe, because the explosion of the inner core of planets would destroy the whole Universe.

In relaxation studies of NaCl crystals the rate of current decrease is stepped. On first differentiation current is expressed by series of straight lines of decreased magnitude and on second differentiation by a series of decreasing well separated symmetrical pulses described in the Appendix. Journals of Physics informed me twice that readers are not interested in oscillations of little particles. And these little quanta oscillations measured by the author 45 years ago and published later displayed the most common source of the energy in the Universe, which is responsible for its longevity and for gravitation between sun and planets. Furthermore water has been long considered a necessity of all living matter. The author attributed it to its very high relative permittivity increasing the speed of reactions by about 10^5 times and **he identified water to be the engine of life.** Now **the engine of life needs energy and that energy is provided of the collapse of protons,** which convert their mass into radiations in locations where the probability is highest due to space geometry. Some chemists attribute some exothermic processes (those generating heat) to conversion of mass to energy without specifying details. Such process induces radial radiations in all directions, which after interactions within the volume of rounded Universe volume has only one way out representing total dispersion to infinity at negligible density obeying the Second Law of thermodynamics.

The energy 1×10^{-10} J is related to probability 1×10^{-10} and it expresses 10 quantum positions in one plane of the Bohr atom with two more positions not interacting representing the hours of the click. The energy in each is 10 J, which for three orthogonal positions generate 30 J exactly. Two pulses relativity corrected in wave function format generate $\sqrt{2}$ lager vector value:

$$(23.4a)$$

$$(T_{300}^2 \times \sqrt{\sigma_s}) \times (T_{300}^2 \times \sqrt{\sigma_s}) \times 1.111\ldots^{1/2} \equiv \sqrt{2} \times 300^2 \times \sqrt{(5 \times 10^{-8})} = 30.00000000 \text{ J}$$

Please note that the accuracy is limited only by the accuracy of the calculator. The classical expression for power emission, $\sigma_s T_{300}^4$, is not negated. The power from earth $\sigma_s T_{300}^4 = 5 \times 10^{-8} \times 300^4 = 405$ W/m^2 has to be added to sun constant to derive the total power generated from the sun $405 + 1372.5 = 1777.5$ W/m^2. Twice that value times E_{searth} and correction $1.04^{2/25}$ derive exactly the gravitation force value $2\,E_{searth}$ x $1777.5/1.04^{2/25} = 3.5438$ x 10^{22} N. Square root of 405 W that is $(\sigma_s T_{300}{}^4)^{1/2}$ with correction derive energy of 20 J:

$$\text{Energy in two dimensions} = (\sigma_s T_{300}{}^4)^{1/2}/[1 + 1/(8\text{ x}20)] = 19.9996142\text{ J} \approx 20\text{ J} \qquad (23.5)$$

It is proposed that the loss of proton structure is besides being the source of gravitation forces may be also the source of energy of other radiations, which are hard to be justified. This view **expresses numerically the Principle of Minimum Entropy Production** [64, 65] and prevails in chemistry. This is the energy that feeds events in the Universe and disappears in the infinity.

Life of the Universe in seconds=13.751 x1x10^9x31.557x10^6x1.1547= 5.0107 x 10^{17} s
$$\equiv 5/(\mu_{os}\mathcal{E}_{os}) \equiv E_{sspace}\text{ J or s, because with power 1 W at 300 K, Energy = 1 x Time.}$$
$$(23.5a)$$
Life per quantum orientation is 1 x 10^{17} s. On conversion from SI to atomic units involving two pulses, above value is decreased by 10 x10 to 10^{15} s $\equiv 10^{15}$ J or W. The interacting mass of sun 1 x 10^{30} kg generates 1 W/kg-K expressed by 1 x 10^{30} W. That power expresses the four formative pulses each $(10^{15})^{1/2}$ J or W:

$$\text{Sun power/K: } 1\text{x}10^{30}\text{ W} = (10^{15})^{1/2}\text{ x }(10^{15})^{1/2}\text{ x }(10^{15})^{1/2}\text{ x }(10^{15})^{1/2} \qquad (23.5b)$$
$$\text{And each }(10^{15})^{1/2}\text{ W} \equiv 3\text{ x }10^8\text{ x}1.111\ldots^{1/2} = (\sqrt{3}\text{ x }10000\text{ K x }1.111\ldots^{1/4})^{1/2} \qquad (23.5c)$$

Mass of proton 10^{-27} kg in one direction, which is also the force times duration of Universe 10^{17} s equals the proton radiation impulse 1 x 10^{-10} kg-s. Longevity of one kg of protons is 1 kg/1x10^{-10} kg-s $\equiv$ 1 x 10^{10} years, because unit of time in space is one year. Life of the Universe expresses time taken to dissipate in one direction energy stored in protons after converting them to SI

$$10\text{ x }(1/10^{-27})^{1/3} = 1\text{ x }10^{10}\text{ years.}$$

Thus duration of the Universe expresses longevity of all its protons, which rotate around other body in mutual inductance relationship and by doing so discharge the spin energy h_s.

In **integer cubic SI units collapse of proton expresses Planck Radiation 1st term**

$$m_{ps}\text{ x }c_s^{\,2}\text{ x }1.111\ldots\text{ x }1/3\,c_s = 1\text{ x }10^{-10}\text{ x }1\text{ x }10^8 = 0.01\text{ J} \qquad (23.6)$$

this at atomic level in integral cubic SI expresses 0.001 J. Referring to 300 K in integer SI units derives 3 J for three oscillators and 2 J, if the electron is not absorbed; energy for three planes with relativity correction is 10 J. Converted to atomic level 0.001 J expresses Planck constant h_s =1 x 10^{-33} J.

The above important conclusion signifies that the unity vector pulse at 300 K of 1 J or 1 eV per atomic event appearing per quantum orientation releases vectorial energy contained in the proton expressed in its mass 1 x 10^{-27} J in one direction. That mass has to be referred to 300 K in three directions relativity corrected, expressing now energy 1x10^{-24} J. On vectorization by 1 x 10^{-3} generates back 1 x 10^{-27} J value, which expresses energy of 10 muo10s$_s$. Thus radiation vector of energy or of power at 1 x 10^{-27} J or W is most likely to be the result of the collapse of the proton structure, which occurs at the rate of one per year per kilogram of the mass. That rate normally increases with the temperature squared, with the need of two pulses increasing the power to four or even to eight. The inherent signal pulse rate per kelvin has been identified as Boltzmann k_s, while on earth surface the elementary energy pulse without electron contribution is Planck radiation term $2/3 \times \frac{1}{2} k_s T_{300}/e_s = 0.01$ J. In NaCl, vacancy particle stimulating H atom energy appears $\sqrt{3}$ larger as shown in the Appendix. This is 1.2 less (due the no reacting component) than the theoretical term expressed in integer relativity units $k_s T_{300} \times \ln 2/e_s \approx 0.0208$ J.

The sun in its rotation radiates in four directions, so that probability in any one is ¼. The mass of 1 kg referred to 6000 K infers rate 6000, which with probability ¼ times relativity corrector 1.1547 over quantum corrector deliver energy 10 J or eV with variance of 40 k_s, which is k_s per direction:

$$1 \text{ kg} \times 6000^{1/4} \times 1.154700/(1+1/60) = (10 - 0.004) \text{ eV} \qquad (23.7)$$

Static $1/2$ k_s variance expresses hard to explain presence of 80 DOF. Kinetic $1/2$ k_s displays 120 DOF suggesting presence of two protons or H atoms and a two stepped process generating velocity of light. Alternatively from two planes each of 4 directions with 5 quantum orientations the probability is 1/40 with 2/3 to segregate electron; with relativity corrector 1 kg generates:

$$1 \text{ kg} \times 6000^{1/20} \times 1.111 \ldots /1.1547 \times 2/3 = (1 - 0.00893) \text{ eV} \qquad (23.8)$$

with very acceptable variance per orientation $89.3/20 = 2.977$ $k_s \equiv 3$ k_s, requiring another step. Reflection at the earth causes the decrease of probability to 1/40:

$$1 \text{ kg} \times 300^{1/40} \times 1.154700(1 + 1/800) = 1.000000 \text{ eV} \qquad (23.9)$$

All these energies are the result of a natural, regular conversion of proton mass to energy. Alternative processes may occur and plausibility of their occurrence remains still uncertain.

Universal value 10^{10}. The period of stability of protons 10^{10} years has been identified as the age of the Universe; it is a relativity related to NASA estimate of 1.373×10^{10} years. Invert of 10^{10} is also the international gravity constant for hot bodies and $2/3 \times 10^{-10}$ for bodies in the solar system. It is the probability of twelve quantum orientations reduced to 10 interacting to generate velocity of light in any orientation of their plane. In SI units those are 1 W radiation of 10 such values per plane. In atomic terms 1 W becomes 1×10^{-10} W. Space permittivity ε_{os} is the probability that 10^{-10} W pulse will appear in any the specified orientations: $\varepsilon_{os} = 10^{-10} \times 1/10$ and Stephan-Boltzmann constant $6_s \equiv 6000$ K/1.2 $\varepsilon_{os} = 5 \times 10^{-8}$ W/m². The linear probability is 1×10^{-10} and 1×10^{-10} m is the diameter of hydrogen atom. The sun power in SI units and one direction is $(1 \times 10^{10})^{10}$. It is suggested that gravitation constant $G_s = 1 \times 10^{-10}$ is the invert of the age of Universe 1×10^{10} years and also the invert of the longevity of proton of years 1×10^{10} years meaning its conversion to radiation expresses 1×10^{-10} probability per second of the occurrence of the unity reference event. It is also: Bohr space diameter $a_{os} = 1 \times 10^{-10}$ m, $(\varepsilon_{os}/\mu_{os})^2 \equiv 1.0 \times 10^{-10}$ J² so that $a_{os} \equiv (\varepsilon_{os}/\mu_{os})^2$, frequency $= 1 \times 10^{10}$ Hz at 1 K as in Equation 16.11, energy of proton mass at velocity of light 1 **x** 10^{-27} **x** 9 **x** 10^{16} **x** 1.1111=1 **x** 10^{10} J and in Equation 16.60 Coulomb constant =1 **x** 10^{10}.

Not all hypotheses proposed here are necessarily true, but evidence presented is overwhelming. Scientists may be unwilling to admit for being eluded by nature so long. The model postulated is not undermining validity of the physics based on the classical SI units. It is adding a huge amount of new information about the physics and about the Universe.

The model postulates the presence of local velocities of light, being independent orthogonal components generating velocity of light c_s. These components require relativistic and quantum corrections, the bases of the integer cubic model. The orthogonal component light velocity is square of sun temperature $(10000 \text{ K})^2 \equiv 1 \text{ x } 10^8$ m/s, while the temperature 10000 K is the pulse rate created by vectorially adding three 6000 K pulses in orthogonal directions expressed by 10000 K = $\sqrt{3}$ x 6000 K/1.04. The velocity of light is $1/3^3$ of the fourth power of earth temperature $1/3^3$x $(300 \text{ K})^4 \equiv 3 \text{x} 10^8$ m/s. All these concepts necessitate introduction of a new parallel set of integer cubic physical constants derived directly from Planck units. It is suggested that the classical SI set of physical constants have imbedded relativity and quantum corrections, which hide the gravitation features of the Universe and its basic probabilities of specific events.

Presently the mass of any atomic or celestial body has been identified with the number N_e of electron, N_e = 1x10^{30} per kg mass. This is unifying the mechanical and the electrical sciences. Mechanical forces can be represented by electrical forces and vice-versa. **Several masses of planets can be treated as electrical charges their motion creating current generating electromagnetic forces inducing gravity** repulsion and attraction according to the classical rules.

The above corrections and additions did not end the difficulties of applying the practical classical International System to atomic problems. Presently Science recognizes that **the world is probabilistic**, with numerical value applied to the density of charge around nucleus; this is not event probability postulated in the book. The classical SI system appears to work almost perfectly, but inconsistencies needed to be explained. Numerical aspects of probability were uncovered in the eleventh year of the research. **Most dimensional inconsistencies are the result of trying to explain in static SI units kinetic events having different units**.

Max Born was first (1928) to discover that the square of quantum wave equation predicts the probability of finding the electron [55]. He deals with the probabilities of transitions and mainly with the integral of the square of quantum wave over volume of the spherical space generating unity values for all shapes of charge distributions postulated by Schrondinger. Thus except for the unity value, his probability concepts have no common features with the probability model postulated now. According to Cassidy [58, p. 110, reworded] Heisenberg believed in the probabilistic world because of H. u. causal connection between the present and future is lost and quantum mechanics becomes probabilistic or statistical in nature. It is suggested now that **all events are probabilities of the reference event occurrence;** the space map of such probabilities creates the picture of the world that we see. The probability concept applies both to the integer relativistic SI model and to the classic SI physics. For earth the sun gravity is the corrected square of bodies' separation, Equations 16.37-40. The forces of sun gravitation in one orientation over the reactive sun mass express probability $1/3 \times 3.6 \times 10^{22}$ N/(1×10^{30} kg/1.2) = 1×10^{-8} N/kg, which multiplied by local velocity of light 1×10^8 m/s delivers the unity W/kg reference pulse. The integer model, besides relativity and quantum corrections, deals only with the probability of occurrence of the mean value of the parameter in relation to the reference.

In the previous book [46] the electron and its radiation was perceived as the ultimate source of everything. momenta of electron radiation may probably recombine forming protons, Bohr atoms and hydrogen. The measurers of electron frequencies [10], derived now, were awarded Nobel Price. The frequency 5×10^{15} Hz postulated presently is directly related to the value measured by Weitz et ales [10]. Presently it is suggested that most likely the world is essentially built on signals from groups of 2001 electrons forming Bohr atom or 4002 electrons forming hydrogen molecule. The signal perceives length of 1×10^{-10} m, mass 1×10^{-27} kg, and time in one of five of quantum pathways $5/(5 \times 10^{15})$ seconds. The pulse energy of 2 eV or at 300 K converted to vector form expresses the Planck Radiation Law starting with 0.01 W. Further research indicated that features attributed in classical physics to an electron apply to an equivalent electron formed by the (proton+electron) pair and the view that the world appeared suddenly is equally plausible. Its protons carry energy per oscillator, which is continuously discharged at 1 W per kg at 300 K creating relatively corrected energy 30 eV from Lyman wavelength 400 nm and local velocity 1×10^8 m/s deriving an exact value $\lambda \times 1/3 \ c_s/1.154700538^2 = 400 \times 1 \times 10^{-9} \times 1 \times 10^8 Hz/1.154700538^2 = 30.000000000$ J. That is the energy of the collapse of the proton and hence of the hydrogen atom and other structures. That collapse is the source of all energy in the Universe the energy generated all the time and used to create other structures and planets. Energy of proton takes 10^{10} years to be dissipated putting limit on the longevity of the Universe. One joule per meter pulses are needed to create velocity of light generating at infrared 1×10^{12} Hz earth pulse $2 \times 400 \times 1 \times 10^{-9} \times 1 \times 10^{12} Hz (1+1/80) = 1.000000000 \times 10^6$ W.

Permittivity of space ε_{os} expresses Planck radiation energy times linear probability over local velocity of light $1/3 \ c_s$, $\varepsilon_{os} = (0.01$ W $\times 1/10) \times 1/10^8 \equiv 1 \times 10^{-11}$ J/m^3. The linear probability structure of 1/10 and local velocity of light $1/3 \ c_s$, converts the 2 eV signal (3 eV at high temperatures) of one plane into energy density of space radiation ε_{os}. Conversion of 1 J pulse signal expressing both the classical SI system and the integer SI model to SI per degree level/K requires division by 300 Kx3x1.111 . . . generating per orientation $\mu_{os} = (\varepsilon_{os}/10)^{1/2} = 0.000001$ J. Converting to atomic level times 10^{-30} derives Planck constant $h_s = 1 \times 10^{-33}$ J-s, which is interpreted as collapse of Bohr structure of energy 30 eV in all directions with 1 J per direction at 300 K expressed by k_s.

Event probabilities at atomic natural level converted from model SI values are consecutive or concurrent probabilities of one step space quantization 1/10 taking value 10^{-10}; three orthogonal vectors combine into a probability value 10^{-3}; while repeated for three planes they become 10^{-9}; this value derives now electron charge $10^{-9} \times 10^{-10}$, which with the minimum energy 10^{-14} J defines Planck constant $h_s = 10^{-19} \times 10^{-14}$ J-s and with invert probability 10^3 derives the electron mass 10^{-30} kg. Both model integer physical constants and the classical SI are obtained from these and other similar probabilities creating valid laws of physics. The integer physical constants are derivatives of the integer atomic muo10$_s$ mass of 1×10^{-28} kg $\equiv (10^{-14}$ J$)^2$ generating a muo10$_s$'s force 1×10^{-28} N. The squared input of ten energies 10^{-14} J obtains the Planck constant for electrons $h_{selec} = (10^{-13}$ J$)^2/10^{10}$ $= 10^{-36}$ J-s. The research has identified presence of three oscillators in Bohr atom and it has identified the radiation energy unbalance between two celestial bodies caused by different frequencies generated by the hot and the cold masses.

The primary source of signal in space is the two dimensional (rotational) radiation oscillation of space $(1 \times 10^{-14})^2$ expressing energy of the muo10$_s$ 1×10^{-28} J/m^3 per pulse which from the minimum mass 1×10^{-14} J, times local frequency 1×10^{15} Hz and one step quantization delivers the energy $(1 \times 10^{-28}) \times 1/2 \times 1 \times 10^{15} \times 1/10 = 1$ W/m^3. The elementary power 1 W/m^3 times volume probability 10^{-30} and step quantization for three orthogonal directions $1/10 \times 1/10 \times 1/10$ derives the model Planck constant $h_s = 1 \times 10^{-33}$ J-s. Display of energy h_s expresses natural expiration of a Bohr structure. Similar radiation interactions between electrons are assumed to generate protons and Bohr atoms. Bohr atoms congregate into domains of clouds. The almost negligible radiation from 1 kilogram of mass (or 1×10^{-30} kg at the atomic level) at 1 K becomes a sufficient force to initiate process of congregating atoms into huge high temperature masses forming stars. Hydrogen atoms combine into helium producing gravitation pulses of sun attraction; this involves mass to energy conversion inside the sun. The final step of the processes involves mass to iron conversion, which is considered to end the sun life.

Orthogonal directions of space along which radiation discharge are taken to be totally independent. The pressure on the sun gives Bohr atoms the cubic shapes. **Average random positions** of distributed pulses in six orientations can be represented by a cubic shape, in which only one dimensional pulse is considered and pulses directed at right angle to the reference are nulled, because they add to zero. This presentation with the orthogonal geometry of space generates a set of space physical constants having integer values. The Special Relativity derived a range of sun-earth forces of gravitation displaying value 3.0000×10^{22} N. The Relativity theory predicts that observers moving with different fractions of the velocity of light attribute to the same event various values of time duration and of distance. Both sun and earth supply one half of the above forces each value backed by a separate source of energy. Earth circulates around the sun and sun circulates around the earth so that repulsive force applies to both. Sun photon radiation increases the earth temperature and hence the rate of the natural collapse of Bohr structures displayed as 1 N per kg per K or at atomic level by Planck constant h_s. It took eleven years to formulate and explain in Equations 17.50-51 the balanced gravitation forces of sun and of earth based on temperature.

The existence of the second set of physical constants is attributed to the second value of electron charge e_s derived now from Milliken experiment. Bohr atom was identified to have 2001 electrons, so that the moving mass of a planet can be represented by a number of electrons generating a current inducing a physical force. Equation 19.13 will now be so updated by using integer values. The earth current $I_{earth} = 30000$ m/s x 6 $\times 10^{24}$ kg/1 $\times 10^{-30}$ kg x $1 \times 10^{-30} = 1.8000 \times 10^{29}$ A. The force of earth repulsion is now $F_{Gr} = 30000$ m/s x $(1.8000 \times 10^{29}$ A$)/1.5 \times 10^{11}$ m $= 3.6000 \times 10^{22}$ N. Divided by quantum corrector $1.04^{2/5}$ the result 3.5440×10^{22} N is within the accuracy of the reference.

Since the gravitational force decreases as the square of the distance, the resultant linear gradient of mass representing the gradient of force dF_{Gs}/dx is inversely proportional to the sun distance L_{eath} to earth. It produces the relations detailed in Equation 16.20 deriving at 1 K the Boltzmann constant k_s.

The Universe must obey the second law of thermodynamics suggesting that the Universe ends as a huge space with radiation spreading outwards. If however the space is curved as suggested by Einstein the infinity may close on itself. Einstein general theory of relativity treats gravitation not as a force, but as a curvature of space and of time around a body. Einstein's view is not contested here. It may be matter of scale with curvature appearing over long distances.

The starting point of the present radiation model are three 10 eV or J in forward orthogonal directions generating energy $\sqrt{3} \times 10 \times 1.15470538 = 19.99999999$ J $= 20$ J. To such pulses in one plane corrected for relativity and quantum derive integer cubic SI value $\sqrt{2} \times 20 \times 1.11 \ldots^{1/2} \times 1.04^{1/6} = 30.00977$ eV expressing variance of 97.9 $k_s/1.5\ k_s = 65 + 1/7.5$ which compares well with 64 value of Equation 22.76, while 30 eV is Hartley energy 27.2 eV $\approx 30/1.1547^{2/3}$. Maxwell attributes radiation to accelerated charges. Bessel [34] suggests that atoms discharge energy without stimulation. While these views are likely to be true, the model derived the radiation to be h_s expressing the angular momentum of reduced electron spin; the equipartition of energy attributes h_s to all sources. It is possible that electrons form protons. In the reversed process it has been demonstrated that protons disperse into electrons. However this is not the requirement for the other features displayed by the integer model. Accumulation of electrons and of protons generate electromagnetic forces, which may compete in the black holes with gravity forces.

The force of attraction of one body is the force of repulsion of the other body, which generates the same force of its own attraction on the first body. The gradient of power is the impulse of the gravitation constant 1×10^{-10} of SI units expressing unity probability at nature atomic scale decreased by the ratio of atomic to SI units. The energy density in one dimension corrected for relativity, space quantization and quantum derives energy of gravitation constant:

$$2 \times (1 \times 10^{-28})^{1/3} \times 1.1547^{1/2} \times 1/10 \times 1.04^{29/50} = 1.00003 \times 10^{-10} \text{ J} \qquad (23.10)$$

The sun long beam of gravitation force acts only in the plane of earth rotation. With ten equally probable optional quantum orientations the probability for the 1st sun pulse to discharging along the source-receiver axes towards the earth is 1/10. The probability for the 2nd pulse to be also so directed is $1/10^2$, for the 3rd pulse it becomes $1/10^3$ and for all the 10 pulses the probability is $1/10^{10}$ expressing the gravitation constant 1×10^{-10} N-m²- kg⁻² for bodies at equal temperature. For sun-to-earth pair the above value is reduced to 2/3 time quantum corrector $1.04^{1/36}$ deriving earth gravitation constant $G = 0.66739 \times 10^{-11}$ N-m²-kg⁻². This deviates from the Coda 2002 value $0.66742(10) \times 10^{-11}$ N-m²-kg⁻² by 0.0003×10^{-11} N-m²- kg⁻² identified to be 3 x the minimum energy value 1×10^{-14} J, which is ~1/3 of the standard uncertainty 0.00010×10^{-11} of CODATA 2002. CODATA 2006 data by Mohr et ales are given on WEB as $G = 0.667428(67) \times 10^{-11}$ N-m²-kg⁻².

Product of minimum pulse energy per cycle of local time at atomic level, 1×10^{-44} J, and of volume probability derives the integer space SI value 1×10^{-44} J $\times 1 \times 10^{30} = 1 \times 10^{-14}$ J. The SI minimum energy pulse is just minimum radiation energy space in 2 dimensions $(1 \times 10^{-28})^{1/2} = 1 \times 10^{-14}$ J.

How is the muo10_s value 1×10^{-28} J/m³ related to 1×10^{-27} J radiation pulse? Energy 1×10^{-27} J must be referred to all ten orientations, in the same way as 10 N/ kg to 1 N/kg.

Magnetic permeability $\mu_{os} = 10^{-6}$ N⁻² and dielectric permittivity ε_{os} of space are related in by $\varepsilon_{os} = 10\mu_{os}^{2}$.

Sun shares equal forces of attraction with all the planets not generating their own energy based solely on the square of their distance and their mass. The sun gravitation force of attraction is clear in mathematical terms. The force of attraction of the planets is expressed in terms of their mass and their squared distance to sun. All planets except Saturn and Neptune generate the same force of attraction as generated by the sun and expressed by Equation 20.2. The centrifugal force of rotation generates with correction the gravitation force for most the planets with Pluto, Saturn and Venus requiring minor corrections.

The common parameter for all the planet-sun pairs, except Jupiter, is their individual pair gravitation energy $E_{g3} = 3 \times 10^{16}$ J/K with respect to the sun derived from the geometric mass of each couple. Their product E_{g3} and sun pulse 10^6 J derives Special Relativity gravity force of earth $F_{Gs} = 3 \times 10^{22}$ N. Two planets add their own energy for balance.

The most likely cause of rotational radiation is the discharge sequence in four orthogonal directions Right! Right! Right! Right! generating spinning squares, and converting protons of earth $2 \times (6 \times 10^{54})^{1/4}$ J into its repulsion energy 6×10^{14} J. Consecutive spins in other two planes create a cube generated from simple probability rule. A small fraction of the Universe has probably developed the left spin Left! Left! Left! Left! generating a world of antiparticles, which annihilate the right turn particles in interaction.

Discharge progress, which is the velocity of light, is the consequence of three events of radiation occurring (on average) in orthogonal axes at the right angle to each other. Velocity of light was derived in several ways, with Equation 10.22 being the most significant since it describes the wave function process.

There is no doubt that thermonuclear processes in the sun generate thermal flux, which according to the model is converted by the probability of the orthogonal space into vector form expressions. The rate of generating 1 J pulses increase from 300 K with T^4, and the processes on sun are likely to be still the product of the collapse of Bohr proton structure.

The centrifugal force induced by the acceleration of the planet around the sun generates the repulsive force balancing sun gravitation in one plane; the sun has been shown to induce the same value of the centrifugal force while circulating around the earth.

The independent gravity forces are present on the whole surfaces of earth and of planets and as demonstrated in Equation 13.1 equate cross-section area of protons of earth with those of the sun radiation. From earth these radiations are derived from the collapse of the structure of protons. For sun there may be a staged process. To find pulse energy generated by the collapse of the mass of proton the age of the Universe derived by the author in years in Equation 21.54 requires multiplier 31.557×10^6 s/year, requires conversion 10^{10} from atomic to SI level, requires relativity 1.1547 and proton mass 2×10^{-27}:

Energy per pulse $= 13.7546 \times 10^9 \times 31.557 \times 10^6 \times 10^{10} \times 1.1547 \times 2 \times 10^{-27} \equiv 10.0240$ J   (23.11)

The anticipated result required by the integer model is 10.0256 J while the formative velocity process corrected probability from $1/4^4 = 1/256$ to $1/128$ expressing in six directions corrected by probability $6 \times (6 \times 10^{23})^{1/128} \times 1.1547^{2/3} = 10.0056$ J indicating variance $6k_s - 1/3k_s$. The rotation for two pulses in wave function format shows their sequential phase positions, $\leftarrow, \uparrow, \rightarrow, \downarrow$, increasing their velocity by $\sqrt{2}$. The event occurs with probability $1/4^2$. The radiation event has to be repeated to generate $\frac{1}{2}$ velocity of light with total probability $1/256$. The resultant 10.0256 J has now to appear in three orthogonal axes doubling velocity by $\sqrt{3} \times 1.1547 \approx 2$, with output of ~30 J pulse.

Converting mass of proton to energy generates $m_{ps} c_s^2 = 2 \times 10^{-27} \times (3 \times 10^8)^2 = 2 \times 10^{-10}$ J Above energy is decreased by total probability and relativity by $4^4 \times 3^6 \times 6 = 186624$ Generating:

2×10^{-10} J$/(186624 \times 1.1547^{1/2}) = 0.9973 \times 10^{-15}$ J $= (1 - 0.0037) \times 10^{-15}$ J as in Equation 2.36
(23.12)

Uncertainty expresses $36 = 3 \times 12$ quantum positions of e_s momentum plus uncertainty e_s. At 300 K $\times 3 \times 1.111 \ldots$ energy 10^{-15} J becomes energy per cycle 1 J$/1 \times 10^{12}$ Hz $= \mu_{os}^2$.

Now all energies have found their place. Understanding has reached the stage that predictions are confirmed by experimental finding and the integer relativity model becomes comprehensive. The output of 1 J requires the input of $6 \times 186624/(1.111 \ldots^{1/2} \times 1.1547^{1/19})$ $= 1.00016 \times 10^6$ J with variance 150 J, $\frac{1}{2}$ J per kelvin.

The absence of independent electrons oscillations introduces the factor $2/3$, which with two relativity corrections convert energy of 36 quantum positions into 27 positions with surprising accurate variances both for the proton and the electron:

$36 \times 2/3 \times 1.111 \ldots \times 1.0328^{1/2} = 27.10044 = 32 + 1/10 + 0.00044$ (23.12a)
the latter variance expresses $\sim \frac{1}{2} k_{selectr}$. This conversion will be used subsequently.

All sun rays are radiated in 1D because this is the requirement of velocity of light moving in one direction without dissipating energy. The rays are condensed from the vicinity of earth by earth gravitation and all enter earth radially inwards. The total power expresses the minimum generation of entropy 1 W/m² considerably incremented by temperature and mass of the sun. Power of 1 W/m² refers to 300 K and expresses one oscillator, while Planck radiation term 0.01 W $\times$ 300 = 3 W includes independent electron oscillations. These powers derive the cross section area of the protons of earth in Chapter 13. These powers express gravity forces, 2 W/m², for cool earth. The great majority of radiations suffer multiple reflections which diffuses (redirects) in all six directions without imposing any momenta to earth. Their power is expressed by sun constant. These twig processes are well represented by relaxation of dipole impurities of NaCl crystal and of epoxy. In NaCl each step has constant charge, in epoxy recorded charge decreases quickly, as shown in the Appendix.

Spacetime. The book would not be complete without correlating its findings with Einstein's spacetime concept and its equation mc^2. The equation has been split by an international group of scientists into elements of high energy physics indicating that mass equals energy. Ted Jacobson has shown in 1995 that Einstein Equation expresses the law of thermodynamics and correctly describes gravitation. The relation was confirmed by Eric Verlinke, who in 2004 also identified gravitation with laws of thermodynamics.

In terms of the integer relativistic model energy mc^s_2 expresses wavefunction format of two interacting vector pulses increasing velocity of light by $\sqrt{2}$ in each of four repetitions to c_s. Interacting mass of proton 1×10^{-27} kg converted by relativity at velocity c_s^2 derives the energy density of space ε_{os}:

$$m_p c_s^2 \times 1.1111 = 1 \times 10^{-27} \times 9 \times 10^{16} \times 1.1111 = 1 \times 10^{-10} \text{ J} \equiv 10 \, \varepsilon_{os} \equiv G^s \qquad 23.12b)$$

Above energy expressed by two interacting pulses in wave format display their velocity to be increased by $\sqrt{2}$ on vector addition and generate c_s by four repetitions:

$$m_p c_s^2 \times 1.1111 = (\sqrt{1 \times 10^{-27}}/1.1111^{1/2} \times 3 \times 10^8) X (\sqrt{1 \times 10^{-27}}/1.1111^{1/2} \times 3 \times 10^8) \times 1.1111^2 \equiv G^s$$
$$= (3 \times 10^{-14} \times 3 \times 10^8) X (3 \times 10^{-14} \times 3 \times 10^8) \times 1.1111^2 \equiv 1 \times 10^{-10} \text{ J} \equiv 10 \, \varepsilon_{os} \qquad (23.12c)$$

Lengthy discussions on WEB pages were not helpful until on May 3/2013 it became suddenly apparent that the postulated model and spacetime concept share one important feature of having no independent time and space dimensions in several relations. Spacetime concept replaces variables of space and of time by their product called **spacetime**. The integer model and Einstein's theory have no space or time variables during events within atomic structures. Presently minimum production of entropy at 1 K per 1 m^2 is taken at $h_s \equiv 1 \times 10^{-33}$ J. At atomic level in 1D that value on vectorization becomes 1×10^{-43} J x 1/10 W/m^2. Value quoted by some researchers is 10^{-44} W/cm^2, while the author's model derives density 10^{-44} W/m^2. Vectorized energy 10^{-44} W/m^2 is taken to flow in six orthogonal directions with velocities 1/3 c_s in three forward forces generating energy $10^{-44} \times 10^8 = 1 \times 10^{-36}$ W/m$^2 \equiv h_{select}$ with three such pulses generating added sum in one direction 3×10^{-36} W/m^2. This value is consistent with the new proposed electron Planck constant hselect. To obtain energy 10^{-44} W/m^2 at c_s requires four steps of wave function format of energy $\varepsilon_{os} = 10^{-11}$ W/m^2:

$$\varepsilon_{os} X \varepsilon_{os} X \varepsilon_{os} X \varepsilon_{os} = 1 \times 10^{-44} \text{ W/m}^2 \quad (23.12d)$$

Usually proton moves with velocity 1/4 c_s and requires four steps to generate c_s

$$1/4 \, c_s \times (\sqrt{2})^4 = c_s \text{ with energy added to } 4 \times 10^{-44} \text{ W/m}^2$$

Now spacetme = distance x time t. Since distance in 1D is 1/3 c_s x t.

Spacetme = 1/3 c_s x t^2 and in the last step expresses power demsity:

$$[\sqrt{(1/3 \, c_s)} \times t] X [\sqrt{(1/3 \, c_s)} \times t] = (10^4 t) X (10^4 t) = 1 \times 10^{-44} \text{ W/m}^2$$

$$(23.12e)$$

Thus $10^4 t = 1 \times 10^{-22}$ s and $t = 10^{-20}$ s $\equiv$ muo10_s x $1/3$ c$_s$ (23.12f)

Earth space (m) around sun times time taken (s) is the product of special relativistic gravitation and $\mu_{os}^{1/2}$ with correction $2\pi \times 1.5 \times 10^{11}$ m x 31.557 s/(1 +1/115) =3.00004$\times 10^{22}$ N x $\mu_{os}^{1/2}$ (23.12g)

Similar space x time around Bohr atom $=2\pi \times 1/2 \times 10^{-10}$m x $2\pi \times 1/2 \times 10^{-10}$m/$1 \times 10^7$ m/s $\equiv \pi^2$ x h_s/μ_{os} J (23.12h)

Their ratio (**discloses multiple findings**) $=3 \times 10^{22}$ N x $\mu_{os}^{1/2}/(\pi^2 \times h_s) = 3 \times 10^{54}$ x $\mu_{os}^{1/2} = \mu_{os}^{1/2}$ x1/2 of earth protons indicating that radiations in three orthogonal directions from each earth proton convert h_s to vectors by1/1000 probability identified as $\mu_{os}^{1/2}$. They pass radially through earth mass and μ_{os} becomes the probability of two atoms generating radiation and accelerating term in wave function format. Above equation derives that relativistic gravitation 3×10^{22} N = 6×10^{54} x $\pi^2/2$ x h_s ≈Number of earth protons 1/10 x $h_s(1+1/75)$ or 5 h_s per plane. **Thus $\mu_{os}^{1/2}$, μ_{os}, $\mu_{os}^{3/2}$ are just vectorization probabilities 10^{-3}, 10^{-6}, 10^{-9}**; the useless minimum entropy product h_s appeared on 14 AUGUST 2913 to be responsible for gravitation and **able to transmit energy through the mass of earth; NO! It is h_s localized in its specific plane and vectorised by 1/10 that generates these events.** Its vectorized value is as given below or expressed in four terms $(10^{-3}x\sqrt{\varepsilon_{os}})^4$: **$1/10$x h_s = 1×10^{-34} J = $(1 \times 10^{-17})^2$ $(\mu_{os}x\varepsilon_{os})^2$ of Equation 2.2b $\equiv (10^{-6}x\varepsilon_{os})$ X $(10^{-6}x\varepsilon_{os})$**

(23.12i)

These spacextime rotating pathways expressed in proposed model units extend the approved science and validify the model and the Einstein concept of spacetime. Pulse 1D $(\mu_{os}x\varepsilon_{os})^{1/2}$=3Jx1.111 . . . $^{1/2}$x **probability 10^{-9}.**

The elementary energy of dipoles measured in NaCl experiments was 0.0173 eV or J which expresses 17 degrees of freedom in space of 0.001 J or eV/event well documented in Equation 2.64. Eighteen H atoms need to be converted to helium to free sufficient input energy decremented by 1.24 and by 2/3 relativity in the absence of independent electron oscillations:

$4 \times 10^{-44} \times 10^8 \times 18 \times 10^{30} \times 10 \times 1.2^4 \times 2/3 \times 1/1.1547 = 0.0009953$ W/m² ≈ 0.001 W/m² (23.12g) with variance 1 k_s per H atom +2/3 k_s.

Integer relativistic space model and Einstein theory (as interpreted by Verlinke) identified gravitation with laws of thermodynamics that is minimum energy of entropy at 1 W/m² per Kelvin or emmition of Planck constant energy h or relativity corrected values.

If values in Equation 17.51 T=1 K and A_{earth} = 1 m² and relativity corrections are neglected, F =0.001 N is indicated by power 1 W/m² vectorized in three orientations. Thus after long considerations both the integer relativistic space mode and Einstein Theory derive that gravitation is expressed by the thermal electrodynamics law of vectorization of the minimum energy production in the Universe expressed in SI units by h and in integer units by h_s. Exhaustion of the energy stored in protons derives the longevity of the Universe.

The research of the author uncovered several well displayed details of physics, remaining usually well hidden in the classical SI physics:

1. Particles of the radiation pulse move within the atom, molecule with local velocities of ½ of the light velocity c_s for electrons, ¼ c_s for protons and 1/3 c_s for join movement. Local velocities are components of light velocity and they are physically significant because of the independence of events occurrence in six space directions or real structural velocities within the atom, molecule or wave function. Electrons rotate with velocity 1/30 c_s. Consequently masses of radiations require relativity corrections. For ½ c_s the correction factor is:

$$K_{rel} = 1/[1-(\text{½}c/c)^2]^{1/2} = 1.154\ 700\ 538 \qquad (23.13)$$

Components of light velocity are not recognized in the current SI system of physical constants leading to measurements, and physical constants having unspecified corrections imbedded in their values.

2. Bohr atom can be represented by three oscillators having usually equal energies. Atom has 32 DOF instead of 30 displayed in radiations creating a mass surplus of 2/30 = 0.0666 . . . That value expressed per each of 6 space directions, over relativity and quantum correction derives essentially the excess of mass over unity as given in WEB data. The electron oscillates with mean frequency of 5×10^{15} Hz, or 10^{15} Hz per orientation, while proton with the mean frequency of 5×10^{12} Hz, or 10^{12} Hz per orientation and the pair with mean frequency of ~5.0025×10^{12} Hz. The latter two form, an unresolved for its sources, couplet in the spectrum. At 300 K only two oscillators carry energy.

3. Electron charge e_s in atomic terms is the proton momentum expressing its radiation mass 1×10^{-27} kg moving at the average local component velocity 1/3 c_s of the radiation group. The model electron charge $e_s = 1\times10^{-19}$ C multiplied by the cube of relativity correction K_{rel}^3 derives the SI value $e = 1.602\times10^{-19}$ C, or more accurately by Equation 1.10. These two values of charges related by K_{rel}^3 lead to the discovery of a second set of physical constants having integer values. Both the classical SI system and the integer space cubic converted to SI system are derived from the same Universe structure displaying the same probabilistic rules. Neither can properly explain some details of physical processes because both use units of length and time absent in the formative stages of radiation.

4. The extremely high pressure on the surface of the sun compresses the spherical shape of the atom into a cubic shape. The nodes of the standing waves of the infrared radiation on the sun are located between the planes of the cube and not along the electron circular footpaths. The directions of radiation from such cubes are in six orthogonal directions. Three pulses in six orthogonal directions or two pulses in two such directions add to generate a resultant. Over a short period of time the resultant may take different orientation and the cubic structure becomes the spherical one or possibly circular paraboloid. Thus the two structures are not exclusive. Signals from three orthogonal orientations taking place within the atom are nulled except for one directional signal between the source and the receiver.

Close to the surface of the sun such groups of cubic atoms are likely to coalesce forming larger cubes and domains of atoms; they take positions minimizing their energy by pointing their long axes at the center of the sun. Domain of 10^3 or 12^3 of Bohr atoms would generate the radiation expressing the measured sun constant of 1372 W/m². Domain of 27 of Bohr atoms (3^3) would retransmit 10 eV signal of the first step excitation generating space quantization 1 eV (or 1 W/m²) vector signal, the source of gravitation.

Unconstrained geometry of the Bohr atom on earth becomes probably spherical, but the mathematics of the cubic geometry displays new hidden details about physics and about gravitation.

5. Unity probability forming reality requires the use of non-measurable components of probability of physical parameters of energy, mass, momentum, velocity and charge. A radiation event from sun, planet, molecule, atom or a chemical structure is expressed by the energy of the number of degrees of freedom in the quantized orientation requiring equipartition of energy rule. The laws of physics and of the orthogonal structure of space generate the event directed in the specified direction. Energies are converted to vectors. The radiated momenta along the three orthogonal directions are unity vectors, which on addition increase the resultant velocity by $\sqrt{3}$; further corrected by relativity, they exactly double the input. Sequential addition of vectors increases the local velocity needing a domain of atoms to generate energy of particles moving with the velocity of light. The probability of radiations taking place at the same time in the same direction is very low. Space quantization has 10 positions with five in one direction. The temperature of 6000 K is arbitrarily split in six directions; temperature is subjected sequentially three times to 1/10 probability generating 1 K ≡ 1 eV ≡ 1 J/m². Kinetics of small particles has no units of time or space creating dimensional inequality with expressions expressing kinetic atomic interactions.

6. True relations are expressed in atomic probabilistic units of nature; they have to be converted to SI units. Natural probabilistic units have no dimensions, they just expressing their number of degrees of freedom. DOF is the total sum of unity momenta. The model suggests that probability of an event is proportional to its linear dimension. In the cubic integer space system the ratio between the SI units of one meter of length to the Bohr cube is 10^{10}. This makes the probability of a linear atomic event per unit time 10^{-10} smaller than in SI units. The ratio between volume probabilities becomes then 10^{-30}. The linear, surface and volume probabilities are truly independent parameters. With 6×10^{54} electrons forming the mass 6×10^{24} kg of earth the electron mass in kilogram expresses in atomic terms volume probability $1 \times 10^{-30} = 1 \times 10^{-10} \times 1 \times 10^{-10} \times 1 \times 10^{-10}$.

7. Generally the multiple reflections occurring on interaction between the radiations and the planets justify treating most planets as black bodies. Corrections needed for two planets indicate presence of some reflections.

8. Proton is just the summation of moments of 2000 electrons creating its mass and nulling sum of its moments. Protons comprise spinning electrons nulling their moments. On becoming unstable the proton converts to radiation expressed by 2001 electrons moving with local velocity of light 1/3 c_s and velocity converting to c_s. This process involves apparent mass to energy conversion, believed to be a reversible process in nature, which probably created stars and planets from initial electron mist.

9. For earth, Venus and Mars the rest mass of the planet and their attractive gravitation force expresses their content of the number of electrons. If that planet body moves with velocity v_e the current generated $I = m_{bodymass}/m_{electronmass} \times v_e \times e_s$ induces magnetic effects consistent with the rules of electromagnetism. The current produced by the motion of planet expresses repulsion of the sun gravity and it is assessed as 2/3 of the force specified by the reference current producing the old International Paris reference force of 2×10^{-7} N per meter length between two thin parallel conductors one meter apart. For Mercury displaying strong magnetic field the force of reflection is beside other relativity and quantum corrections increased by a factor of 3.

For Saturn, Neptune and Jupiter which are further away from the sun their repulsive gravitation force is decreased by factors 1/30, 1/25 and 1/10 of the value derived for earth, Venus and Mars. Uranus and Pluto generate forces with factors 1/781.36 and 1/232.58, which esencially express factors $1/30^2$ and $1/30 \times 1/8$.

10. The atomic radiation is subject to the statistics of the six sided dice game. This means that the probability needed to generate sequentially the same display in a number of throws is the product of probabilities of sequential throws. Furthermore the four displays of the dice rotating along one axes show the four directions of space in one plane of rotation.

11. The interpretation of the Franck-Hertz experiment is now modified with the energy of electrons decreasing in steps of 4.88 eV. The total quantum excitation energy of mercury vapor is that taken by the number of all independent oscillators plus one. Atomic number of Hg is 80 with 14 groups of electrons and one nucleus making a total of 95. The vapor is diatomic increasing by two the total energy to: $2 \times 95 \times kT = 4.807$ eV. An extra $kT \approx 0.025$ eV is required to free the atoms and energy of $2kT$ to balance the created moment; this adds to a total of 4.882 eV, which compares well with the original 4.88 eV.

12. There is no independent force of gravitation. It is a probabilistic, 1×10^{-10}, derivative of SI parameter of 1 W/m^3. The integer SI gravity constant expresses 4 SI muo10$_s$. The muo10$_s$ mass is 1/10 of the proton radiation mass in one orientation and the mass density of radiation pulse corrected by relativity and quantum.

13. The international constant of gravity in cubic space units converted to SI, G_s, has value ranging from 1×10^{-10} N-m^2/kg^2 for bodies at equal temperatures to $2/3 \times 10^{-10}$ for sun to earth system in which the cold earth generates negligible electron radiation. With sun mass generating on earth force per kg m_{sun}/L_{earth}^2 force acting on earth $m_{sun}m_{earth}/L_{earth}^2$ has to be converted by linear probability, $G_s \equiv 10^{-10}$, to natural atomic units, expressing product of ten pathway probabilities 1/10 of discharging in the correct orientation. It is also expressed in Equation 6.1. G_s is the square of $\varepsilon_{os}/\mu_{os}$. The power of five applies to the overall probability of five space quantizations occurring sequentially in the same direction to initial probability expressed by Planck's Law. It is identified now with the model gravitation constant for masses at the same temperature:

Gravitation constant $G_s \equiv [\varepsilon_{os}/\mu_{os}]^2 \equiv (0.01)^5 = 1 \times 10^{-10}$ N-m^2/kg^2 is becoming for the sun-to-earth system effectively $2/3 \times 10^{-10}$ N-m^2/kg^2.

G_s value does not apply for Neptune and Jupiter, the planets generating part or most of their own energy for gravitation.

14. Ninety six expressions specifying correct values of the apparent gravitation forces have been derived for the sun-to-earth pair in terms of various SI and cubic space physical constants. They express both the force of attraction, the force of repulsion or pure energy and are difficult to be identified as such. The gravitation expresses both the force of attraction and the force of repulsion. They **are better described as release of energy and the absorption of energy.** Really they express **the same force of 10 N/kg per K generated by every mass in all directions and converted to 1 N/kg in linear interactions.** The force 1 N/kg appears to be specific to the sun-to-earth couple and to the inside of the Bohr atom, but it does not impose a gravitational model valid all over the Universe.

15. **The planets are the satellites of the sun** producing stable rotation. The sum of satellite's linear momenta and of the momenta delivered by the sun generates their circular path. Relativity increases particle mass and decreases particle velocity generating the needed relativity corrected circular radius.

16. The basic interaction between the source and the receiver is one dimensional in which the oscillations can occur solely in the direction of the propagation because oscillations in other orientations are nulled. Linear variations of velocity generate angular spin of moments exchanging energy within atomic dimensions and keeping total moment in 3D constant.

17. The model frequency of 5×10^{12} Hz is validated by the radiation frequency of free space measured as wavelength of 0.21 m. Converting from 300 K to 2.725 K and multiplying by 2, since bi-atomic hydrogen is twice as heavy as Bohr, and by $\sqrt{3}$ derives:

Space frequency $v_{space} = 5 \times 10^{12} \times (2.725/300)^2 \times 2 \times \sqrt{3} = 1.429 \times 10^9$ Hz $\qquad$ (23.14)
of wavelength $\lambda = 3 \times 10^8 / 1.429 \times 10^9 = 0.2099$ m.

The velocity of light is derived from an event being expressed by the probability of unity.

18. The dissociation energy of the Bohr atom is given by Equation 15 of the Appendix. The Bohr atom is simulated by the void-impurity pair in a NaCl crystal. That energy W_{max} expressed in integer cubic SI space units with $\tau = 1536$ s and $v_{Lms} = 1 \times 10^{12}$ Hz at 300 K is:

$W_{max} = k_s T_{300} \ln (\tau \times v_{Lms}) = 1 \times 10^{-23} \times 300 \ln (1536 \times 1 \times 10^{12}) = (1 - 0.0490)$ J or eV/event.
$\qquad$ (23.14a)

The variance/kelvin is 0.000165 J $\equiv 0.165 \times 10^{-22}$ J corrected $1/3$ k_s. Energy W_{max} is the experimentally found thermal energy needed to break the bond in one orientation between the two charges. That value in atomic terms $m_{ps}c_s^2$ converted linearly to SI by 10^{10} and by relativity is 2 J, because it includes the reflected pulse:

Energy of radiation = $2 \times 10^{-27} \times 9 \times 10^{16} \times 10^{10} \times 1.111 \ldots = 2.000 \ldots$ J or eV/event
(23.15).

As mentioned earlier the squared velocity is correct, but misleading because it omits the wave function process of velocity of light formation.

Energy in hydrogen atom is treated by SI as a negative value because energy is radiated. The integer model treats energy as a positive value because it supplies most of the needs of the Universe. It is according to the author the exothermic process of heat and energy generation needed in creation of all atomic elements in the Universe. At high temperature hydrogen on the sun converts sequentially to iron. At low temperature heavy radioactive elements, created by energy from decaying protons, and decaying protons display their energy by radiation. Although clouds of electron are present in the Universe it is not possible to relate them firmly as the source of protons.

19. The elementary quantum vector sun signal in 1D display of 1 W is identified to be 1/10 of the space energy momentum per kelvin of the sun temperature expressed also in 1D:
$$1/10 \times \varepsilon_{os}^{1/3}/T_{sun} \times c_s/(1.1547^{1/2} \times 1.04^{1/16}) \equiv 1.0000097 \text{ W} \qquad (23.15a)$$

20. Sun gravitation. **The gravity force of sun attraction is consequence of the release of energy in the pulses of the sun.** This release can be described in different ways. One way is to start with the sun temperature of 1000 K discharging in six orientations a total of 6000 K. Temperature 1000 K time three steps of space quantization discharges along the source-receiver axis the power of 1 W per pulse:

$$1000 \text{ W} \times 1/10 \times 1/10 \times 1/10 = 1 \text{ W per pulse} \equiv 1 \text{ W pulse on earth at 300 K} \qquad (23.16)$$

Actually two orthogonal 1000 K pulses in the plane of earth rotation add vectorially and, while subjected to quantum corrector $1.04^{3/4}$, generate the sun constant $1000 \text{ W} \times \sqrt{2} \times 1.04^{3/4} = 1373.2$ W. Thus 1 W on earth per oscillator is

$$1373.2 \text{ W}/\sqrt{2} \times 1/10 \times 1/10 \times 1/10 \times 1.04^{3/4} = 0.99999 \text{ W or eV} \qquad (23.17)$$

The gravity force in terms of sun gravitation on earth represents gravity force per kilogram
(23.18)
$$F_G \text{ (N)/Mass of earth (kg)} = 3.5438 \times 10^{22}/5.9742 \times 10^{24} = 0.0059318 \text{ N/kg} \approx 0.006 \text{ N/kg}$$

The above value referred to 300 K over $\sqrt{3}$ time 1.04 derives similarly 1 N/kg
$$0.006 \times 300/(\sqrt{3} \times 1.04) = 0.999 \approx 1 \text{ N/kg} \qquad (23.19)$$

This result reinforces the earlier finding, that **the mass of earth and of sun generates a force of 10 N/kg per degree kelvin.** This is confirmed in the calculation of the force of sun gravitation decreasing with the square of the distance converted to reference value by linear probability, single step quantization, $\sqrt{3}$ and quantum corrector. The agreement to within 1 part in 35000 is a very good evidence of the validity of the above proposition:

Sun gravitation force   (23.20)

$$=2x10^{30}x6000/(1.5\ x10^{11})^2x10^{10}x10\ \text{N}\ x\ 1.1547/(\sqrt{3}x1.04^{1/12}) \equiv 3.5439x10^{22}\ \text{N}$$

The cubic space gravitation force of $3.5x10^{22}$ N in atomic terms of nature is:
$$3.5\ x10^{22}\ x\ 10^{-20}/(1+1/6)\ \text{generates}\ 300\ \text{N} \equiv 300\ \text{K} \qquad (23.21)$$

21. The long search uncovered the following identity indicating that earth mass generates also the force of 1 N per kg (1/10 N vector) and also 2/3 N per square meter of surface area per unit linear velocity around the sun:

$$4xA_{cirearth}\ x\ \text{Linear integral earth velocity}\ x\ 2/3\ x\ 1/1.04^{3/5} \equiv$$
$$4\ x1.278\ x\ 10^{14}x\ 30000\ x\ 2/3\ x\ 1/1.04^{8/5} = 0.998\ x\ 10^{19} \equiv 1/e_s$$

Earth mass x 300 K/velocity of light x cor. $\equiv 6\ x\ 10^{24}x300\ \text{K}/(2/3x3x10^8)\ x1.2/1.1111$
$$\equiv 1.001\ x\ 10^{19} \equiv 1/e_s \qquad (23.22)$$

It is suggested that $1/e_s$ for both expressions expresses the rate of appearance of radiation momenta per meter, of mass $1\ x10^{-27}$ kg, moving with local velocity of $1/3\ c_s$.

The above equations express the features of jet engine and of radiation propulsion. The momenta generated are proportional to the area of the source, to the temperature of the source and to the local velocity of the exhaust.

22. The sun generates in one plane a force of one newton per kilogram of its mass, radiation particles moving with local the velocity of light $x1/3\ x3\ x10^8$:

$$(F_{Gs}/mass_{sun})/\sqrt{3}\ x1/3\ x3\ x10^8/1.04^{1/2} = 1.0015\ \text{N/kg} \qquad (23.23)$$

The generation of the force of one newton per kilogram mass is derived using cubic space constants also from **the product of the inverse of the fine constant α = 1/137.035 and the generation of the force 0.006 N/kg of earth mass** of Equations 2.51 with corrections:

$$137.035\ x\ 0.006\ \text{N/kg}\ x\ 1.11111^2/1.04^{3/5} = 1.00024\ \text{N/kg} \qquad (23.24)$$

The variance is H. u. because per charge of 6 directions it is $0.4\ x1.1111^21x10^{-23} \approx 1/2\ k_s$.

23. The product of radiation particle mass of 10^{-27} kg and square of the local velocity $1/3\ c_s$ derives $\varepsilon_{os} = 10^{-27}x(1x10^8)^2 = 1x10^{-11}$ J/m; that value converted by $(300\ \text{K})^2\ x\ 1.1111 \equiv 1x10^5$ W becomes a pulse equal u_{os}, $1x10^{-6}$ NA^{-2}. For expressions comprising three steps of 1/10 quantization the quantum unity terms for 1 K on Figures 2-4 becomes 300 K for that temperature. Hence μ_{os} converted to 300 K requires factor $(300\ \text{K})^3$:

$$\text{Total energy} = 1.1111\ x\ 1\ x\ 10^{-6}\ x\ (300\ \text{K})^3 \equiv 30\ \text{eV or J} \qquad (23.25)$$

Energy of 30 eV is displayed in Lyman frequency $2.467\ x\ 10^{15}$ Hz quoted in Table 3 by the application of Equation 3.9.

Force of 1 N time √3, times quantum correction 1.04 and corrected for temperature $(300^3)^3$ now to power nine derives the earth numerical force of gravitation:

Earth force of gravitation $= 1$ N $x\sqrt{3}$ $x1.04$ x $(300^3)^3 \approx 3.5456$ $x10^{22}$ N   (23.26)

Power nine expresses three sets of three pulses in orthogonal orientations generating one result.

The variance from the reference $3.5438x10^{22}$ N is $2x10^{19}$ N$x1x10^{-19} \equiv 2$ eV, which expresses $3k_s$ x 300 x $1.1111 \approx 0.1e_s$ is suggesting these are the momenta of the photon and (proton+electron) pair as explained earlier. The quantum value puts a limit on the accuracy of gravitation force measurement, an optimistic value being $3.5438(1)x10^{22}$ N, expressing 28 parts per 10^6, in broad agreement with accuracies of Equations 17.50-51.

Above method derives the linear velocity of earth from the base value of 1 m/s to:

$$1 \text{ m/s } (300 \text{ K})^3 \text{ } x \text{ } (1/10)^3 \text{ } x1.1111 = 30\ 000 \text{ m/s} \qquad (23.27)$$

And it converts velocity of light c_s at 300 K to the elementary level of 1 m/s

$$3 \text{ } x \text{ } 10^8 \text{ } x \text{ } (1/10)^3 \text{ } x \text{ } (1/10)^3 \text{ } x \text{ } (1/10)^3/(3 \text{ } x1.1111) = 1.00 \text{ m/s.}$$

Or vice versa 1.00 m/s x $(300$ K$)^3$ $x1.1111$ x $10 = 3$ x 10^8 m/s.

The energy radiated by the sun using the product of Stephan-Einstein constant and of T^4, times $6A_{cirsun}$, over the mass of the sun of $1.0x10^{30}$ kg discharged in one direction derives the force per kg/K:

                                                                                    (23.28)

$10x\sigma_s xT^4x\,6\,xA_{cirsun}/1x10^{30} \equiv 10x10^{-8}x6000^4x6x1.5x10^{18}/(1x10^{30}\ x1.04^{17/25})=0.005999$ N/(kg-K).

The unit of value 0.006 is N/(kg-K) because $\sigma_s = 5$ x 10^{-8} W-m^{-2} T^{-4} expresses four pulse probabilities of $(\sigma_s)^{1/4}=0.009969 \equiv 0.01$ J, the 1st term of Planck Radiation.

24. Most of proposed relations between constants are dealt in Chapter 2 with some additional relations presented here. The minimum pulse energy of space is derived from integer permittivity constant of space: ε_{os} $x1/10^3$ $=1x10^{-14}$ J/m^3. The gravitation constant can be expressed as product of integer permittivity of space and earth pulse: $G_s = 1/(\varepsilon_{os}$ x $E_{searthSI}) \equiv 1/(1x10^{-11}x1x10^{21}) = 1$ x 10^{-10} N-m^2/kg^2. The velocity of light over minimum energy pulse derives Special Relativity gravitation force: $3x10^8/1x10^{-14} \equiv 3x10^{22}$ N. The gravitation constant between the sun and earth to power ½ corrected for relativity and quantum delivers 10^5 J energy pulse: $(2/3$ $G_s)^{1/2}x$ $1.11111/1.04^{1/5}=100014.4$ J$\approx 10^5$ J. Integer consistent of permittivity of space can be expressed in terms of the constant of gravitation, h_s and $h_{selectr}$ and frequencies 10^{12} Hz and 10^{15} Hz:

$$\varepsilon_{os} \equiv h_s \text{ } x \text{ } 10^{12} \text{ } x10^{10} = h_{selectr} \text{ } x \text{ } 10^{15} \text{ } x10^{10} =1 \text{ } x \text{ } 10^{-11} \text{ J/m}^3 \qquad (23.29)$$

Integer Boltzmann constant is Planck constant converted by linear converter to SI

$$k_s = h_s \; x \; 10^{10} = 1x10^{-23} \, J \tag{23.31}$$

Similarly
$$k_{selectron} = h_{selectron} \, x \, 10^{10} = 1 \; x \; 10^{-26} \, J$$

Integer Boltzmann constant is given by: $1/e_s = E_{searthSI} \; x \; 0.01$ and

$$\varepsilon_{os}/G_s \; x \; E_{searthSI} \equiv 1 \; x \; 10^{-11}/10^{10} \; x \; 1 \; x \; 10^{21} = 1 \, J \tag{23.32}$$

25. Interactions and processes in space are subject to vector formation, size conversions, and plane conversions, probability of appearance in the right direction and temperature corrections; the presence of the number of degrees of freedom at the base level establishing energy due to energy equipartition rule. The sun radiating in one plane in 4 orientations generates pulse $1/4x6000 \, K \equiv 1500 \, W/m^2$ which appears in 5 quantum orientations as 300 W pulses decreasing to 100 W per plane expressing the hydrogen molecule on the outside of the sun surface. Other atomic, molecular or chemical structures generate a wide range of energies, which are subjected to the same listed above rule of interactions considered in great detail for the hydrogen. It's the reference value for one plane of $1 \, W/m^2$ at 300 K converts into a vector directed along the source-receiver axis by two step quantization $1/10 \; x1/10$ and becomes the vector 0.01 W of Planck Radiation Law. The different energy values around us generated by the wide range of sources create the beautiful world around us, full of color and extended to invisible range short and long waves creating the Universe. Thus with 17 DOF and $T= 6000 \, K$ the energy of $1 \, W/m^2$ in one quantum orientation and wave function format becomes:

$$(17^2 \; x \; 6000^2)X(17^2 \; x \; 6000^2) = 1.08243 \; x \; 10^{20} \, W/m^2 \tag{23.33}$$

Above energy times size converter for rotation to atomic terms, 10^{-20}, derives:

Atomic energy $= 1.08243/(1.111\ldots^{1/2} \; x1.04^{0.7}) = 0.999574 \approx 1 \, eV/m^2 \tag{23.34}$

The variance expressing $4 \, k_s$ for four directions.

26. The mass of earth $6x10^{24}$ kg corrected for relativity and quantum generate a reference force of attraction of the sun 1 N/kg for each of 5 local pathways:

$$5 \; x \; 6 \; x10^{24} \, kg \; x1 \, N \; x \; 1.111\ldots^{3/2} \; x \; 1.04^{11/50} = 3.5440 \; x10^{22} \, N \tag{23.35}$$

The calculation follows the argument of § 64 in each of three orthogonal directions particle has four orientations making a total of twelve. Two are at right angle to the progression are mutually nulled and do not count. Five are reflected and cancel the long train converting it into a radiation pulse.

27. Velocity of light in 1D and other dimensions displays generates several parameters discussed in Chapter 15. Integer mass of earth in 1D display times $\sqrt{3}$ expresses sum of

3 vectors in orthogonal orientations and with relativity and quantum correctors derives velocity of light:

$$c_s \equiv \sqrt{3} \text{ x } (6 \text{ x } 10^{24} \text{ kg})^{1/3} \text{x } 1/1.111 \ldots ^{1/2} \text{x } 1.04^{1/8} \equiv 3.000508 \text{x} 10^8 \text{ m/s} \qquad (23.36)$$

One D display of light velocity over quantum corrector derives power density generated by earth or thermal power density of the sun on earth deriving in Equation 5.13 balancing gravity power densities:

$$\text{Power density of gravitation} \equiv c_s^{1/3} \text{x} 1.04^{1.2} \equiv 701.65 \text{ W/m}^2 \qquad (23.37)$$

The pulse energy per plane at 300 K is expressed by the product of the velocity of light, temperature, relativity correction for 1/3 c_s and the model gravitation constant that is the graviton energy:

$$3 \text{ x } 10^8 \text{ x } 300 \text{ x} 1.111 \ldots \text{ x } 1 \text{ x } 10^{-10} \equiv 1 \text{ J or eV} \qquad (23.38)$$

In SI units 1 J is per kg or 1W/m^2, while at atomic level energy 1 eV is per orientation event.

28. The number of oscillators in a complex atom molecule is the atomic number plus number of independent electron groups plus one, all squared.

29. The cubic integer constants are derived from the Planck constants in the same way as the classical SI constants and most generate directly integer values. The space is hiding its true nature by displaying physical constants, which incorporate unrecognized up to now relativity and quantum corrections. The relativity correction derives for the 3D display of e_s:

$$e_{3Ds} = \sqrt{3} \text{x} 1.1547 \text{ x } 1 \text{ x } 10^{-19} = 1.999999 = 2.0 \text{ x } 10^{-19} \text{ C} \qquad (23.39)$$

30. While classical SI system works very well for the source and the receiver being at the same temperatures, the system needs adjusting, when these temperatures are significantly different. The radiations frequencies generated by the lower temperature control the radiation absorbed according to the rules of mass spectrograph. Since only 2/3 of energy is absorbed, a different gravity physical constant is required to describe this situation. The research derived over 100 accurate expressions for the gravitation force of earth and of the planets.

Total energy 3 eV in one plane, that is 1 eV, multiplied by the volume probability 10^{-30} and by the invert of space quantization 10^3 derives the SI radiation energy and power in joules and in watts representing the proton radiation mass of 10^{-27} kg. In converting graviton **1.66 . . .** 10^{-35} J to **1** eV of integer cubic space unit the multipliers are: volume probability 10^{30}, space quantization 10^3 and the factor for not interacting radiation: $1.66 \ldots 10^{-35} \text{ x} 10^{30} \text{ x} 10^3 \text{x } 1.2 \equiv \mathbf{0.999\ 999\ 999}$ **eV** $\qquad (23.40)$

31. The derivation of the sun gravity force on earth per square meter of N/m^2 requires relativity correction for 1/3 c_s, quantum correction and converter to SI units with ¼ c_s expressing velocity of proton. It confirms that proton infrared frequency is responsible for the gravitation. One newton/m² is not a small force:

$$3.5438 \, x10^{22} \text{ N} \, x \, 1.1111 \, /(1.04^{7/10}A_{cirearth} \, x \, \tfrac{1}{4} \, c_s) = 1.0015 \text{ N/m}^2 \approx 1.000 \text{ N/m}^2 \qquad (23.41)$$

32. In Chapter 16 the force of attraction within Bohr atom was successfully explained in terms of the force of 1 N/kg being emitted by the proton and by the electron. Such force was likely also emitted by the electron comprising the primeval fog of space. That force is very minute, but over long time it is sufficient to direct radiations centrally and bring condensation of electrons into protons with the formation of hydrogen atoms. If the Universe is created from very small volume, electrons must be condensed into protons carrying positive energy available for Universe events during its existence.

33. Formation of radiation pulse comprises adding vectorially component velocities and creating products of probabilities generating the pulse in the correct direction decreasing very significantly chances of that event. Two proton velocities moving at $\tfrac{1}{4} \, c_s$ along axes Y and X add to value $\sqrt{3} \, x \, 1.154700$. Two such values from the plane YX and the plane -XZ add generating light velocity cs. The resultant probability is the produce of four probabilities $(1/6 \, x \, 1/6)^2 = 1/36^2 = 1/1296$

34. The mass of the sun generates at distance the force of 1 N and the momentum expressed in atomic probability terms at 1 N-s/kg at 300 K. The identity converts the mass of sun to force by dividing it by the mean of particle's velocities $1/3 \, c_s$ and it displays the equivalence of the centrifugal force and of the force of 1 N-s/kg at 300 K:

The sun gravitation force = $2x10^{30}$ [kg] x1 [N-s-kg^{-1}] $/(1/3x3 \, x10^8)$ [m-s^{-1}] $x\sqrt{3}x1.04^{10/17}$
$$\equiv 3.5438 \, x10^{22} \text{ N-s}^2/\text{m} + 1/4 \, x \, k_sT_{6000}/e_s \qquad (23.42)$$

Since the force 2 N/kg has been used in Equation 16.10 that value is taken to be generated by the side of the sun of $1x10^{30}$ kg facing the earth.

The force 1 N/kg = 10 N x1/10 per kg mass of earth around the earth surface with three step space quantization generate the geometry of attraction in all orientations

$$6 \, x10^{24} \text{ [kg]} \, x10 \text{ [N-s-kg}^{-1}] \, x1/10x \, 1/10x1/10x1/\sqrt{3}x1.04^{10/17} \equiv 3.5448x10^{22} \text{ N-s} \qquad (23.43)$$

These two equations identify the generated parameter from the mass to be an impulse representing sum of momenta at atomic level. This is consistent with the earlier author's experimental observation that to separate dipoles an impulse has to be applied comprising a specific minimum field for a specific time. With radiation (proton+electron) mass oscillating each at $5x10^{12}$ and $5x10^{15}$ Hz in 5 pathways with multipliers T^2 for pulse rate and T for conversion and one for relativity derives Planck Energy and on multiplication with $1/3 \, c_s$ the sun pulse:

$$5(1x10^{-27}x5x10^{12}+1x10^{-30}x5 \, x10^{15}) \, x6000^2x6000/1.04^2 = 0.009985 \text{ J} \rightarrow x \, 1x10^8 \text{m/s} =1x10^6 \text{ J}$$
$$(23.44)$$

35. Fingers of the human hand and toes of foot are ancient indicators of the preferential space field orientations for the first land vertebra animals.

36. The probability to generate events at velocity of light in specific orientation derived from its local components and based on quantized structure of space was uncovered to be based on one reference pulse of energy 1 J or eV.

The energy variance between values generated by the SI velocity of light and the velocity of light of the integer SI space model:

$$2\,m_{es}(c_s - c)^2/1.1547 = 1\times10^{-30} \times (3\times10^8 - 2.99792\times10^8)^2/1.1547 = 0.9975\times1\times10^{-19} \approx 1\text{ eV}$$
(23.45)

represent the minimum energy quantum value at 300 K, which probably is deducted in the measurement. The variance 0.0025 J referred to 300 K equal 0.78 J is taken to represent 2/3 J in view of the absence of electron oscillations.

37. Multiplying the gravitation force 3.5438×10^{22} N by 1.111 . . . x $1.0328^{1/2}$ x X_L derives $6\times10^{33} \approx 6\times10^{33}$ the invert $1/6\times1\times10^{-33}$ identified to be 1/6 of Planck constant h_s of Table 2. Hence forces are derived from atomic probabilities arising out of 11 DOF per plane of hydrogen atom.

38. The uncertainty of measurement according to the integer SI model fitted the velocity of light uncertainty measured in 2012 by CERN. The average value of Higg's bosom derived in the model fitted also the value measured in 2012 by CERN.

39. Equation 10.39 establishes magnetic permeability $\mu_{os} = 10^{-6}$ NA^{-2} to be proton momentum of minimum energy. Hence $\mu_{os}\times10^{12} \equiv 10^6$ J pulse is the proton impulse per second and $\mu_{os}\times10^{15} \equiv 10^9$ J pulse is the electron impulse per second of Equations 2.23.

40. Special Relativity Theory predicts divers, but accurate values of times and of distances for observers moving with different local velocities of light. Several values calculated using space cubic physical constants indicate that these results carry relativity corrections, which makes lengths contracted and times dilated.

With mass being identified as energy in the sentence following Equation 22.123a three pulses from sun mass 1×10^{30} kg over local velocity $1/3\ c_s$ generate Special Relativity gravitation force of $3\times1\times10^{30}/1\times10^8 = 3\times10^{22}$ N. The same force is generated by the mass of earth (now claimed to be equivalent to energy) in one direction vectorized in two dimensions 6×10^{24} kg x ½ $\times1/10\times1/10 = 3\times10^{22}$ N.

41. The energies derived from both, SI units and from integer cubic SI space units vary by relativity and quantum corrections. Energy of Bohr atom of 10 eV discharging in three orthogonal directions together generates 30 eV in comparison to 27.21 eV of Hartree energy in SI units. Two Bohr's atoms display energy $\sqrt{3}$ x 30/1.04 = 49.963 eV $\approx$ 50 eV. While moments of two 10 eV energy pulses displaced by 90° added vectorially generate 10 x$\sqrt{2}$/1.04 =13.60 eV. Derived energies may refer to pulse, be per wave or per second; they may be in classical SI units, integer SI units, joules or electron-volts, and they may be in atomic terms referring to a specific temperature expressing electron, proton or radiation energy. The signal of 30 eV is displaced in Lyman frequency 2.467×10^{15} Hz applied to Equation 3.9.

42. All physical constants are derived from the unit of length, unit of mass and unit of time plus electrical unit like charge, current or permittivity. Equally valid fourth parameter deriving all others, with three of the above, is Boltzmann constant k_s expressing 6 x $1.04^{1/10}$/(Avogadro number):

$$k_s \equiv 6.02357/(6.00000 \times 10^{23} \times 1.04^{1/10}) = 0.9999986 \times 10^{-23} \text{ J} \qquad (23.46)$$

The variance of $<1 \times 10^{-27}$ J is less than the radiation pulse energy.

Linear probability, temperature and relativity corrections define relations between some physical constants: $3k_s \times 10^{-10} = 3h_s$, at 300K $3 \times 300k_s \times 1.1111 = 10^{-30} \equiv 10^{-30}$ kg $\equiv m_{es}$.

43. The chain of pulses of Equation 10.38 displays 32 interacting components moving at the local light velocity, which also oscillate 4 positions forward and four positions backwards. They display wavelength $\lambda = 1/2 \times 10^{-10}/8 = 8 \times 10^{-10}$ m generating wave expressing pulse rate $1 \times 10^8/(8 \times 10^{-10} \times 1.1111^2) = 1 \times 10^{17}$ waves/second, which carry $1 \times 10^{17} \times 1/8 \times 1/[1.111111^2(1+1/80)] \equiv 1.000000 \ldots \times 10^{16}$ J. Ten local pulses add the rate to generate light velocity.

44. The International Gravitation Constant in integer cubic space units converted to SI is $G_s = 1 \times 10^{-10}$. Converted again to atomic eV term by multiplying by 1×10^{-19} C it becomes 1×10^{-29} J. The 1×10^{-29} eV value divided by the sun pulse of 1 MeV or N or J derives Planck constant $h_s = 10^{-33}$ J-s.

G_s at$=1 \times 10^{-10} \times 1 \times 10^{-19} = 1 \times 10^{-29}$ eV and 10000 $K_s \equiv \sqrt{3} \times 6000$ K x $1/1.04^{0.98} = 10\,000.4$ K. Coulomb constant $= 8.333333333 \times 10^9 \times 1.2 = 1 \times 10^{10}$ is the exact invert of the gravitation constant. It is also the relativity unconverted age of the Universe and **life of proton in years.** Multiplying by linear probability 1×10^{-10} converts the classical SI units into true atomic values expressing lower probability of the event of smaller dimensions. For energy and momentum in three directions the multiplier derives 1 kg $\times 10^{-30} \equiv 1$ mass of electron.

The gravity constant G_s is the integer SI force pressure 1 N/m^2 converted to nature probability atomic terms by linear probability 1×10^{-10}.

45. The velocity of light is derived by the same expression as in classical SI units:

$$\text{Velocity of light } c_s = 1/\sqrt{(\varepsilon_{os}\mu_{os}/1/1.1111\ldots)} = 1/\sqrt{(1 \times 10^{-6} \times 1 \times 10^{-11}/1.1111\ldots)} = 3 \times 10^8 \text{ m/s}$$
$$(23.47)$$

Probability of an event in atomic terms uses size conversion 1×10^{-10} times linear probability of 1 occurrence out of 10 in one plane becoming $P_{10\uparrow-9} = 1 \times 10^{-10} \times 1/10 = 1 \times 10^{-9}$

With ten local velocities, 10^8, probability $P_{10\uparrow-9}$ creates the unity probability event:

$P_{10\uparrow-9}$ **x 10 x 1 x 10^8** = 1.00 unity probability event attributing local velocity value 1×10^8 m/s becoming 3×10^8 m/s for three planes.

46. The most probable radiation from the sun surface or its vicinity is coming from a coalescence of 27 hydrogen atoms forming a cube. This is the conclusion of Chapter 11. The radiation forms sum of three orthogonal oriented pulses generated from the centers of three surfaces of the cube facing the earth. Each surface has 9 atoms (3x3), inclusive of atoms located along edges, discharging in two orientations the whole of mass 54 **x10**$^{-27}$ kg in opposite directions. Twenty seven atoms discharging together create a probability index 1/27. With correction for relativity, quantum and linear probability discharge derives Planck Radiation power:

$$(27 \text{ x}10^{-27} \text{ kg})^{1/27}/[1.111 \ldots \text{ x } (1+1/60) \text{ x } 1/10] \equiv 0.0100018 \text{ W} \qquad (23.48)$$

It is suggested that at 300K power 0.01 W becomes 3 W. The uncertainty 1.8 **x** 10^{-25} **x** 300 **x** 1.1111 = 6 k_s is 1/5 k_s per atom.

47. Acceleration on earth in the model integer SI units is 10 m/s.

48. Equipartition of energy in a semi-stable system requires the total energy in one dimension to be 6000 K and to be expressed, after corrections (Equation 2.64), by square root of 2 times the cube of seventeen degrees of freedom of space system per K (10 space quantization, 3 planes, 3 orientation of space and 1 spin):

$$\textbf{6000 K} \equiv \sqrt{2} \textbf{ x } 17^3/(1.1547 \text{ } 1/04^{7/100})=\textbf{6000.67 K} \qquad (23.49)$$

Earth gravitation induces force attributed to temperature 600 K = 2x300 K created by vector addition and 17 x 600 represents number of degrees of freedom each expressed by unity of energy; Boltzmann constant, using the converter 1x10^{-19} to integer SI units, has been found to express the probability required to have in all degrees of freedom energy 1 J $\equiv$ 1 eV to generate in one direction energy 1x10^{-23} J:

Probability= converter/(total energy) x corrector=1x10^{-19}/(17x600)x1.04$^{1/2}$=0.999808 x10^{-23}
$$\equiv 1\text{x}10^{-23} \text{ J } (23.50)$$

Boltzmann constant per unit local velocity 1/3 c_s in probability product, with the converter 1 x 10^{30} to integer SI units, derives the pulse of 1 J:

$$[k_s \text{ x } (1/3 \text{ } c_s)]\text{X}[k_s \text{ x } (1/3 \text{ } c_s)] \text{ x } 1 \text{ x } 10^{30}= 1 \text{ J} \qquad (23.51)$$

49. In assessing the credibility of the concepts proposed and the feasibility of radiation particles moving locally with fractions 1/10, 1/4, 1/3 and ½ of the velocity of light it is necessary to consider in detail the accuracy of the results obtained. Table 18 in the first book [45, p.161] has shown that uncertainties of the CODATA 2005 physical constants fit well the anticipated Heisenberg uncertainty values of the integer cubic values of the model. Several other parameter values arbitrarily corrected by relativity and quantum derive either an integer value or have a residual, which can be expressed by a fraction of the quantum parameter $q=hv=kT/e$. It is claimed that when properly analyzed the physical parameters expressed in cubic space SI converted units derive more accurate data than the SI classical units. This note does not apply to the gravity forces in Tables 5-14, where corrections were not made or several relations may be deceptive.

50. To obtain acceptable uncertainties expressed by Boltzmann constant k_s or its simple derivative per degree of freedom calculated results had to be multiplied by 300 K. It raised the likelihood that SI masses in general express values per kelvin. Referring ½ mass of sun and the mass of earth, with their probabilities 1/32 and 1/34, to temperatures 1000 K and 300 K and identifying the resultant probability to be power generated derives for both masses 10 W with uncertainties $3k_s$ and ½ k_s. This is a strong indication that SI sun and earth masses express value per kelvin and require adjusting to obtain correct results:

$$(1000 \text{ K x 1 x } 10^{30} \text{ kg})^{1/32}/1.154700^{1/2} \equiv 10.000282 \qquad (23.52)$$
$$(300 \text{ x } 6\text{x}10^{24})^{1/34} \text{ x}\sqrt{3}/1.11111111(1+1/80) \text{ x } 1.0004 \equiv (10.00000-0.000053) \text{ W}$$

Velocity of light is again indicated as being the factor increasing the rate of pulses. Combining it with other two factors generates the energy 20 J or eV with uncertainty ~ $k_s T_{300}$:

$$(1000 \text{ Kx1x}10^{30} \text{ kg x3 x}10^8)^{1/32}/(1+1/80) \equiv 20.0242/(1+1/80) \text{ W, J or eV} \qquad (23.53)$$

While 1000 Kx1x10^{30} kg $^{1/32}/(1+1/80) \equiv (500 -0.036)$ J or eV with variance $1.5k_s$.
Signal from the proton mass referred to 300 K, which probably expresses energy of mass generates Planck's 1st Radiation signal with uncertainty k_s:

$$(300 \text{ x } 2\text{x}10^{-27} \text{ kg})^{1/32}/(\sqrt{3} \text{ x } 1/10) \equiv (0 \; 01 + 0.000104) \text{ W} \qquad (23.54)$$

Signal from electron mass referred to 300 K gives error of 4.1 x 10^{-25} W, which multiplied by 300 K derives uncertainty 12 k_s:

$$(300 \text{ x } 1\text{x}10^{-30} \text{ kg})^{1/32}/(1.1547^2 \text{ x } 1.04^{22/25}) \equiv 0.0999959 \text{ W} \qquad (23.55)$$

An unexpected finding includes the identity of 6000 K to power $1/4^3=1/64$ deriving accurately energy 1 eV or J and the mass of sun to power 1/64 derives accurately energy 3 eV or J :

$$(6000 \text{ K})^{1/64} \text{ x } 1.1547 \text{ x } 1.04^{1/5} \equiv 0.999933 \text{ eV with variance 2/3 } k_s$$
And $$(2 \text{ x } 10^{30})^{1/64} \text{ x } 1.04^{1/5} \equiv 3.0032 \text{ J with variance 1/3 x } k_s \qquad (23.56)$$

Temperature 1000 K generates creates: $1000^{1/6}/1.1111\ldots^{1/2} \equiv 3.0000\ldots$ J or eV $\qquad$ (23.57)

The results indicate that cubic geometry with probability 1/64 is derived as sequence of four orientations of discharges in three planes. In the classical SI system it is represented by the spherical geometry with probability 1/60; both the integer probability model and the classical SI systems derive correct energy values and identities indicating that cubic integer system is very real and that it equally represents the spherical system.

The attractive force of sun gravitation is equal, for all planets with two minor corrections, to the force of attraction F = mass x acceleration induced by the rotation of the planet mass. That force in atomic terms expresses kinetic interactions, which have no units of length and time. It is suggested that consequently mass in kinetics is identified with energy leading to identities in Equations 22.101-22.122. Furthermore kelvin temperature to power four is identified with the momentum of 1 kg of mass:

For sun $(6000 \text{ K})^4 \equiv 1 \text{ kg} \times 2 \times 3 \times 10^8 \text{ m/s} = 1 \text{ kg} \times 6 \text{ directions of space} \times 10^8 \text{ m/s}$.

Product of two moments generates velocity of light and expresses, uncorrected by relativity, sun of mass to be $1.67916 \times 10^{30} \text{ kg}$:

$$(6000 \text{ K})^4 \text{ X } (6000 \text{ K})^4 \times \text{cor.} \equiv 1 \text{ kg}^2 \times 1.67916 \times 10^{30} \times 1.1547^{3/2}/1.04^{1.05} = \qquad (23.58)$$
$1.99999428 \times 10^{30} = (2-0.0000058) \times 10^{30} \text{ kg}$, the variance $6 \times 10^{24} \text{ kg}$ is the mass of earth.

Since the sun interacts with only one beam the above procedure for earth produces different outcome:
$$300^4 \equiv 10 \text{ kg} \times 3^3 \times 3 \times 10^8 \qquad (23.59)$$

On earth two 300 K pulses generate 600 K pulse and

$600^4 \equiv 1\text{kg} \times 1.296 \times 10^{11}$ while $1.11111^3 \times 1.0328^{3/2}/\varepsilon_{os} = 1.295804 \times 10^{11}$ indicating identity $600^4 \varepsilon_{os} \times \text{correction} \approx 1 \text{ J, W or eV}$.

While $600^4 \text{ X } 600^4 \times \sqrt{3} \times 1.0328 \equiv 3.00460136 \times 10^{22} \text{ N}$ (23.60)

Temperature 1000 K generates very clear result: $1000^{1/6}/1.1111\ldots^{1/2} = 3.0000\ldots \text{ J}$ or eV

Plank's 1st radiation energy term power is very simply expressed in integer units by:

$$10\varepsilon_{os} \times 1/3 \times c_s = 0.01 \text{ J} \qquad (23.61)$$

The same form of relation should be produced by the classical SI units, but there is now need to use relativity and quantum corrections and local velocity of light suggesting this to be evidence of the presence of local velocities $1/3 \times c_s$:

$$10\varepsilon_o \times 1/3 \times c_s \times 1.111\ldots \times 1.04^{15/36} = 8.85418 \times 10^{-11} \times 1 \times 10^8 \times 1.111\ldots \times 1.04^{15/36}$$
$$= 0.01000007 \text{ J} \qquad (23.62)$$

It is interesting to note that probabilities $1/10$, 3^4 with relativity correction squared derive the first Plank Radiation energy $0.01 \text{ J} \equiv 3 \text{ J}$ or eV at 300 K:

$$1/10 \times 1/(3^4)/(1.111\ldots^2) = 1/10 \times 1/81/(1.111\ldots^2) = 0.01234567900/(1.111\ldots^2)$$
$$\equiv 0.009999999999 \text{ J} \qquad (23.63)$$

The corrected inverted probability $3^8 = 6561$ defines with corrections accurately the temperature of 6000 K:

$$6561/1.1547 \times 1.111111^{1/2} \times 1.04^{1/22} = 6000.035 \text{ K} \qquad (23.64)$$

Thus temperature of 6000.035 K is just probability of 1/699.035= 0.00016665694≈ 0.001 x 1/10 x 1.1547005$^{3.5}$ x 1.04$^{0.19}$=0.00166678 (23.65)

Some of the calculations could be considered unjustified manipulations with figures, but the nature is the activator and the author uncovers its ways.

51. There is no missing invisible mass in the Universe; there are cold masses of stars, which generate much smaller power of gravitation than visible suns at 6000 K. They effectively do not rotate around other bodies and do not accelerate and do not convert protons into radiation, the main gravitation process.

52. The atomic source of the centrifugal force induced by acceleration remained puzzling for many years. To distant observer a unidirectional gravity force of attraction cannot be balanced by the vectorial forces of thermal flux radiated in all directions from the planets. That balance obtained is by the centrifugal force induced by planet rotation. The thermal balance between the sun and earth is created by locating earth at a position where the earth temperature and the energy released as the consequence of the normal rate per K of collapse of proton structure is equal to the sun rate of moments around earth. The sun rate is function of its temperature, the distance to earth and the magnetic lens effect. At atomic level the kinetic interactions eliminate acceleration term in $F = ma$ so that force $\equiv$ mass. On earth two 300 K generate the 600 K pulse and the force becomes $600x6x10^{24}$ units=$3.6x10^{27}$ units. With $\mu_{os}/\varepsilon_{os}=v_o/v_{Hls}= 10^{17}/10^{12}\equiv1.0x10^5$ **J** the repulsion force generated by earth is

$$3.6 \ x \ 10^{27} \ \text{units}/1.0x10^5 \equiv 3.6 \ x \ 10^{22} \ \text{N} \qquad (23.66)$$

In Einstein view planets float around the sun with no expenditure of energy. The alternative is to consider equal momenta and vectorized energy supplied by the sun and by the planets interacting with each other. Elementary radiations from inherent instability of protons are rotating in opposite orientations and probably nulling the total momenta. The same process establishes the age of the Universe.

53. Because of equality between dimensions of power and energy (Equation 23.1) the huge value 6x10100 treated initially as power now can be considered energy of $6x10^{100}$ J expended by the sun during it total longevity. Elementary estimate of the age of the sun derive it to be the longevity of the Universe considered contraversary.

Sun total longevity is $6x10^{100}$ J/$3.9x10^{29}$ W x1/(31.6 10^7 s/year) = $2.92x10^{63}$ years (23.67)

Longevity in 1D: $(2.92x10^{63})$1/6 = 37.8 x10^9 years (23.68)

Relativity corrected longevity per plane :

37.x10^9 tearsx1/3x1.111 . . . =14x10^9 years ≈13.73x10^9 years (23.69)

Creation of the sun system took millions of years. This allowed to sustain the equality of energies in all degrees of freedom for the whole of our limited Universe according to the requirement to maximize entropy.

54. The Ampere Law predicts a central force at right angle to the circular current. Observer moving linearly with velocity 1/2 of the circular velocity observes also presence of a force of repulsion between the particle in circular motion and the centre. The validity of the calculation may be doubted, but the switchgear design is based on this force principle to push and to extinguish the arc.

55. The balance, between the sun gravitation and the planet repulsion induced by rotation, is semi-stable. If energy is added to the system the new equilibrium position is shifted closer to the stationary sun. This semi-stability is attributed for very slow movement of the galaxy mass to the center of the rotating system of masses creating huge stars. Such stars collapse creating black holes. Planets rotating around the sun and masses of suns rotating around the center of galaxies are not neutral, but comprise electrons contained in their protons. Their linear velocity displays considerable magnetic flux and forces generated by the Ampere Law. **It is suggested that the gravity force at black hole apexes are overcome by the forces of the Ampere Law due to rotation of electrons and of protons. The Ampere forces generate extremely strong radiations beams in both axial directions extending far into the space.**

56. Ten invert values of fine structure constant $10/\alpha_s$ $x1.04^{1/10}$ =1371.5 are now identified to represent the solar constant 1371.5 W/m². The integer Coulomb constant $1x10^{10}$ is the exact invert of the gravitation constant.

57. Corrected by relativity six circular areas of surface area of sun 1.5000 x 10^{18} x 6 x 1.1111 = 1 x 10^{19} m² are an exact inverts of electron charge e_s=1 x 10^{-19} C.

Integer sun radius (without quantum cor.) is $(15 x10^{18}/\pi)^{1/2}/1.1547$ $x1.04^{1/11}$ =6.008 $x10^8$ m.
Earth diameter corrected for relativity and quantum is 1.000 $x10^7$ m.
Circumference of earth corrected for relativity is 40 $x10^6$ m as proposed by French Academy of Science in 1791.

58. Three consistent estimates of mass density $5x10^8$ kg/m² radiated to earth by sun in Equations 3.17-3.18 using different data confirmed by evaluation of the correct gravitation force in Equation 3.20 form a strong support for the validity of the model.

59. A quantum value of a parameter in nature is the smallest **average** amount that can be measured after a semi-stable balance has been reached locating equal energies in all degrees of freedom. This definition of a quantum value applies to energy and to momenta and possibly to mass, when split. It expresses equal probabilities of events in all space quantization orientations. Cubic quantum value 3/2 x $k_s x T_{6000}/e_s$ x1.11111 = 1.0 eV ≈ $\sqrt{3}/1.2^3$. Quantum value applies essentially to atomic particles; it minimizes and equalizes energy between particles and maximizes entropy areas of space extending over light years in size causing generation of common quantum energy value.

60. Although the results presented display an excellent agreement it is difficult to differentiate between the presence of kT/e particles and an incorrect quantum corrector. There may minor calculating errors, incorrect individual event assessments in the proposed text or incorrect dimensions, but they do not discredit the whole theory. All basic proposed processes are very well supported by evidence. The velocity of light has local atomic components requiring individual relativity corrections. There are three oscillators in Bohr atom. Sun interact with earth with band of 5×10^{12} frequencies. Quantum correctors are obviously missed in the classical SI system. Integer physical constants created directly from Planck units represent the geometry of space. Geometric masses of sun-planet couples define gravitation energy. There is no independent gravity force and the naked reality is expressed by the probability factors in the atomic units of nature. The kinetics interaction rules within the atom have no dimensions of time and of distance; consequently energy is expressed by mass while time expresses velocity. The product of velocity, of mass and of power of four of kelvin temperature expresses the pulse rate; the elementary signal from the sun of 50 J of Equation 2.15a is then the product of the rate with 1/3 probability in one orientation, the power of proton pulse of Equation 2.36 and corrections:

$$(1000 \text{ K}^4 \text{ x3 } \times 10^8 \text{ m/s } \times 1 \times 10^{30} \text{ kg})^{1/3}/(\sqrt{2}) \times 1.111 \ldots ^{1/2} \text{ x } 1 \times 10^{-15} \text{ J} \equiv (50 - 0.1035) \text{ J} \quad (23.70)$$

Lack of dimensions of time and distance makes it difficult or impossible to explain processes inside the atom using SI units.
The elementary pulse from earth is 1 J, W or eV:

$$(6 \times 10^{24} \text{ x } 600^4 \text{ x } 3 \times 10^8)^{1/3} \text{ x} \sqrt{3} \times 1/(1.1547^{1/2} \times 1.04^{15/128}) \text{ x } 1 \times 10^{-15} = (1 - 0.000195) \text{ J, W or eV}$$

with uncertainty $2\,k_s$.

Man created signals obey the nature laws, but they may not appear in nature.

61. Radiation shown on Fig. 4 is represented by 30 quantum positions each being a source of 1 J. This representation applies to spherical sun and with restrictions to earth. The structure of the spherical H atom attributed in Chapter 10 to 32 quantum positions on the spherical surface with wave of 2x4 particles oscillating between the outer surface and the spherical centre. These particles change their phase position by 90° at each interlinked half wave while circulating energy of 128 units within the sphere. This description has to accommodate unequal energy density distribution within the sphere absent in the cubic structure. The half wave of cubic structure is replaced by 128 of interlinked cubic structures discharging energy each in 8 corners making a total of 128x8 = 1024 units, which divided by quantum corrector derives $1024/1.04^{3/5} = 1000.184$ J, displaying uncertainty 0.184/8 x 300/1.1547= 6 W, J or eV. At 1 K the uncertainty $0.023/300 \approx \frac{3}{4}\,k_s$ has an acceptable value.

A cubic H atomic structure may be taken to have 10x10x10=1000 units of average linear dimensions 10^{-11} m $\equiv \varepsilon_{os}$ identified numerically with dielectric permittivity. Each of 1000 units contains two interlinked energy components of electron mass 1×10^{-30} kg forming a loop around each other and cubes interlinked forming a chain converting phase of components something like dx/dt → dy/dt → dz/dt → and back.
The following relations give support to above optional description:

$$\text{Mass of proton } 2 \times 10^{-30} \text{ x} 1000 = 2 \times 10^{-27} \quad (23.71)$$

The linear dimension of the electron cube $\varepsilon_{os} \equiv 1 \times 10^{-11}$ m;
The volume of the cube $h_s \equiv \varepsilon_{os}^3 \equiv (1 \times 10^{-11})^3 = 1 \times 10^{-33}$ m^3;

Boltzmann constant k_s defined=electron mass x atomic spin velocity=$1x10^{-30}x10^7=1x10^{-23}$ J; Energy of the loop (Eq.10.42) $3x10^{-12}x1x10^{-19}=300 x10^{-33}$ or $1.0\ h_s$ per kelvin per loop. With volume $1 x 10^{-33}$ m³ energy density per 1 m³ is 1 W, J eV (23.72)

62. The classical h and the integer model h_s SI Planck constants relate to infrared frequencies. Infrared radiation involves probably conversion of proton mass to energy on being accelerated to velocity of light. This creates high power, but on being converted to vectors by the probability of space quantization $(1/10\ x1/10)^3$ the energy of 1 MeV photon pulse is reduced to 1 eV vector. While the converted energy is referred to 300 K, the relativity corrected proton, subjected to velocity c_s generates the energy 30 J ≡ 30 eV calculated below for three planes; it really expresses the product of momenta of two local energies with corrections speeding at the velocity of light times 300 K:

$$3/2\ x\ m_{ps}\ c_s^2 x\ 1.1111^2 xc_s\ x300\ K = 3/2\ x\ 2x10^{-27}x\ (3x10^8)^3 x\ 1.1111^2 x300\ K= 30\ J$$
$$= (\sqrt{3}x\ 3x10^{-14}x1.11111^{1/2}x3x10^8)^2 x\ 1.1111^2 xc_s\ x300\ K = 30\ J \qquad (23.72a)$$

63. *Magic numbers of the ancients.* Geometry of space and independence of six orientations create probabilities and magic numbers in the quantum theory of the Universe. Restricting the quantum limitation to proton discharge of 5 or 10 directions per plane or per direction has to be reconsidered. While these values are not incorrect, properties of physic must have a simpler basic source of behavior. It is suggested that generation of circular rotation requires four sequential discharges in Z plane in directions Y, X, -Y, -X. These discharges take place in three independent orthogonal orientations generating the maximum resultant shown in Figure 5. In each of three orientations the rotating radiated particle has four independent directions. Hence the number of options is 3 x 4 =12. The generated rotating pulse creates events expressed by irregularly displaced hours of a clock generating distributions neglected in the model. It is suggested that twelve is the Magic number of physics. The others are three, four and ten because the components at right angle to the propagation direction are summed to zero reducing twelve components to 10. There are 12 months in the year; it is the number of hour per day, number of cards in a suit. There are 12 hours per day. The number twelve has a long history. It expresses wholeness in the Bible. There were 12 apostles. There are twelve hours per day and per night. Twelve has religious significance for Hans, Egyptians and Hindus. There are twelve signs of Zodiac. There are 12 Northern and 12 Southern stars. There were twelve tribes of Israel. For three planes 10 becomes 30. There are 30 days per month. There are several derivatives of 12. The product $12x30x1.04^{3/8} = 365.33$ days per year is very close to actual 365.242 days per year. Earth temperature 300 K is 30x10. Sun temperature is 300x2x10 = 6000 K. Thirty probability pathways creates probability index 1/30 indicating strangely that most sources deliver unity signal. Consequently a star and the hydrogen molecule from monstreous sized hydrogen cloud deliver to earth at 300 K photon elementary quantum energy values 3 eV, which per K becomes in SI units 0.01 J. Actually pulses from sun and stars contains often about 20 such photons. The most important number 10^{10} has been discussed earlier.

Thus the above "Magic numbers" are inverts of various probabilities. The mass of sun with probability 1/10, corrected for relativity and quantum derives its own temperature per direction with variance of $1/5\ k_sT$ in:

$$(\text{Sun mass kg})^{1/10} \times 1.1547 \times 1.04^{1/12} \equiv 999.82 \text{ K} = 1000 \text{ K} - 1/5 \, k_s T \times 1.1111 \qquad (23.72b)$$

64. Essential factors generating the integer model of reality in which probability replaces model SI units

Units of the classical SI system view the world as static particles generating radiations moving with the velocity of light c. In the integer model atomic **interactions are purely elastic. They are kinetic such as observed between balls on a pool table and they express true reality having no dimensions of time or distance.** One unity event appears as the reference and other events express solely the probability of that occurrence. Elastic kinnetic interactions between radiation particles and within the atom and between radiation particles in space bring identities **Mass = Force,** and **Time = Velocity**, which have to be adopted into SI system. In the integer model SI units the radiation pulses are assumed to move with average local velocities of $1/3 \, c_s = 1 \times 10^8$ m/s in three orthogonal orientations with the classical relativity correction 1.1111 . . . adding vectorially and generating c_s. Local light velocity 10^8 m/s in probability units expresses Planck radiation energy 0.01 eV converted to SI units $0.01 \times 1 \times 10^{10} = 10^8$ J. All conversions take place in one location, so space has no function in the formation of integer probability model. Space disperses the density of the signal. At the source in specific directions, the energy is subjected to various structure requirements expressing input probability. At the receiver the absorbed energy is expressed as the probability of output. Independent orthogonal directions of space with two more velocities and relativity corrections create optional output vectors probability from input reference moments or energies.

Over most of earth separation the sun attracts earth with forces directed only along the rotational plane of earth. **With three relativity corrected pulses generating each pulse of four orientations per plane of rotation, as shown for sun in Figure 14, there are 12 pulses.** Components of these pulses in one plane are shown in Figure 2. **Two of twelve pulses are not interactive** because they discharge in the plane at right angle to the propagation orientation. Thus **there remain 10 elementary quantum pulses orientations per plane having each independent sequential probability of occurrence of 1/10.** Generating one pulse of 1 J in a chosen orientation, an event of unity probability is repeated ten times with resultant probability $1/10 \times 1/10 \times 1/10 \times 1/10 \times 1/10 \times 1/10 \times 1/10 \times 1/10 \times 1/10 \times 1/10 = 10^{-10}$ expressing the linear probability of the integer model, and establishing the volume probability 10^{-30}. For three planes the probability is 10^{-30} with each kilogram of the input of 1×10^{30} kg of the sun generating 1 J. Any departure from the indicated sequence and probability effectively destroys the signal to insignificancy.

Equipartition of quantum energy in space may be rapid or it **may take millions** of years. With three planes there are $3 \times 12 = 36$ radiation orientations and pulses, but with 30 interacting with the receiver the momentum transferred is decreased by 1.2 for one plane, 1.2^2 for two planes and 1.2^3 for three planes of interactions.

Rotation has four independent orientations with 1/10 probability per orientation generating probability of 10^{-4} for 1D signal of specific orientation. Each of the signals of the four orientations is formed from three unity events (signals) in orthogonal directions generated with 10^{-3} signal probability. These probabilities generate sequential probability 10^{-12}; with above 10^{-4} value one obtains the resultant probability 1×10^{-16} identified with invert of the squared local light velocity $(1/3 \, c_s)^2 \equiv 1/(1 \times 10^{-16})$. At atomic level speed is unity and the pulse is expressed by the squared wave function value $(1/3 \ c_s)^2$. The product of two such probabilities from two planes displaced by 90^0 generate resultant probability called wave function format $(1 \times 10^{-16}) \times (1 \times 10^{-16}) = 1 \times 10^{-32}$ with signal value $\sqrt{2}(1 \times 10^{-16})$ x 1.1547 x $1.04^{1/2} = 1 \times 10^{-15}$ J $-1/5 \times 10^{-19}$ J, expressing an acceptable uncertainty. To convert to SI units 1×10^{-32} has to be multiplied by volume probability 1×10^{30} generating 0.01 J, the Planck radiation term.

The discovery of the space pulse $E_{sspaceSI} = (\varepsilon_{os}\mu_{os})^{-1} \equiv 1 \times 10^{17}$ J and of vector space frequency $v_{ovector} = 1 \times 10^{17}$ Hz puts a limit on the accuracy of measurement of energy per cycle as 1 J, which with liner probability becomes the gravitation constant $G_s = 1$ x 10^{-10} J. Consequently some classical SI uncertainties may be over-optimistic with their measurement accuracy.

Physical constants expressing energy in classical SI units are normalized and so are the uncertainties of the mean probability in the model integer SI units. The integer model SI uncertainty in one dimension (1D-direction) per kelvin expressed by Planck radiation term is:

$$k_s \ \textbf{x} \ E_{searthSI} = \textbf{1x10}^{-23}\textbf{J} \ \textbf{x} \ \textbf{1x10}^{21} \ \textbf{J-s} = \textbf{0.01 J}^2\textbf{-s} \qquad (23.73)$$

This is the radiation term of Equations 3.14 or 3.14a representing the energy quantum, which is normalized per kelvin hence generating at 300 K energy of 3 J for 3 planes or 1 J or 1 eV per plane.

The uncertainty of the mean probability in the model integer SI units in one dimension in the specified orientation per kelvin and per $E_{searthSI}$ for each of ten quantum pathways (expressing 1/10 probability of pulse formation) is:

$$k_s \ \textbf{x} \ (E_{spaceSI}) \equiv k_s \ /(\varepsilon_o \ \mu_o) = \textbf{10}^{-23} \ \textbf{x} \ \textbf{10}^{17} = k_s \ c_s^2 \ \textbf{x1.1111} = \mu_{os} = \textbf{1 x10}^{-6} \ \textbf{J-s} \qquad (23.74)$$

The uncertainty summing values for ten pathways and two oscillators becomes 20 $\mu_{os,}$ or

the input energy 2×10^{-13} J x $1 \times 10^{8} = 2 \times 10^{-5}$ W $\qquad$ (23.74a)

Accuracies in the model express quanta values; the author discussed in [46] that accuracies in the classical SI system express the measurements, which inaccurately describe the physical problem. It is suggested that **the uncertainty of the energy measurement expressed in the model SI units and in the classical SI units are both 2 x10^{-5} J/Hz.** That value is very close to the value announced on 22 September 2011 by CERN for the measured excess of the velocity of light for neutrinos.

The value **2 x10^{-5} J/Hz** multiplied by the linear probability 1×10^{-10} and 3×10^{-17} Hz to express it in the atomic level derives the uncertainty of:

Energy uncertainty at the atomic level = 60h_s = 60 x 10^{-33} J-s $\qquad$ (23.75)

Above uncertainty represents two Bohr structures needed for c_s wave function formation each having 30 quantum orientations with an uncertainty 2 x ½ h_s per orientation.

Alternative calculation starts with 1 DOF at 300 K, linear probability, 1.111 . . . relativity corrector and 2/3 multiplier expressing presence of solely infrared frequency components:

The uncertainty at the atomic level is given by $6 \times 1/2 \, k_s T_{300} \times 1 \times 10^{-10} \times 2/3 \times 1/10$ $\equiv 60 \times 10^{-33}$ J-s$=60 \, h_s$ because $k_s \times 10^{-10} \equiv h_s$. Ten comes from linear probability of ten pulses.

Radiations from protons or neutrino have 30 quantum radial orientations for three planes, each quantum attributing $h_s = 1 \times 10^{-33}$ J-s to uncertainty. Reflections in each orientation convert the wave output into a pulse with double energy and two $\frac{1}{2} \, k_s T$ or three $1/3 \, k_s T$ degrees of freedom adding to $k_s T$ and increasing the total number of DOF (degrees of freedom) to 60 with uncertainty $60 h_s$. The energy derived in Equation 8.46 is 20 eV. With 60 DOF the temperature of 6000 K/60 derives pulse of 100 J, which referred to 300 K decreases the pulse to 5 J. Now 5 J $= 3$ J $\times 1.1547^3 \times 1.0328^2 \times 1.04^{1/3} \approx 1$ J or 1 eV per orientation.

For earth and sun the probability P derived from three orthogonal orientations is taken as   (23.76)

$$P = [1/60 \times 1/10^2] \, [1/60 \times 1/10^2] \, [1/60 \times 1/10^2] \times 1.04^2 = [1/6000]^3 \times 1.04^2 = 5.0074 \times 10^{-12}$$

The surface re-radiation is derived most probably from two protons of unspecified molecule of unity mass oscillating at 1×10^{12} Hz and generating power 1 kg $\times v_{Lms}^2 = 2 \times 10^{24}$ J becoming for 3 planes 6×10^{24} J $\equiv$ mass of earth, since Mass = Energy. The resultant power using P converted to one charges derives μ_{os}:

$$\mu_{os} \equiv 2 \times 10^{24} \times 5.00 \times 10^{-12} \times 10^{-19} = 1 \times 10^{-6} \quad (23.77)$$

Thus the constant of space permeability of the model μ_{os} is derived as the signal from the infrared oscillations of the two protons. For balance similar power pulses from two protons of the hydrogen molecules are probably coming from the sun. The Planck radiation term (Equations 3.13-14a) in three orthogonal orientations generates with relativity correction probability of the same value of μ_{os}:

$$\mu_{os} = (0.01 \times 0.01 \times 0.01) = (0.001)^2 = (1/10 \times 1/10 \times 1/10)^2 = 1 \times 10^{-6} \text{ NA}^{-2} \quad (23.78)$$

Please note that $2 \times [1/6000]^3 \times 1.04^2 \times 1 \times 10^{12}$ Hz $= 10$ W is numerically twice the above 5 J.

In model SI units space permittivity ε_{os} can be expressed by linear probability and in the wave function format with each pulse generated by three orientations; it expresses also the space energy density (J/m^3) :   (23.79)

$$\varepsilon_{os} \equiv (0.01)^3 \times (0.01)^3 \times 10 = 1 \times 10^{-11} \text{ Fm}^{-1} \text{ converts at SI atomic level to } 1 \times 10^{-12} \text{ Jm}^{-3}$$

which at 1×10^{12} Hz delivers 1 W/m^{-3}; times volume probability delivers pulse energy density 1×10^{-12} Jm^{-3} $\times 1 \times 10^{-30} = 1 \times 10^{-42}$ Jm^{-3}, which times $1/10 \times 1/10 \times 1/10$ generate the vectorized energy 1 x 10^{-45} Jm^{-3}; that value to power 1/3 derives earlier quoted energy 1 x 10^{-15} J of Equation 2.36.

Probability postulate requires displaying ε_{os} in terms of 60 DOF. Two single ε_{os} pulses are generated initially in both opposing orientations and with relativity and arbitrary quantum corrections: (23.80)

$$2\varepsilon_{os} \equiv (1/60)^3 X(1/60)^3 / 1.1547^{1/2} x 1.04^{0.068} = 1.999936 \times 10^{-11} \text{ J/m}^3$$

In term of 30 DOF, probability expresses permittivity μ_{os} corrected by relativity and by three steps of probability occurrences:

Probability = Unity probability signal x $(1/30)^3 X(1/30)^3 \equiv \mu_{os}$ x $(1/10 \text{ x} 1/10 \text{ x} 1/10) = 1 \times 10^{-9}$
(23.81)

Physicists at CERN should be congratulated at measuring velocity of light to such great accuracy. They asked for comment. Their measurements according to the author do not support speed higher than velocity of light. Their results support the validity of the integer values of physical constants proposed by the author. The upgrading of physics is based on particles moving with local velocity of 1/3 c_s, because velocity is also the rate of pulse delivery with three components generating c_s.

The above uncertainty value 2×10^{-5} has to be referred to 300 K. At 300 K over first relativity correction, 1.111 . . .$^{3/2}$, times the above 2 x10^{-5} J/Hz, times charge increment factor, 1.2^3, and times second relativity correction, 1.1547 and quantum correction $1.04^{28/50}$ derives 0.00999975 J $\approx$ 0.01 J expressing the 1st Planck radiation term:

$$1 \text{ kg at } 300 \text{ K}/(1.111 \ldots^{3/2}) \text{ x } 2 \text{ x} 10^{-5} \text{ x} 1.2^3 \text{ x} 1.1547/1.04^{28/50} \equiv 0.00999975 \text{ J} \quad (23.82)$$

Please note that the fractional variance from unity probability of 0.01 J is 2.5 x10^{-5} J.

Energy $k_s c_s = 10^{-23}$ x 3 x $10^8 = 3 \times 10^{-15}$ J-m/s times volume probability10^{-30} derives elementary pulse energy 3 x10^{-45} Jat of the Equation 2.32.

Earth balance using different wording. Thermal pulses from earth are generated in radial directions from the whole volumetric surface. In the region around earth the resultant radial vector forces form fan shaped field geometry. It is claimed that vector forces oriented at right angle to the source receiver axis play no role as interacting rotational forces between the sun and earth. The two forces displaced by 90° from two protons are in wave function formation. They produce √2 larger pulse most likely with their resultant probability being the product of the input probabilities as indicated earlier. As shown on Figure 16 they generate a resultant 1/10x1/10. This has to be repeated in three planes each pulses occurring with probability 1/30. With this process being repeated at the end of each of three vectors the overall probability of deriving a pulse along the receiver-source direction becomes: (23.83)

$$[1/30 \text{ x} 1/10^2] X[1/30 \text{ x} 1/10^2] X[1/30 \text{ x} 1/10^2] = [1/3000]^3 = 3.7037037 \ldots \text{ x } 10^{-11}$$

Divided by relativity correction 1.111 . . .^{2} obtains $3\times10^{-11} \equiv 3\varepsilon_{os}$ **identifying triple permittivity $3\varepsilon_{os}$ with the probability of the formation of the pulse along the earth-sun axis.** With the input μ_{os} the resultant is **$3\varepsilon_{os} \times \mu_{os} \equiv 3 \times 10^{17}$ J^{-1}**. With input energy of 3×10^{-15} J as of Equations 2.36-37 the output after dividing by relativity correction is

$$3 \times 10^{-15} \times 3 \times 10^{-11} \times 1.1111 = 1 \times 10^{-27} \text{ J, the value of Equations 2.81-2} \qquad (23.84)$$

On the other hand product ε_{os} of space permittivity and of the infrared proton frequency per plane 1×10^{12} Hz with linear probability derives the unity pulse 1 J or 1 eV

$$1/10 \times \varepsilon_{os} \times v_{Lhs} = 1 \times 10^{-11} \times 1 \times 10^{12} \times 1/10 \equiv 1 \text{ J or 1 eV} \qquad (23.85)$$

The 1 J output requires the sun power input of 1×10^{30} W, which cannot have a smaller values and so it is in sense a quantum value.

Equipartition of discharge momenta in six orthogonal orientations of space generates the probability factors of the integer model. It implies that the Bohr atom is locally in the state of maximized entropy

To be accelerated to the velocity of light protons are subjected twice to probability conversion displayed on Figure 4. It is suggested that the vectors in the three orthogonal orientations are each the resultant of the same probabilistic process all orthogonal orientations being the longest path across the cubic vector structure at ¼ unity velocity. A specific number of elementary quantum units of energy have to be attributed to the processes to make it feasible. Since each of the three orthogonal vectors is the sum of three vectors the proposed total number of units including the factor of two for reflections is 18. Inspection of Chapter 11 shows that the closest possible number of atomic aggregates is 27 Bohr atoms. Fraction of energy appearing in 1D signal is decreased by 1.2^3 factor time relativity correction for velocity ½ c_s, 1.1547, over presumed quantum correction $1.04^{3/50}$ generating:

$$\text{Number of atomic units} = 27/(1.2^3 \times 1.1547) \times 1/1.04^{3/50} = 17.9998 \text{ units} \approx 18 \qquad (23.86)$$

Although the result appears very reasonable, slightly modified numbers generates similar accuracy:

$$27/(1.2^2 \times 1.1111^{1/2}) \times 1.04^{1/4} + 2/3 \, k_s \, T_{6000} = 17.94 \text{ units} + 0.04 \qquad (23.87)$$

Please note that 30/27=1.11111 . . . is the exact value for the relativity correction for 1/3 c_s. It is suggested that that the first equation expresses radiations from earth spreading proton vectorized radiations in three orientations from earth surface as suggested earlier and the second are equipartition energies of the thermal radiations from the sun moving in one plane with local average velocity 1/3 c_s, which oppose the first bringing balance.

Larger number of discharging atoms such as 12^3 atoms appears possible especially since they are an integer multiplier of the smaller group $12^3/3^3 = 64$.

The old International Standard in Paris defined the current of one ampere by generated force of 2×10^{-7} N per meter length, between thin and long parallel conductors. The model proposed attributes this force, after applying the relativity corrector 1.111 . . . to be generated by the local velocity of light, 1/3 c_s, moving in the orthogonal orientations shown in Figure 5 and forming a twice larger value. The atomic force acting in one plane between the electron and the proton expressed in the model SI units is quoted in Table 1 as 1×10^{-7} N. This is the elementary force of 1 N/kg at 300 K reduced by linear probability by 1×10^{-10} times, with the input energy increased by a factor 10^3 to generate the vector force. According to the author the Paris Standard converts SI units into the equivalent natural atomic by the linear factor of 10^{-10}, the energy multiplier 10^3 and gains for each conductor $\sqrt{2}$ by the geometry of adding two vectors displaced by 90° generating for both conductors the total force of 2×10^{-7} N.

The expansion of the Universe is now again examined in terms of the probabilistic source of the basic radiation signal coming from the edge of the Universe. That signal as explained in Chapter 11 is coming from H_2 molecules generating in each of six orthogonal orientations signals of two proton-electron pairs, two protons and two electrons. The probability of the signal to appear in wave function format as two pulses displaced by 90° is expressed in integer model SI units as permeability of vacuum corrected for relativity:

$$(1/6 \times 1/6 \times 1/6 \times 1/6) X (1/6 \times 1/6 \times 1/6 \times 1/6) \times 1.2^3/1.04^{0.7} = (1/1296)(1/1296) \times 1.2^3/1.04^{0.7}$$
$$= 1.00095 \times 10^{-6} \approx \mu_{os} \qquad (23.88)$$

The variance is 1 J per cycle of space frequency at 1 K of $\frac{1}{2} \times 10^9$ Hz (Equation 4.5 of [45]). Product of two $\sqrt{3} \mu_{os}$ pulses at 300 K is attributed to come from He or four H atoms. Multiplied by surface probability 10^{-20} creates SI pulse energy in three planes:

$$(\sqrt{3} \mu_{os} \times 300)^2 \times 10^{-20} \times 1.1111 \approx 3 \times 10^{-27} J \qquad (23.89)$$

The above decrement of the momentum means that the source charge facing the receiver is moving away from the receiver. The velocity of the receiver charge expanding from its centre of mass towards the source is likely to be same as the source charge. With the mass of the receiver taken as the reference any observed infrared Doppler Effect would be due to the mass of the source moving away from the receiver. The mass of the receiver seems to be the correct reference and hence the high speed of the Universe expansion due that effect may be illusional. Nevertheless proposed instability of protons generates radiations, which all proceed outside the Universe, causing **its expansion at velocity of light becoming an inherent feature of the observation**, unless Einstein theory changes it.

In the above argument radiation is represented by the orthogonal geometry of space in which the four pulses of rotation were each formed from discharges in three orthogonal directions.

This is a research book, which deals with the topics as they are disclosed, so earlier chapters may not include implications of the discovered probability formulation of the reality in conclusion 64. There may be some misconceptions, errors of judgments, dimensional inconsistencies, simple errors, repetitions and huge spurious data, the source of final conclusions. This is the cost of advancement.

65. The book upgraded the classical SI system comprising features of the Universe recognized by our perception by investigating consequences of the presence of independent components of velocities of light along orthogonal orientations and applying to them relativity and quantum corrections. At the end of the project it was realized that interactions of radiations in space and probably within the atom obeys rules of kinetics having no units of time and length; this brings havoc to dimensions of equations expressed in SI units. Importance of probability was realized after working nine years on the project with the finding that all events express probability imposed in the orthogonal space and the need to generate one dimensional oscillation vectors. The book considers only the infrared frequency range, because it has been surprisingly uncovered that h_s applies to that range.

With no dimensions of length or time the number of interactions with SI functions names overlap leading to generating dozens of correct expressions for the force of gravitation. Improved understanding of physics and of gravitation has to be considered together. **There is no need to change the practical classical SI system, but it is useful to know classical SI system limitations. The new structure of the physical events derived from the simple concepts of probability is very consistent, but several aspects of science remains unknown in detail.** Both physics and gravitation needed updating; young unbiased people will have to improve the views postulated.

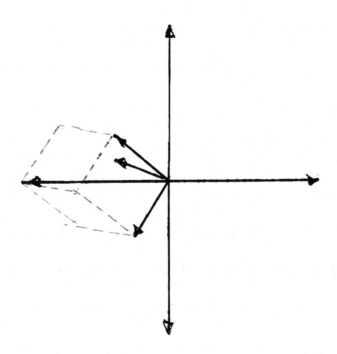

Figure 14. Radiation pulse created from sun by four rotational orientations each genera- ted by three relativity corrected pulses at velocity 1/3 c_s; the created probability of vector formation is 1×10^{-120} so that sun generates pulse $(1 \times 10^{30}$ kg$)^{1/12}/\sqrt{3}$ $\times 1.04^{13/20} \equiv 1.0008$ J

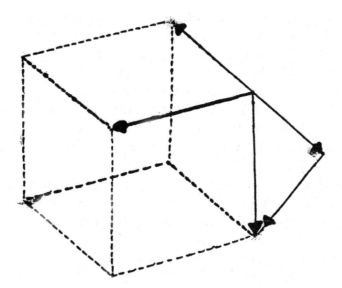

Figure 15. Hypothetical radiation pulse created from earth by three orthogonal orientations each generated by two relativity corrected pulses at velocity 1/2 c_s; two above pulses are needed to create probability=μ_{os}^2=$(0.01)^3$X$(0.01)^3$=1x10^{-12}≡$1/10^{12}$Hz

66. *Addendum*

The final overview of the book uncovered important relations given below and needed reminders.

Boltzman constant is the angular rotation momentum of electron= m_{es} x υ_{es} = 1x10^{-30} kg x1x10^7 m/s = 1x10^{-23} kg-m/s. It is also derivarive of squared proton frequency k_s = $1/10$ x $1/(10^{12})2$ = 1x10^{-23} J and

$h_s \equiv k_s$ x10^{-10}

Geometric mass of proton-electron couple = $\sqrt{(m_{ps} m_{es})}/1.111 \ldots ^{1/2}$ = 3x10^{-29} kg

Geometric mass energy =$\sqrt{(10^{-27}}$x$10^{-30})(3$x$10^8)2$ x$1.111 \ldots$ = 3x10^{-12} J exactly.

The power at 3x10^{12} Hz is 3 W in agreement with 2 W in the absence of electron oscillations.

Energies and moments in 1D with velocity 10^8 m/s:

Electron: 10^{-30} kg x $(10^8$ m/s$)^2$ = 1x10^{-14} J;

Proton: 10^{-27} kg x $(10^8$ m/s$)^2$ = 1x10^{-11} J = ε_{os}

Linear momentum of electron: 10^{-30} kg x 10^8 m/s =1x10^{-22} kg-m /s

Linear momentum of proton: 10^{-27} kg x 10^8 m/s =1x10^{-19} kg-m/s ≡ es≡ electron charge

Relativistic forces of gravity of the sun is expressed my the interacting mass of sun, three planes, local velocity 1/3 cs, and quantum ratio 36/30

$$F_{Gs} = 3\text{x}1\text{x}10^8/10^8 \text{ x}1.2 = 3.6 \text{ x}10^{22} \text{ N} \qquad \text{(see Equations 6.21, 6.19, 6.22)}$$

Ionization energy of Bohr atom in 1 D is 10 J or eV. The energy carried in the sun pulse, of 50 J or eV ≡ 60 J in relativistic terms, has 40 J or 50 J of moments structurally nulled as explained on Fig. 10 of reference 45 p.95. Energy 60 J applies to H2 molecule, while 30 J (eV) = 3x10 is the Bohr atom energy.

Generation of 2 W sun pulse; increasing proton velocity from ¼ c_s to c_s in two steps of 1/4 c_s x $\sqrt{3}$x1.1547 x2= c_s and nine component h_s energies times 1.1111 express energy 10^{-32} J. Radiation needs two such interacting pulses of energy 2×10^{-32} J. At 6000 K power is 2×10^{-32} x 6000^4/[1.111 . . .^{2} x (1+1/6)] = 2×10^{-21} becoming 2×10^{-20} W on SI level. Divided by probability 10^{-10}x10^{-10} derives 2 W/kg. Such process has no energy or momentum loss. Only a small fraction of no coherent power of 2 W has probability to be converted into the gravity force.

Pulses or waves. The direction of radiation is established following the failure of atomic potential barrier. The forward wave is cancelled by the reflected wave from the back of atom with a delay which allows generating the pulse and deriving correct value of hydrogen atom diameter (Equaton15.25).

Quantum correction, 1.04 is actually the ratio of two relativity corrections 1.1547/1.111

Range of application. Model deals mainly with events in physics taking place outside the atomic spherical structure. It does not deal with High Energy Physics, expressing internal events in atoms. Cubic structure appeared best to express independent orientation of space.

System International SI is perfect for practical applications, but is inadequate for representing atomic events. Current SI system of physics is significantly incomplete.

24. References

1. ZOLEDZIOWSKI, S. and SOAR, S.: 'Life curves of epoxy resin under impulses and breakdown parameters', *IEEE Trans. Electr. Insul.* EI-7, 84-99 (1976)
2. ZOLEDZIOWSKI, S.:'Quantized thermal depolarization of solid dielectric', *IEEE Trans, Die.El.Ins.* 2, 12-26 (1995)
3. ZOLEDZIOWSKI, S.:'Spectrometry of quantized thermal relaxations of polarizations in a solid', *IEE Proc.-Sci. Meas. Technol.* 146, 257-262 (1999)
4. ZOLEDZIOWSKI, S.:'Empirical treatment of high field conduction', *IEE 7th Inter. Conf. Die. Mat. Meas. Appl.* 430, 82-85 (1996)
5. D.Lide ed. Handbook of Chemistry and Physics 85th Ed. 2004-05, CRS Press
6. Fundamental Physical Constants <http://physics.nist.gov/constants> (2005)
7. TONOMURA, J., ENDO, A., MATSUDA, T., & KAWASAKI, T.: 'Demonstration of single electron built-up of an interference pattern', *Amer. J. Phys.*, 57, 117-120 (1989)
8. SMOLIN, L. 'Quantum Gravity', p.157, (Basic books, Perseus Group, 2001)
9. JAIN, G., 'Properties of electrical engineering materials'Harper International, 1966, N.Y.
10. WEITZ, M. *et al.* 'Precision measurement of the 1S ground state Lamb shift in atomic and deuterium by frequency comparison', *Ph. Rev. A*, 52, 2664-2681 (1995)
11. Palmer, R., Stein, D., Abrams, E. and Anderson, P.: 'Models of hierarchically constrained dynamics for glassy relaxations', *Phys. Rev. Lett.*, 1984. 53, 958-961
12. WHITHAM, W. and CALDERWOOD, J. H.: 'Electrode behavior and the determination of defect energies from measurements of ionic conductivity in NaCl crystals', *IEEE Trans. Electr. Insul.,)*, EI-8, 60-68 (1973)
13. MARMET, Paul WEB Pages, Over 100 contributions
14. GREEN, Bian, The fabric of the Cosmos, Vintage Books, New York 2004
15. PAGE L. & ADAMS N. Electrodynamics, New York, Van Nostrand, 1040, Chap. 3.
16. PEEBLES, P.J. Principles of Physic Cosmology, Princeton University Press, 1993, p.56, 417.
17. WHITT T., Our undiscovered Universe Aridian Publishing, Melbourne 2007
18. SMOLUCHOWSKI R., The Solar System, Scientific American Library, 983. p.12
19. WOIT, Peter, Not even GNOWR, The failure of the String Theory and . . , Basic Books, 2006, New York
20. D'Abro A., The rise of the new physics, vol. 1&2, Dover Publ., New York, 1951.
21. WEAVER J. ed. The world of Physics, 3 vol., Simon & Schuster New York, 1986
22. SILVERMAN M. More than one Mystery (Quantum interference), Springer-Verlag, New York, 1995
23. ROSSSER, W., Classical Electromagnetism Via Relativity Butterworths London, 1968
24. SMOLIN, Lee, The trouble with Physics, Houghton Mifflin Co, New York 2006
25. ASPEDEN, H., Physics without Einstein, Camelot Press, London 1969
26. McGEHEE, B., New Universe Theory, Authorhouse, Bloomington, 2005
27. BARTUSIAK M.. Einstein's Unfinished Symposium, Berkley Books, New York, 2000
28. MICHAUD A., Electromagnetic fields, Kazan State University, 2(28), v.13, 2007
29. FERMI Enrico, Elementary Particles, Yale University Press, 1961
30. INGRAM, D.J., Radiation and quantum physics, Oxford Physics Series, 1973, p.40.
31. BARNES, T.G., Foundation of Electricity & Magnetism, Heath & Co, Boston 1965.

32. SERWAY, R., MOSES, C. & MOYER, C., Modern Physics, Books. Cole, London 1997
33. EISBERG, R., Fundamentals of modern Physics, Jon Wiley, New York, 1961.
34. BEISER, A., Concepts of Modern Physics, MaGRAW-HILL, London 1973
35. BLUMICH, B., The incredible shrinking scanner, Scientific American, Nov.2009, p 92.
36. SCHWARZ, P. www.superstringtheory.com
37. MARMET, Paul, International IFNA ANS Journal : Problems in Nonlinear Analyses in Engineering Systems" N3 (19), vol. 9, 2003
38. TEDENSTIG O. Matter Unified HTML Basic Research Stockholm Sweden 2000
39. Probability Theory, Wikipedia Encyclopedia, WEB
40. LA FRENIERE, G. Matter is made of Wave, WEB pages and books (2002-2008).
41. MARMET Paul, The GPS and the constant velocity of light. WEB
42. BORN, Max, Atomic Physics, Blackie & Son, London, 1961.
43. BAKER, Joanne, 50 physics ideas, Quercus, London 2007.
44. CRILLY, Tony, 50 mathematical ideas, Quercus, London 2007
45. ZOLEDZIOWSKI, Stan M., Probabilistic, Exact Physics and Secrets of Gravitation, PayMaster, 200 p,. Edmonton 2009.
46. ZOLEDZIOWSKI, Stan M., Updated Gravitation and Physics in the Probability Model of the Digital Universe, PayMaster, 95 p,. Edmonton 2010
47. ZOLEDZIOWSKI, Stan M., Research reports for the U.K. Atomic Energy Authority (four), 1968-1970.
48. WAHLIN, Lars, The Collapsing Universe, Johnson Publishing, Co, Boulder 1985 and Deadbeat Universe, Colutron Research Corporation, Boulder 1999
49. WEIDNER, R. and SELLS, R., Elementary Physics, Allyn and Bacon, Boston, 1973.
50. BRADLEY, Reymond, Does God play dice with Universe? hhtp://www.sfu.ca/ philosophy/Bradley/does_god_play_pdf
51. MCMANON, David, Quantum Mechanics, amozon.co.uk
52. PAIN, H. J., The Physics of Vibrations and Waves, J.Wiley, London 1970
53. DEKKER, Adrianus, Solid State Physics, McMillan, London 1965
54. LISI, G and WEATHERHALL, O., A Geometric Theory of Everything, Scientific American, December 2010
55. WEIDNER, R. and SELLS R., Elementary Modern Physics, Allyn and Bacon, Boston 1973.
56. 'Max Born Explaining Born Probability Waves', WEB
57. ZOLEDZIOWSKI, S., Prace Naukowe Politechniki Warszawskiej, *Elektryka, 80, 43*-67, 1985
58. CASSIDY, D., Heisenberg, Uncertainty and the Quantum Revolution, Scientific American, May 1992, 106-112.
59. GUO, T. and GUO W.: "A transient-state theory of dielectricic relaxation and the Curie Schwiedler law", *J. Phys. C, Solid State Phys.*, 1083, pp. 1955-60
60. von HIPPEL, A., Dielectrics and Waves, J.Wiley, New York 1959
61. von HIPPEL, A. et ales, Molecular Science and Molecular Engineering ibid.
62. ZOLEDZIOWSKI S. Probabilistic World of Physics and Gravitation, With Relativistic Integer Physical Constants, PageMaster, p, 174-188, reference 45 written in 2002.
63. HAKING S. et ales, The dreams that stuff is made of, 1071 p., Runnning Press Book Publishers, Philadelphia 2011.

64. KITEL, C., KNIGHT, W., RUDERRMAN, M., Mechanics, Educational Services In. 1965 USA California, Lib. Congress No 64-66016 31.579-601↓ 65. JAYNES, E.T., The Minimum Entropi Production Principle, AnnRevPhysChem 1980 66. PRIGOGILI, I., Time structure and fluctuations, Nobel Lecture, Dec. 1977.

25. Appendix

Sun energy

Total sun power cannot be reliably calculated, so results below are only the research considerations. Number of units of minimum energy $1x10^{-14}$ J in 1 kg of mass moving at velocity 1/3 cs:

1 kg x $(1/3\ c_s)^2/1x10^{-14}$ = 1 x10^{40} per kg so that in the sun of $1x10^{30}$ kg the generated power is 1 x10^{70} WSI time volume probability 10^{-30} making it really 1 x10^{40} Wat.

The sun power in SI of 1 x10^{70} W is per kelvin. With sun circular area 1.5 x 10^{18} m^2 and power generated per square meter $\sigma_s T^4$ = $5x10^{-8}$ x (6000^4)= 6.48 x 10^{10} W/m^2 sun power is 1 x10^{70} x 1.5 x 10^{18} x 6.48 x 10^{10} = 9.72 x 10^{98} WSI. Increasing input by 10 for linear probability and quantum corrector $1.04^{7/10}$ derives sun power $1.00007x10^{100}$ WSI. The sun generated power per orientation or sun momentum in terms of natural units is:

$$\textbf{Sun power} = \textbf{1x10}^{100}\ \textbf{W}_{SI} \tag{25.1}$$

Sun temperature 1000 K in three directions and in 10 orientation and invert of linear probability generates power:

$$\text{Sun power per orientation } (1000^3)^{10}\text{x } 10^{10} = 0.999999999 \text{ x}10^{100}\ W_{SI} \tag{25.2}$$

Further analyses indicated that power 10^{100} WSI is closely interrelated with the integer space physical constants indicating that the nature of the Universe is probabilistic:

$$\text{Planck constant } h_s = (1/10^{100})^{1/3} \equiv 1 \times 10^{-33} \tag{25.3}$$

With 30 orientations of space in three dimensions the derived sun pulse is:

$$\text{Sub pulse to earth } = (1/10^{100})^{1/30} \equiv 2000 \text{ eV} \tag{25.4}$$

Radiation pulse in three dimension presentation $(1/10^{-27})^3$ converts to sun power in one orientation, discussed again in Equation 17.40 at

$$(1/10^{-27})^3\text{x}1/10^{-19} \equiv 10^{100} \text{ JSI} \tag{25.5}$$

The total sun SI momentum per plane is derived from two corrected pulses squared, each expressed by the product of interacting mass of sun and the squared light velocity. Such squared energy is equal the product of energy parameter 1 J-m, number of such parameters and the atomic force expressing J/m making it dimensionally correct:

$$[(1x10^{30}x3x10^8)^2]^2x1.154700x1.111\ldots/(1x10^{-7}x1.04\ x10) = 1.0074x10^{100} \text{ number of events} \tag{25.6}$$

The variance based on the incremental SI factor $10^{80} = 10^{30}$ x 10^{30} x 10^{20} expresses

$$7.4 \times 10^{97}/(10^{80} \times 1.111\ldots^2) =\approx 5.994 \times 10^{17} \approx 6 \times 10^{17} = 6\, h_{sspaceSI} \qquad (25.7)$$

Radiation pulse in three dimensions presentation $(1/10^{-27})^3$ converts to sun pulse of 500 eV with probability 1/30

$$[(1/10^{-27})^3]^{1/30} \equiv 500 \text{ eV} \qquad (25.8)$$

Atomic spherical size is the most efficient package for energy and it represents reality better. Cubic shape explains independendance of generating pulses in six directions, accordingly to probability rules. There is no doubt that the gravity on the sun creates a pressure within the range 10^5 - 10^6 greater than that the earth compressing the hydrogen atoms into cubes. Consequently the force acting on the earth surface on a mass of 100 kg is 1000 N inducing the same acceleration, 10 m/s. So the equation is identical in form to that of the sun-earth couple, except that the radius of earth is replaced by the distance from the sun to earth.

Sun attraction = Sun distance/earth radius $\times c_s^2 \times$ factors

$$= 1.5 \times 10^{11}/3 \times 10^6 \times 1.5(\sqrt{3})^3 \times 1.04^{1/4} = 3.5420 \times 10^{22} \text{N} \qquad (25.9)$$

The product of probability 1×10^{-100} and energy 1×10^{100} JSI of Equations 25.1, 25.5 or 25.9 generate 1 eV or 1 J.

Consquently the product of the mass of earth over the squared radius time 1 kg and time gravity constant G_s with relativity and quantum corrections derive the gravity force at the surface of earth. The expression is correct dimensionally with 10 N generating an acceleration of 10 m/s on the mass of 1 kg

$$6 \times 10^{24} [\text{kg}] \times 1 \, [\text{N/kg}]/[(6000 \times 10^3)^2 \, \text{m}^2]/(1.1547^2 \times 1.111\ldots^2 \times 1.04^{1/3} = 9.993 \text{ N}$$
$$(25.10)$$

Hence if the output event from the sun-earth pair is considered to be of unity probability, expressed by energy 1 J, the input must be 1×10^{100} units expressing 1×10^{100} J per direction as explained in Chapter 17.

Finding and trying to comprehend the mysteries of such relations created this book. It took a long time to justify the conversion of the SI cubic space size of the Bohr atom of 1×10^{-10} m to classical SI value of $\sim 1.05835 \times 10^{-10}$ m using the identity:

$$1 \times 10^{-10} \times 1.111\ldots^{1/2} \times 1.04^{1/10} \equiv 1.05823 \times 10^{-10} \text{ m} \qquad (25.11)$$

Expressing 0.01% accuracy, the difference between the two value taken to be 0.00012eV $\times 1 \times 10^{-19} = 1.2 \times 10^{-23}$ J representing ~Boltzmann constant that is H. u. per degree K. The relativity correction $1.111\ldots$ for $1/3\, c_s$ is needed because that is the speed of oscillating particles. Special Relativity theory predicts that distances apparent to different observers may vary with the velocity of the observers and the above difference follows that theory.

The earth moves around the circumference of space volume $4\pi/3(1.4961x10^{11})^3 \approx$ 1.403 x 10^{34} m^3, which time 1.11111 h_s c_s^2 x $1.04^{1.1}$ derives numerical circular area of the sun expressing thus its density at 1 W/m^2:

$$4\pi/3 \text{ x } (1.5x10^{11})^3 x1.1547 \text{ } h_s \text{ } c_s^2 x \text{ } 1.04^{1.2} = 1.4983 \text{ x } 10^{18} \approx 1.500 \text{ x } 10^{18} \text{ W} \qquad (25.12)$$

Above power per $E_{sspaceSI}$ J-s in two directions derives the Hartree energy 30 eV

$$1.500 \text{ x } 10^{18} \text{ W x 2 x } 10^{-17} \text{ Hz} = 30 \text{ W} \equiv 30 \text{ eV}$$

In rotation 30 eV expresses 2D energy presentation =
$$\sqrt{2}\text{x}30\text{eV}\textbf{x}1.111\ldots^{3/2}\text{x}1.04^{1/6}=50.016 \text{ eV}$$

$$(25.13)$$

The energy of 50 eV in waveform formation generate the 1D $1x10^{12}$ Hz frequency

$$[1/6 \text{ x } (50 \text{ x } 50)\text{X}(50 \text{ x } 50)/1.04]^2 = 1.0032 \text{ x } 10^{12} \text{ Hz} \qquad (25.14)$$

While using vectorized $E_{ssearthSI}$ 1 x 10^{21} J-s derives Planck Radiation term:

$$6 \text{ x } 1.11111 \text{ x } 1.500 \text{ x } 10^{18} \text{ W } /1x10^{21} \text{ J-s} \equiv 0.01 \text{ J}$$

The energy variance 0.016 eV representing Heisenberg uncertainty $\sim 1/\sqrt{2} \text{ x } k_s T_{300}$, the value derived in author's experiments.

Components of thermal radiation

Thermal radiated energy in integral model SI units is

$$\sigma_s T_{6000}^4 = 5 \text{ x } 10^{-8} \text{ W/m}^2 \text{ x}6000^4 = 6.48 \text{ x } 10^8 \text{ W/m}^2 \qquad (25.15)$$

The value equals the probability product of two 90° vectors (1 J x c_s)X(1 J x c_s) generating radiation proceeding at the velocity of light divided by converter to SI units and the correction discussed earlier:

$$(1 \text{ J x } c_s)\text{X}(1 \text{ J x } c_s)/cs \text{ x}1.2^3\text{x}1.1547 \text{ x}1.111\ldots\text{x}1.04^{7/10} = 6.471\text{x } 10^8 \text{ W/m}^2 \text{ }(25.16)$$

Expressions deriving components of the radiation energy:

$$6000 \text{ K } \textbf{x } 10^8/1x10^{-19} \textbf{ x } 10^{-30} \equiv 6 \text{ eV, i.e. 1 eV in six directions}$$

$$\frac{1}{2} \text{ x } 6000 \text{ K } \textbf{x } 3 \text{ x}10^8/1x10^{-19}\text{x}10^{-30} \text{ x}1.2 \equiv 10.000 \text{ eV}$$

$$6000^2 \text{ K } \textbf{x}\sqrt{3}/(1/4 \text{ } c_s) \text{ x}1.1547 \text{ x}1.04^{1.05} \equiv 1.00036 \text{ eV} \qquad (25.17)$$

The Planck Radiation energy of 0.01 eV $\equiv$ 6 $\textbf{x}$ (10 N-s/6000 K) $\qquad (25.18)$

Electron charge e_s appears to be the product of 3/2 Boltzmann constant and 6000 K sun temperature relatively corrected by 1.111 . . . for the average local light velocity 1/3 c_s:

$$e_s = 3/2\ k_s \times 6000 \times 1.111\ldots = 0.99999999999 \times 10^{-19}\ C \qquad (25.19)$$

Six directions with three planes in each and ten possible space quantization pathways in each plane require input temperature of 6000 K to produce output of 1 K in a specific direction:

$$1\ K \times 10 \times 10 \times 10 \times 6 = 6000\ K \qquad (25.20)$$

Squared kelvin time quantum correction derives 1/8 of the velocity of proton

$$(6000\ K)^2 \times 1.04^{25/24} \equiv 1/8 \times 3 \times 10^8 \qquad (25.21)$$

Quadrupled $(\sqrt{2} \times 6000)^4$ kelvin over $1.04^{11/12}$ derives well the electron frequency 5×10^{15} Hz of five pathways:

$$(\sqrt{2} \times 6000)^4 / 1.04^{11/12} \equiv 5.0009 \times 10^{15}\ Hz \approx 5 \times 10^{15}\ Hz \qquad (25.22)$$

Square root of $6000^{1/2}$ K derives energy E_n of 50 eV less $4 \times 1/2xk_s T_{300}/e_s$ representing quantum step of four 13.62 eV dissociation energies of H2 atom:

$$6000^{1/2}/(\sqrt{2} \times 1.111111) \times 1.04^{1/3} \equiv 49.94 = 50\ eV - 4 \times 1/2xk_s T_{300}/e_s \qquad (25.23)$$

And squared $(\sqrt{2} \times 50\ eV)^2 \times 1.11111^2/1.04^{0.7} \equiv 6005.6\ eV \approx 6000\ eV + 6 \times 1.0\ eV$

while $\quad E_n = 50\ eV = h_s v_{Hms}/e_s = 1 \times 10^{-33} \times 5 \times 10^{15}/ 1 \times 10^{-19} = 4 \times 13.62/K_{rel}^{1/2} \qquad (25.25)$
And $\quad E_n = 50\ eV \equiv 1/\sqrt{3} \times 6000^{1/2}\ K \times 1.111111 \times 1.04^{1/6} \qquad (25.26)$

The temperature of 6000 K in one of 6 orthogonal directions converted to SI units appears to represent at atomic level the Planck constant:

$h_s \equiv 1/6 \times 6000\ K \times e_s/c_s^2/1.11111 = 1/6 \times 6000\ K \times 1 \times 10^{-19}/9 \times 10^{16} \times 1/1.11111 = 1 \times 10^{-33}\ J\text{-}s \quad (25.27)$
with $\qquad\qquad\qquad\qquad \sigma_s \times c_s \equiv 15 \qquad (25.28)$

Electron frequencies 5×10^{15} Hz and proton frequencies 5×10^{12} Hz express energies at 300 K. When they are referred to 6000 K and to 1 K the frequencies vary proportionally to temperature to express correctly that relationship, with the elementary energy 1 eV or 1 J. Energies generated with reference to 6000 K:

At 5×10^{15} Hz $\times$ 20, i.e. electron energy in integer space SI units $(1 \times 10^{17})^2 \times 10^{-30} \times 1 \times 10^{-19}$
$$\equiv 1 \times 10^{-15}\ J \equiv 10000\ eV \qquad (25.29)$$
At 5×10^{12} Hz $\times$ 20, proton energy is $(1 \times 10^{14})^2 \times 10^{-27} \equiv 10\ eV \equiv 10\ J \quad (25.30)$

Energy of the sun pulse $3 \times 10^{-17} \times 6000\ K \times 1.111111 = 2 \times 10^{-13}\ J \equiv 10$ of 1×10^{-14} J in two planes
$$(25.31)$$

294

Energies generated energies with reference to 1 K:

At $5x10^{15}$ Hz/300K =1.666 . . . $x10^{13}$ Hz time vol. probability $10^{-30} \approx 1.67x10^{-17}$ J. pulse in integer SI units (25.32)

At $5x10^{12}$ Hz/300K =1.666 . . . $x10^{10}$ Hz, in absence of electron interactions, that value x 2/3 $=1x10^{10}$ expressed numerically the gravitational constant G_s (25.33)

Wien Law discloses that hotter objects emit their radiations at shorter, bluer wavelength, and cooler objects emit their significantly decreased in intensity radiations at longer, redder wavelength. Using Wien Law derives constant peak wavelength λ_p=2.898 $x10^{-3}$/K. At 6000 K with quantum correction and one step space quantization the energy density:

$$2.898 \text{ x } 10^{-3}x6000 \text{ K}/\sqrt{3} \text{ x}1/1.04^{1/10}x1/10 \equiv 0.99997 \approx 1 \text{ J or W/m}^2 \text{ per plane} \qquad (25.34)$$

At 300 K power and energy $8.1 \text{ x } 10^9 = 1 \text{ x } 300^4 \equiv 27 \text{ eV x } 3 \text{ x } 10^8$ (25.35)
$$. 27 \text{ eV x } 1.11111 \equiv 30 \text{ eV}$$

Hertz Displacement Law in terms of frequency expresses also thermal energy generated from earth in 6 orientations

$$v_{max} = 6 \text{ x}1 \text{ x } 300^4 \text{ x } 1.11111^2 = 6 \text{ x } 10^{10} \text{ Hz/K} \qquad (25.36)$$

Expressions using velocity term with temperature require square of temperature.
Planck constant h_s is per kelvin. With earth radiating 300 K in three planes:

The effective power generated = $\sqrt{3}$ x 1.1547 x 300W $\equiv$ 600 W (25.37)

Stephan-Boltzmann classical constant σ can be corrected for relativity and quantum with accuracy of one pulse by Planck radiation term 0.01 increased 50% in incorporating electron energy at higher temperatures and power of four:

$$\sigma \equiv (0.01 \text{ x } 3/2)^4 \text{ x } 1.1547/1.0328 \text{ x } 1.04^{1/23} = 5.66967x10^{-8} \text{ W-m}^{-2}. \qquad (25.37a)$$

The temperature has been identified with the product of (pulse rate x pulse magnitude). The temperature of 300 K is likely to represent the product of thirty radiation orientations of 10 J. The temperature rate is interpreted to mean that the sun generates the rate of 1000 rate in six orientations transmitted in its beams. That rate has to be decreased by the relativity factor 1.111 . . . to 900. **On reaching earth the 1D pulse rate of 900 is distributed to 3D in elastic exchanges with gaseous atmosphere of the earth generating rate 300 represented by 300 K.** Similar process has been observed in relaxations of energies in plastics. The temperature of 3000 K represents probably the product of thirty radiation orientations of 100 J sun pulses generated by H2. The energy rate generated by the corrected product of three ($\sqrt{3}$ x 300 K) pulses equals 1 J delivered at the local velocity of light $1/2$ c_s:

$$\text{Power} = (\sqrt{3} \times 300 \text{ K})^3 \times 1.1547^{1/2}/1.04^{1/8} = 1.5002 \times 10^8 \text{ W} \equiv 1 \text{ J} \times 1/2 \, c_s \qquad (25.37b)$$

Radiation, corrected for relativity and quantum, from the sun on earth per each of 34 quantum orientations is:

$$6000^{1/34}/(1.111\ldots^2 \times 1.104^{7/6}) \equiv 1.0000 \text{ J(eV)} -0.00061 = 1 \text{ eV} - 6 \, k_s$$
$$(25.37c)$$

Allowing for the not interacting component 1.2 and two relativity corrections the observed sun temperature of 6000 K is likely to be different at the source, because that value is subjected to event probability 1/3 and it derives the twenty joules (or eve) pulse:

$$6000 \times 1.154700^2 = 7999.9954 \text{ K and } (8000 \text{ K})^{1/3} \equiv 20 \text{ J or eV,}$$

$$6000 \times 1.2 \times 1.154700^2 \times 1.04^{2/5} = 9999.66 \text{ K and } 6000 \times 1.2^2 \times 1.04^{1.04} = 8999.7 \approx 9000 \text{ K}$$

With variance $=30 \, k_s$. Single formula requires several backings to be validated $\qquad$ (25.37d).

Sun mass

Sun mass interacting with earth 2×10^{30} kg has imbedded two values of energy related with probability P_R 1/30 and $(1/30)^2$.

Energy of the pulse with P_R of $(1/30)^2 = (2 \times 10^{30})^{1/900} \times 1/1.1547^{1/2} \times 1/1.04^{1/3} \equiv 2.00087$ eV
$$(25.38)$$
Energy of the pulse with P_R of $(1/30) = (2 \times 10^{30})^{1/30} \equiv 10.23$ eV
$$(25.39)$$

Mass, light velocity and force

While the relation between the mass and the generated force has been clarified two identities were uncovered, which may lead other researchers to clarify the meaning of mass in 1D display in terms of light velocity. Light velocity in atomic terms was finally derived in the conclusions. The first relation expresses the velocity of light as product of one dimensional display of earth mass and relativity corrections:

$$(6 \times 10^{24} \text{ kg})^{1/3} \times 1.1547^4/(1.1547^{1/2} \times 1.04^{1/20}) \equiv 3.00036 \times 10^8 \text{ m/s} \qquad (25.40)$$

The second relation expresses the gravitation of earth as a product of a cube of SI classical velocity of light, space quantization and relativity and quantum corrections:

$$(2.99742 \times 10^8)^3 \times 1/10 \times 1/10 \times 1/10 \times 1.1547^2/1.04^{1/3} \equiv 3.5441 \times 10^{22} \text{ N} \qquad (25.41)$$

The last relation between variables expresses simply the force per unit mass times the mass of earth deriving according to the author the true integer gravitation force:

$$6 \times 10^{24} \text{ kg} \times 0.006 \text{ N/kg} = 3.6000 \times 10^{22} \text{ N} \qquad (25.42)$$

With mass of electron being 1×10^{-30} kg and protons being formed from electrons (see Chapter 13), the number of electrons on earth is $6 \times 10^{24}/1 \times 10^{-30} = 6 \times 10^{54}$. Hence the number of electrons per kilogram is 10^{30}, the invert of the energy converter of SI units to atomic units. The gravitation force of earth is expressed by one half of the 1D display (1/3 power index) of the product of the number of electrons forming the mass of earth and the mean frequency of protons at 300 K in one of six directions of space, corrected by relativity and quantum :

$$F_{Rs} \equiv \tfrac{1}{2} \times (1/6 \times 6 \times 10^{54} \times 1 \times 10^{12} \times 300)^{1/3} \times 1.11111^{1/2} \times 1.04^{1/10} = 3.5419 \times 10^{22} \text{ N} \qquad (25.43)$$

Thus oscillations of electrons essentially by proton frequency 1×10^{12} Hz are confirmed.

Two pulses in wave function format $(1 \text{ J} \times 3 \times 10^8)^2 \times 3 \times 10^8$ generate a third pulse $q = \sqrt{2} \times 9 \times 10^{16}$ J moving at c_s. Probability 1/1000 create energy correct orientation: energy q time c_s and divided by the product of relativity and quantum correctors derives the reference sun gravitation force:

$$F_{Gs} = \sqrt{2} \times 9 \times 10^{16}/1000 \times 3 \times 10^8 / (1.15479^{1/2} \times 1.04^{1/10}) = 3.5441 \times 10^{22} \text{ N}.$$
Alternatively

Gravitation force $\equiv (3 \times 10^8)^3 \times 1/10^3 \times 1.33333 \equiv 3.6000 \times 10^{22}$ N $\qquad (25.44)$
In Special Relativity terms $F_{Gs} [\text{N}] = c_s^3 \text{ [m}^3/\text{s}^3] \times 1/10^3 \times 1.111 \ldots \equiv 3 \times 10^{22}$ N

Above differences are the consequence of using different observers. Equation 25.44 indicates that velocity of light is the carrier of unity energy of 1 J, because other terms are dimensionless.

Three 1D displays $c_s^{1/3}$ over the quantum and square of the relativity correction derive sun power due to energy density of 1500 W/m² - $\tfrac{1}{2} k_s T_{6000}$:

$$3c_s^{1/3}/ (1.1547^2 \times 1.04^{1/10}) = 1499.72 \equiv 1500 \text{ W/m}^2 - \tfrac{1}{2} \times k_s T_{6000} \qquad (25.45)$$

The next identity incorporating total Bohr atom energy 30 eV in wave format $(30 \times 30)^8$ to power 8, applied at positions of cube corners of the spherical structure, is interesting, because its symmetry may create opportunity for probability to induce not only velocity of light, but also angular electron momentum creating spin. Both requirements are necessary for creating vectorized radiation beam and probably only spherical and cubic geometries are able to generate right conditions (Eq. 5.16 is similar).

$$1/12 \times (30 \times 30)^8/(1+1/80) = 3.5438 \times 10^{22} \text{ eV} \times 10^{20} \times 10^{-19}/10 \equiv 3.5438 \times 10^{22} \text{ J/m} = 3.5438 \times 10^{22} \text{ N} \qquad (25.46)$$

Rotational representation is the result of vector addition of orthogonal components, which are truly independent. Hence, while representation by wave rotation around a center is perfectly valid, the true source of central waves in six directions uncovers further details of physics hidden in the circular model. With ten radial h values, and their distributions, it is possible to fit ten waves around the circle. Integer model fits the half-wave or its odd multipliers across the center.

The model proposes that protons convert to radiation all the time and emit radiation causing attraction. This is a rare event, but there are very many protons and it describes properly the gravitation and the stability of solids.

Further analyses

The consistent theories in the book were created by examination of hundreds of numerical identities having initially no meaning and incorrect dimensionally. Mean human temperature of 36.6 K +273.2 =309.8 K; divided by $1.04^{4/5}$ derives 300.23 K, which is taken to be 300 K + uncertainty of $8ksT_{300}$. The uncertainty value per K converted to atomic units is $0.23/300 \times 1 \times 10^{-19} = 0.3333 \times 10^{-23} = 1/3$ k_s expressing a very reasonable result. The calculation suggests that humans body temperature evolved the value of 300 K to sustain thermal equilibrium with surroundings at 300K and to absorb most efficiently vectorized 1 N pulses from the earth or the sun ≈ 1373 W/m²/($10^3 \times 1.1547^2$). It is suggested that orientations of cell division and growth in vertebrates which follow DNA replication after meiotic cell reproduction are basically controlled by the 1 N/m forces of earth gravitation.

Volta has discovered and Engels has calculated (Atmospheric Electricity, Wikipedia, WEB) that a proportion of the vertical electrical potential carries constant electrical field of 88 V/m. This value is now deduced to be the consequence of the presence of large amount of water in clouds above and below the measuring range generating the classical field $\varepsilon_{rwater} \times 1$ N/m = 88 V/m. In water environment in the body with sufficient nutrients the reaction rate is increased 10^5-10^6 times. The 1 N/m at SI level, $1 \times 10^{-10} \times 1 \times 10^{10}$, or most probably 88 V/m components create electrical fields at the critical atomic stage of life development. They generate the geometry of the human body comprising four limb orientations with five fingers or toes for each orientation and so express the characteristics of the exciting radiation. Our head, backbone and tail are remnants of the fish background, but it seems that all animals born at one time in free space are vertebrates, who share radiation geometry characteristic of earth and of sun. Smaller organisms born in earth, in water structures or within other structures are not subjected to such high exciting fields and they do not display these features.

Voltage and energy from wave format of two 88 V pulses times SI probability of 1/10 over relativity correction 1.111 . . .: $88 \times 88 \times 1/10 \times 1/111$ . . . $\times 1.04^{1/6} = 701.55$ V in agreement with Equation 5.13. Combining, earth and sun pulses increased by $\sqrt{3}$ of the three cubic components, derives $\sqrt{3} \times 88 \times 2 \equiv 304.8$ V, J, eV or K. divided by $\sqrt{3} \times 88$ derives the 2 eV pulse of Equations 16.9-16.10. Volta's graph of ~ 1000 V over height 30 m derives the space gradient 33.33 V/m divided by 1.111 . . . obtains 30 eV or J and 30 eVx10 $\equiv 300$ K. Radiations from mass generate equal direct and alternating potential gradients. '

The radiation pulse from the sun is largely the consequence of mass to energy conversion of 18 atoms of hydrogen (Conclusion 64) into 4 atoms of helium.

The excess of mass of atomic weights released calculated from any periodic table of atoms is 18.000-16.0164=1.9896 units. Time relativity correction 1.111 . . .$^{1/2}$ derives

2.00011 units expressing mass of two protons 4×10^{-27} kg radiating in six directions each $4/6 \times 10^{-27}$ kg in the SI model units. The term proceeding to earth has the reflection component generating the radiation mass $1.333 \ldots \times 10^{-27}$ kg, which divided by the relativity correction $1.333 \ldots$ derives the mass of the model radiation particle of $m_{ps1D} = 1 \times 10^{-27}$ kg.

Two pulses of voltage gradient 88 V/m displaced by 90° create a wave function format progressing the pulse $\sqrt{2} \times 88$ at the local velocity of light. The input is 10 times higher due to presence of ten optional pulses per phase: $\sqrt{2} \times 88 \times 10/(1.111 \ldots^2 \times 1.04^{1/5})$ $\equiv 1000.17$ pulses/s $\equiv 1000$ W at 1 W/m². The earth atmosphere potential gradient of 1 V/m at 1 K is increased locally by the presence of charges in water in the clouds and in earth by the factor of specific permittivity of water of $\varepsilon_{osr} = 88$ in the same way as it has been found experimentally for NaCL crystal. Clouds are masses of water with air gaps between them expressed also by the space to earth. The voltage gradient in a parallel narrow gap in a dielelectric medium is greater by the dielectric constant ε_{osr} of the dielectric. It is suggested that the experimentally verified incremental value of 3/2 for spherical voids does not apply because the signal is 1D and the breakdown occurs before the side charges have chance to appear lowering the stress. Referred to 300 K the local gradient becomes 26400 V/m. SI energy contributing to radiation originating from sourced from two 88 V/m at right angle (actually 18 atoms) is $\sqrt{2} \times 88 \times 300/(1.111 \ldots^2 \times 1.04^{1/5}) \times 1 \times 10^{-19} = 3.0005 \times 10^{-15}$ W as given in Equation 2.36. It follows as mentioned earlier that several values quoted in the book are calculable components of energies present in the reality.

Two stress parameters 88 V/m displayed by 90° with sufficient other inputs form the wave function pulse progressing at the local velocity of light. Corrected for temperature, relativity and quantum the derived gradient value $E = 300 \times 88^2 \times 1.154700^2/(1.04^{4/5}) = $ **3.002×10^6 V/m expresses well the breakdown stress of air at uniform field at 1 NTP of 30 kV/cm.** Such stress produces zigzag electron streamers hundreds of meter long converting to lightning on thermal ionization of air. Converting to atomic units:

$3 \times 10^6 \times 10^{-19} \equiv 0.001/(3 \times 1.111 \ldots) = \sqrt{\mu os}/(3 \times 1.111 \ldots)$ is attributed now to the equipartition of energy in earth atmosphere. In the absence of units of time and of length, stress is treated as velocity and 1 kg converted to atomic units derives:

$$1 \text{kg} \times 3 \times 10^6 \times 10^{-19} \equiv 10^{13} \text{ J} = 1/(10 \times 10^{-14} \text{ J}) \qquad (25.46a)$$

The relatively corrected energy for three planes is $3 E \times e \times 1.11 \ldots = 3 \times 3.002 \times 10^6 \times 1 \times 10^{-19} \times 1.11 \ldots = 1 \times 10^{-12}$ J. At 1×10^{-12} Hz it generates 1 W per pulse. The relativity corrected product for 3 planes of the potential gradient, the electron charge and the local velocity of light derives the square of the 1st Planck radiation term:

$$3 \times 3 \times 10^6 \times 1.111 \ldots \times 10^8 \times 10^{-19} = 0.0001 = (0.01)^2 \text{ W/m} \qquad (25.47)$$

Here the Planck constant in model SI units is expressed as the product of the mass of proton converted to energy by square of local velocity $1/3\ c_s$, the converter $1/3\ c_s$ from atomic to SI units and the volume probability 10^{-30}:

$$\boldsymbol{h_s = 1 \times 10^{-33} \text{ J-s} \equiv (1/2 \times 2 \times 10^{-27})(1 \times 10^8)^3 \times 10^{-30} = 1 \times 10^{-33} \text{ W}} \qquad (25.48)$$

This relation can be considered as presence of mass to energy conversion; further multiplication by velocity converts energy to power. Several other expressions incorporating power of three or higher applied to a physical parameter illustrate the process. Vectorial addition with wave function formation generating velocity of light necessary for propagation increases the amplitude of the event by square root value, differing by $1.04^{1/120}$ from correction of $1.154700^2 \times 1.111 \ldots^{1/2} \times 1.04^{1/6} = 1.41467400$.

Expressing proton mass in one direction relativity and quantum corrected derives:

(25.49)

$$(2 \times 10^{-27})^{1/3} / (1.111 \ldots^2 \times 1.04^{1/2}) = 1.000718 \times 10^{-9} \text{ kg}$$ times the pulse energy 1 GeJ, Equation 2.23, **derives 1 W** referring to 1 kg and to proton mass in 1 D

Expressing proton mass in two directions relativity and quantum corrected derives:

$$(2 \times 10^{-27})^{1/2} \times 1.111 \ldots \times 1.04^{1/6} = 5.00158 \times 10^{-14} \text{ kg} \tag{25.50}$$

Expressing split proton mass in two dimensions (power ½) and correcting for relativity the other way derives:

$$(1 \times 10^{-27})^{1/2}/1.111 \ldots = 3 \times 10^{-14} \text{ kg exactly} \tag{25.51}$$

The meaning is very clear. **The mass to energy conversion is wave function formation of two moments $(3 \times 10^{-14} \text{ kg} \times 1 \times 10^8 \text{ m/s})^2 = (3 \times 10^{-6})^2 = (3 \, \mu_{os})^2$ generating local velocity of light that needs to be relativity corrected and the pulse is $9/10^{12}$ times $1.111 \ldots \times 10$ J or 10 eV per infrared cycle, hence it is effectively μ_{os}^2/cycle. Thus the common expression, mass x c^2, is valid numerically but it has no physical formative back up.**

The mass of earth in 1D with relativity and no interacting radiation component derives 1000 J pulse which in three planes derives 1 GJ pulse. At frequency 10^{12} Hz it space pulse $E_{sspace} = 1 \times 10^9 \times 1 \times 10^{12} = 1 \times 10^{21}$ J.

(25.52)

Energy of earth mass in 1D $= (6 \times 10^{24} \text{ kg})^{1/3}/(1.2^3 \times 1.111 \ldots^{1/2}) \times 1.04^{1/16} = 1000.0623$ J
And

Special Relativity gravitation $= 1000 \text{ J} \times 300^3 \text{ K}^3 \times 1.111 \ldots 10^{12}$ Hz $= 3 \times 10^{22}$ N

(25.53)

The mass of earth in one plane, corrected for relativity and quantum derive frequency 2×10^{12} Hz:

Earth mass frequency $= (6 \times 10^{24} \text{ kg})^{1/3}/1.111 \ldots^{1/2} \times 1.04^{1/5} \equiv (2-0.00039) \times 10^{12}$ Hz

(25.53)

Applying 1×10^{12} Hz to the wave function product of two pulses of 1 J at 300 K and c_s and converter to atomic level derives:

(25.54)

$(1 \text{W} \times 1 \times 10^{12} \text{Hz} \times 300 \text{ K} \times 3 \times 10^8 \times 1 \times 10^{-30}) \text{X} (1 \text{W} \times 1 \times 10^{12} \text{Hz} \times 300 \text{ K} \times 3 \times 10^8 \times 1 \times 10^{-30}) = 3^4 \times 10^{-14}$ J.

The above represents probability. The inverse of probability is the magnitude of the event; that event expresses also the time it takes to exhaust energy of earth mass and remove ability to sustain gravity balance:

$$1/3^4 \times 10^{-14} \times 1/1.111\ldots^2 \times 1/10 = 1.00000000 \times 10^{10} \text{ J} \equiv 1 \times 10^{10} \text{ years} \qquad (25.55)$$

Alternatively the spin of the equivalent electron h_s, as derived in Equation 22.57 per direction, is lost in rotation lasting in atomic terms one year. The radiation power of ½ of proton mass obtained by multiplying by the square of light velocity, converted by 10^{-30} to atomic terms and vectorized by 10^{-3} and corrected by relativity is:

½ $\times 2 \times 10^{-27} \times 9 \times 10^{16} \times 10^{-30} \times 10^{-3} \times 1.111\ldots = 1 \times 10^{-43}$ W applies also to one year and the ratio $h_s/1 \times 10^{-43} = 1 \times 10^{10}$ years is the stability period of the proton. These powers are in three orthogonal orientations generating a resultant increased by factor $\sqrt{3} \times 1.0328 \times 1/2/(1.1547 \times 1.111\ldots) = 1.371$ making the age of the Universe 13.71 billion years.

$$(25.55a)$$

Power of 1 W is at 300 K created by three 300 K pulses in orthogonal orientation and relativity correction. Converting to atomic level at 1 K derives Planck Radiation constants, which in the integer model an infrared parameter of the equivalent electron:

$$. h_s = 1 \text{ W} \times 1 \times 10^{-30} \times 1/(3 \times 300 \text{ K} \times 1.111\ldots) = 1 \times 10^{-33} \text{ J} \qquad (25.56)$$

The units of the product of $\varepsilon_{os} \times 1/3 \ c_s$ are W/m$^2 \equiv 0.001$; they require multiplication by 10^{-20} on converting to atomic units deriving Boltzmann constant $k_s = 1 \times 10^{-23}$.

Similar conversion of the electron mass into energy predicts a new Planck constant applicable to electrons:

Electron $\boldsymbol{h}_{selec} = 3m_e(c_s)^3 \times 10^{-30} \times 1.111\ldots^2 == \mathbf{1 \times 10^{-36} \text{ J-s}} = \mathbf{1 \times 10^{-3} \times} h_s$

$$(25.57)$$

which is a significant finding considering the year's long controversy in μ_{os} **interpretation.**

With the new electron Planck constant $\boldsymbol{h}_{selec} = \mathbf{1 \times 10^{-36}}$ **J-s** the local energy is $\qquad (25.58)$

Electron energy $= \boldsymbol{h}_{selec} v_{Hms} = \mathbf{1 \times 10^{-36} \times 1 \times 10^{15} = 1 \times 10^{-21} \text{ J-s}} \qquad (25.59)$

It is suggested that Boltzmann constant for true electron is thousand times smaller than k_s
$$k_{selec} = \mathbf{1 \times 10^{-26} \text{ J-s}} \qquad (25.59a)$$
The above implies that 2xn protons oscillate locally at right angle and doing so generate velocity of light with n = 27 units generating sufficient energy (and probability) to produce a radiation pulse. Converting $\boldsymbol{h}_{selec}$ from atomic natural units to SI units requires 1×10^{20} multiplier deriving the gravitation constant of Table 1:

$$G_s \equiv (\mathbf{1 \times 10^{-30} \times 1 \times 10^{20}}) = \mathbf{1 \times 10^{-10} \text{ J-s}} \qquad (25.60)$$

The presence of h_{selec} has been realized by the author some years ago, but it was not postulated until relations to other physical constants have been clarified. This analyses has lead to defining the meaning of SI model Planck constant h_s. It is the inverse of radiation energy arriving per second on earth from the sun from two relativity corrected pulses 3 J $x1.11111^{1/2}x(1x10^8)^2$ $x1/10$ moving at local velocity generating power $[3 J x 1.111...^{1/2} x (1 x 10^8)^2 x1/10]^2 \equiv 1/h_s$. This is an exact relation. Although at the true atomic level the power of sun per direction is $1 x 10^{30}$ W, the power of the sun per orientation in exaggerated integer SI units is given exactly by:

$$[1/h_s \text{ x } (c_s)^2 \text{ x} 1.11111 ...]^2 = [1x10^{33} \text{ x } (3x10^8)^2 \text{ x} 1.11111 ...]^2 = 1 \text{ x } 10^{100} \text{ W}$$
$$. (25.61)$$

Omitted relations and Planck constant

The two distant electrons joining two free protons form hydrogen molecule releasing radiation energy:

$$q_s = \int_{aos}^{\infty} 2e_s/(2\pi\varepsilon_{osos})d \ a_{os} = [2 \ x10^{-19} \text{ x } 4/(6 \ x1x10^{-11}a_{os}^2)]x1x10^{-10} = 4/3 \ x10^{-12} \text{ J} \quad (25.62)$$

Dividing by relativity corrector, 1.1547^2 derives $1x10^{-12}$ J and generates at frequency 1 x 10^{12} Hz power 1 W. Converting 1 W or 1 J to the atomic units using $1x10^{-30}$ multiplier and electron charge $1/1x10^{-19}$ derives $1 \ x10^{-11}$ W or J, which is energy per equivalent electron equal ε_{os} J, representing in SI units energy density per m^3 ε_{os} J/m^3. (25.63)

The conversion of the electron mass in SI integer cubic units derives also $1x10^{-30}$ kg x $(1/3 \ c_s)^2 = 1x10^{-14}$ J, which converted to atomic units by multiplier $1x10^{-30}$ derives energy $1x10^{-44}$ J given near Equation 2.32. In terms of the classical SI units the result is essentially the same after corrections for the relativity and quantum:

$$\text{Classical min. energy} = m_e x1.11111/1.04^{1/4} = 0.91093x10^{-30}x1.11111/1.04^{1/4} = 1.0015 \ x \ 10^{-14} \text{ J}$$
$$(25.64)$$

This calculation supports the presence of the local velocity of light and hence of the need for the relativity and quantum corrections. The radiation mass $1 \ x10^{-27}$ kg with SI integer cubic units moving at the local velocity of light derives energy:

$$1x10^{-27} \text{ kg x}(1/3 \ c_s)^2 = 1 \text{ x } 10^{-11} \text{ J} \equiv \varepsilon_{os} \text{ J/(m}^3\text{-s)} \quad (25.65)$$

Above relation appears to indicate that the space comprises energy of the converted energy of the proton radiating in two directions, filling the space. Similar result is derived from the calculation using atomic units where SI proton momentum is converted to SI units by dividing by local velocity:

$$1x10^{-27} \text{ kg x}[1x10^8 \text{ (m/s)}]^2 = 1x10^{-11} \equiv \varepsilon_{os} \text{ J/(m}^3\text{-s)} \quad (25.66)$$

With pulses coming from three opposite direction the moments would add to Planck uncertainty $h_s = 1x10^{-33}$ expressed in derived earlier relation $\varepsilon_{os}^3 = h_s$. Planck constant is the

SI radiation particle $1x10^{-27}$ kg converted to atomic units time squared velocity of light, volume probability and atomic converter of 10^8: $h_s=1x10^{-27}x(1/3c_s)^3x10^{-30}=10^{-33}$J-s. Planck constant is also Planck radiation term time linear probability 1/10 and $1x10^{-30}$:

$$h_s = 0.01 \times 1/10 \times 10^{-30} = 1x10^{-33} \text{ J-s} \qquad (25.67)$$

Sun radiation of 1 J at earth temperature of 300 K, expresses in six 50 J orientations of uncorrected 1 J.

Temperature and velocity of light expressed as pulse rate

The temperature of 6000 K time relativity correction and the decrement of transmission 1.2 in the source-receiver direction derive input of 10 000 K:

$$6000 \times 1.1547^2 \times 1.2 = 10\ 000 \text{ K} \equiv (1/3\ c_s)^{1/2} \text{ m}^{1/2}/\text{s}^{1/2} \qquad (25.68)$$

$$\text{And } (10\ 000 \text{ K})^{1/2} = 100 \equiv 10 \times 10 \qquad (25.70)$$

The product 10 x 10 expresses input in two planes having each ten optional orientations deduced from five orientations of Stern and Gerlach in 1921 shown in Fig.1. The infrared frequency of $1x10^{12}$ Hz is the exact product of local velocity of light and the temperature 10000 K:

$$(1/3\ c_s) \times 10000 \text{ K} \equiv 1 \times 10^{12} \text{ Hz} \qquad (25.71)$$

Electron oscillation generates a local pulse rate of 1×10^{15}. Thus the sun temperature, proton frequency and local velocity of light form set of values connected logically to other parameters.

The main results

The 2013 model follows the 2009 book and the shorter 2010 edition in which the list of integer physical constants was derived. After three more years of research into physics probability was identified as the best predictor of events. Validity of the mathematics of probability is proved experimentally by hundreds of expressions. Rules of probability are applied to the three independent orthogonal orientations of space carrying radiations arriving on the pixels of the television screen of our eyes. The signal comprises pulses or half-waves of the model have one color and it's the intensity is described by its pulse rate. The microscopic and the macroscopic balance (Equations 12.18-19) were established between the earth and the sun and the space between them. The local velocity of light in 1D is clearly related to temperature:

$$(1 \times 10^8)^{1/3} \times 1.111111^{1/2} \equiv 1/12 \times 6000 \text{ K} + 15\ k_s T_{300} \qquad (25.72)$$

Final corrections uncovered new findings located in arbitrary locations; these findings require the use of powers of three and five. The radiated power vector is generated by the wave function of vectors of voltage and of current each of which is also pulse generated by wave function of two vectors. This process generates $\sqrt{2}$ larger values with the square of probabilities. The probability requirement requires power of two for each, the resultant probability progresses also at velocity of light adding another power making a total of five. Equation 25.73 is an example of local velocity of light needing power of five times linear probability of 1×10^{10} to obtain the momenta of the sun 1×10^{100} J-s derived independently in Equations 25.1, 25.5 and 25.6:

$$[(1 \times 10^8)^5 \times 1 \times 10^{10}]^2 \equiv 1 \times 10^{100} \text{ W} \tag{25.73}$$

Repeated results are needed to confirm validity. The unit's dimensions mismatches were covered by the sign of equivalence: $\equiv$. Dimensional agreement was obtained in some cases by the introduction of unity values in expressions. In the final correction the probability model with kinetics rules allowed to dispense with all units. Some readers may consider this inexcusable. Finally rate of emission of proton derived from space probability established the age of <u>Universe</u> and uncovered the true source of all energy activating processes in the Universe this being inherent instability of protons deriving gravity between celestial bodies and between atoms.

Theory of relaxation of 2002. The draft paper [62] explains the reason for water causing huge acceleration of the rate of processes in organic chemistry. Refusal to publish the paper below induced the author to look closely at physics to update it.

Note of 2013: The value of relative permittivity of water $\varepsilon_{os}=92$ used in the book has been updated to 88. Activation energies of 0.79 eV, the potential barriers, are the dissociation energies of Na ion jumping into the Na vacancy over the average space of 0.29 nm. The integral $1/2 \times 1/ (4\pi\varepsilon_o\varepsilon_r \times 0.29 \times 10^{-9}) = 0.2757$ eV, which multiplied by space quantization of 5, unknown in 2002 and divided by $K_{rel}^2 = 1.15474$ also an unknown factor in 2002 derives 1 eV. Similarly energy 1 eV divided by quantum corrector squared 1.1111^2, derives about 0.79 eV. Dissociation energy of Bohr atom $\sqrt{2}m_e /(4\pi\varepsilon_o \times 10^{-9})$ $(1/0.095-1/0.185)/1.111^3 = 13.67$ eV, where 0.095×10^{-9} m is the Na radius and 0.185×10^{-9} m is the sum of Na and Cl radii. The variance 0.02 eV is the uncertainty$2/3$ kT/ e_s. Measured energy step $0.0174/1.1111/2 = 0.01650$ eV $\approx g_{ev}$. The macroscopic relation (ref. 5, Fig.1) between the measured time constant τ_n and time t of the process, $\tau_n \approx 0.8$ t. Converting it by the square of $(rms/mean)^2 =1.111^2$ derives $\tau_n \approx$ (time t from the beginning of the process). Thus application of correctors derived in the book to the above paper of the year 2002 derived several correct parameters of physics. While the atomic oscillations taking in two dimensions are sinusoidal, the radiations transmitted are 1 D and hence they should be represented by rectangular pulses and not by waves. In 2002 the importance of space quantization was not uncovered and phonons were the proposed activators of charge movements because relaxation intensity increased nearly to power of four. Presently phonons are treated as parameters vectorized by space quantization. The waves in Figure 17 should be rectangular. Above paper explains the reason for water causing a huge acceleration of the rate of processes in organic chemistry facilitating formation of life structures. High permittivity of water, having a dielectric constant $\varepsilon_o \approx 80$, reduces the height of potential barriers in structures to about a quantum value. Refusal to publish this important finding that high permittivity of water is the propellant of life induced the author to look closely at physics to update it. On leaving the oceans to improve the access to oxygen the early live creatures took water inside their bodies because that was the necessity of life.

Present findings affect the earlier paper in the following way:

The consistent theories proposed in the book were created by examination of numerous numerical identities having initially no meaning and being onally. Mean human temperature of 36.6 K +273.2 =309.8 K; divided by $1.04^{4/5}$ derives 300.23 K, which is taken to be 300 K + H. u. of $8ksT$. The uncertainty value per K converted to atomic units is 0.23/300 x 1 x 10^{-19} = 0.3333 x 10^{-23}= 1/3 k_s, a very reasonable value. The calculation indicates that we are tuned to receive 1 J pulses from the sun. The lattice energy of 0.79 eV from numerous mentioned sources 0.79 x 1.1547 x $1.1111111^{1/2}$ x 1.04= 10.00018 eV = 10 eV + $2k_s$ uncertainty, energy 10 eV deriving all important energies of the model: 10 x1/10=1 eV, 10 x3 =30 eV, 10^3=1000 eV, J 6 x1000 = 6000 K ect.

Applying integer cubic SI units to Equation 6 energy per direction quantum corrected is q_s =1/2 x $k_s T_{300}/e_s$ x $ln2$ /1.04 = 0.009997315 eV=0.010-0.000002685 eV.

The variance of 1/4 x10^{-26} J = ¼ $k_{selectron}$ confirms presence of the new physical constant proposed $k_{selectron}$.

There is more than one way of generating vector forces from thermal pulses.

Therefore mathematical feasibility does not prove physical reality.

End note

The total time taken for the research presented including the early studies of quantum adds to some fifteen years. The huge accumulated data distributed in different chapters makes it hard to comprehend the model postulated, although its mathematics is elementary. These last few pages clarify features some of which might have not been clearly stated.

The model and the parallel table of physical constants is based on relativity, which is well recognized in physics, but neglected because of correct, but misleading Einstein's formula mc^2 = energy. Relativity can change the value of the observed force and of observed distance with the product (force x distance) remaining the same for the model and for the SI system. Consequently nearly all energy values derived in the relativity model have the same values as those of the SI system. The small relativistic gravitation force 3.0000×10^{22} N of the model derives a large sun-to-earth distance:

$$3.5438 \times 10^{22}/3.000 \times 10^{22} \times 1.5 \times 10^{11} = 1.7719 \times 10^{11} \text{ m}$$

in comparison to model value 1.5×10^{11} m. The product (force x distance) has the same value 5.316×10^{33} Nm; multiplied by $1.0328^{1/2}_{es}$ derives energy $\sim 5.38 \times 10^{14}$ J of Equation 7.4.

The sun gravitation needs surplus energy from at least 18 atoms of hydrogen converted to helium to generate the minimum 2 J pulse at 300 K around earth; generation of velocity of light c_s from sun by the proton's oscillations at $1/4$ c_s needs a three stepped processes of vector addition in three space orientations. Energy arriving around earth per oscillator is 1 W/kg-m^2 expressing the minimum entropy production for the static state. Equal force is generated by earth by a four stepped process of wave function format each increasing proton and (proton+electro) velocity by $\sqrt{2}$. The force generated by earth acceleration induced by rotation is upsetting stability of protons and converts some masses to energy. The same value of force applies to the force between the proton and electron of the hydrogen atom. At 1 K mass of 1 kg can generate power of 1 W for 1×10^{10} years, which corrected for relativity $1 \times 10^{10} \times 1.15472 \times 1.0328 \equiv 13.77$ billion years expresses the age of the Universe derived by CERN.

The Millican 1923 electron charge corrected by relativity (Equation 1.10) is exactly $e_s = 1 \times 10^{-19}$ C. The charge and other three parameters derive all Planck Units and directly the integer set of SI constants relativity corrected. Applying them to some three dozen laws of physics in Chapter 22 generate the same energies as obtained using classical SI units. At 300 K hydrogen atom has only two oscillators generating power of 2 J. The gravitation pulses are of infrared thermal frequency; they are vectorized by probability $1/1000 \times 1/\sqrt{3}$ and express sun gravitation. **It is suggested that sequences (rates) of oscillatory pulses from sun are converted to waves of earth particles by resonance. Electron pulses from sun are subjected to numerous reflections with protons causing their diffusion in six directions canceling their input momenta.** Classical physics applies usually incorrectly the term electron to equivalent electron, which is the equivalent mass of proton+electron oscillating with infrared frequency. Spin energy released by rotation of equivalent electrons, which are essentially protons $k_sT = 10^{-23}T$ J/oscillator or $h_sT = 10^{-33}T$ J/oscillator are the primary source of energy of all active processes in the Universe. **Two new physical constants are needed for electrons, $k_{select} = 10^{-26}$ J and $h_{select} = 10^{-36}$ J-s.** Both classical SI and relativity model SI units express events of length, intensity and time observed by humans, but they cannot deal **with kinetic events of radiation generation having no units of time or length.** Two eyes identify units of time and space. One eye needs events registration over time and analyses to deal with time and length. Kinetics of particles have no such ability.

Event probability is based on the reference, which might be input or output. The mass of 1 kg has 1×10^{30} electrons making most mechanical and electro-dynamical events equivalent. The integer model enriches the classical SI physics. Atomic receiver detects only amplitude and frequency (rate) and its mathemathics have no units of length or of time.

Infrared thermal pulses from the sun oscillate in 1D within atomic dimensions in both directions and do not transfer any unidirectional resultant force, although they carry energy. Gravity forces are really components of the above energy and of power expressed by the sun constant (137.5 W/m^2) subject to probability 1/34; 34 expresses number of quantum space positions on a spherical surface generating the gravity reference pulse 1 W/m^2. Probability and relativity correction applied to sun constant expresses accurately the power generated including its variance:

$$(137.5 \text{ W/m}^2)^{1/34}/1.111111^2 = 1.001752 \text{ J or eV/event}$$

The variance 17.5×10^{-23} J expresses ($\frac{1}{2} k_s \times 34 + k_s$). In some circumstances $36 = 3\times12$ is used. Thirty four positions are not fixed, they are subject to distributions.

The minimum pulse energy is the minimum pulse per orientation generating with probability 1/3 $(1\times10^{-14})^{1/3} = 1\times10^{-52}$ J. While the observed sun power with which the earth and planet interact is no doubt 2×10^{30} kg, it is still uncertain whether the total sun power is really 6×10^{100} W.

SI identifies radiation with waves, waves groups and waves packets. Waves require constant supply of power to deliver their energy. The integer relativistic model confirms the view that mass of 1 kg generates the force of 1 N at 1 K and accepts the Principle of minimum entropy production involving generation of energy h_s at 300 K per kilogram of mass. The corrected energy generated at velocity of light:

The minimum energy in three orientations $300\times1\times10^{-33} \times 9\times10^{16} \times 1.111\ldots = 3\times10^{-14}$ J.

Discharge of this energy is attributed to the instability of protons deriving also correctly the longevity of the Universe of 13.73 billion years. These values derive also correct gravitation force in Equations 23e and 21.40.

Many of the model considerations deal with radiations involving single events, which may occur in very large or in atomic size volumes, while sustaining individual quantum characteristics of the single event. All such event express the probability of the occurrence of an event derived from the structure of space and based on the reference taken, which is unity and may be input or output.

The energy of the gravitation pulse is radiated in one direction, all the energy of the source being dissipated through the weakness of the potential barrier with multidirectional internal waves being converted into a unidirectional pulse. To generate velocity of light required for radiation the proton and (proton+electron) couple have to increase their velocity from the initial ¼ c_s to c_s. For earth pulses this needs four steps, while their individual components move with velocity 1/3 c_s generating c_s.

In a space radiation path the corrected product of the mass of proton and its velocity ¼ c_s:

$2x10^{-27}x¼x3x10^8/[1.1547^2x1.1111x(1+1/80)] = 1x10^{-19}$ kg-m, the electron charge e_s.

Is Universe subject to large uncertainty? On 8 September 2013 it was still unclear whether radiation is the consequence of generating velocity of light of proton or of equivalent electron. It was finally derived that these two energies vary by uncertainty of measurement expressed in SI terms, in model SI terms and in atomic terms. There is no doubt that that Heisenberg concept of uncertainty is valid. The model SI value of uncertainty in 3D is: $(\varepsilon_{os})^3 = (1x10^{-11})^3 = 1x10^{-33}$ F^{3}m^{-3} ≡ $1x10^{-33}$ J/m^3. With measurements taken at 300 K this value is applied to atomic sizes. The true SI value in 1D is Boltzmann constant $k_s = 1 \times 10^{-33} \times 1 \times 10^{10} = 1 \times 10^{-23}$ J; in 3D it is $1 \times 10^{-33} x1 \times 10^{30} \times 10 = 0.01$ J is the Planck constant. Uncertainty 1×10^{-33} J converted to SI by $1 \times 10^{30} \times 10$ multiplier, divided by relativity corrector and electron presence 3/2 factor derives:

$$1x10^{-33} \; x1 \times 10^{30} \times 10/1.111 \ldots x3/2 = 20 \times 10^{-6} \text{ J}$$

This value according to the CERN press release of June 8/2012 expresses exactly the measured fractional excess velocity of light of neutrinos. The later CERN press release attributes the difference to instrumentation. The model of physics proposed identifies this value with measurable accuracy.

It is suggested that the uncertainty is proportional to mass, because the mass generates a gravity force proportional to mass. Hence the mass of earth $6x10^{24}$ kg x 1 x $10^{-33} = 60x10^{-10}$ J because although energy requirement is 18 atoms, velocity wave function format requires two hydrogen atoms with energy of 60 eV having uncertainty of 1 eV per degree of freedom. The mass of sun $2x10^{30}$ kg x 1 x 10^{-33} x 10 = 0.02 J expresses the sun pulse to be twice the Planck constant.

The author believes that the Universe is real and accurate. Quanta are related to atomic particles and to degrees of freedom. With large number of freedom this may affect macroscopic objects and create a distribution to longevity.

Reminder   Mass of electron =1/100 mass of muo10$_s$

Radiation current = electron charge x velocity of light
=$1x10^{-19}$ kg-m $3x10^8 = 3x10^{-11}$ A ≡ $3\varepsilon_{os}$ Fm^{-1} indicating that ε_{oss} is not a static parameter.
Velocity of light in radiation is generated as the product of two pulses both rotating in phase and displaced at 90° in three planes for sun pulses and displaced at 90° in one plane for earth pulses. The probability of such discharges is 1/6 for each of 6 directions repeated twice for sun pulses:

$$(1/6x1/6x1/6)^2 = 0.000021433$$

increasing proton velocity $\sqrt{3}x1.1`547x2 = 3.999998 ≈ 4$ times from ¼ c_{uss} to c_{uss}. Earth pulses are added in one plane while displaced by 90° increasing velocity in vector addition by $\sqrt{2}$. To increase velocity of protons from $1/4\ c_s$ to c_s four steps are needed generating total probability of $(1/6)^4 = 0.0000771604$, which is 36.0008 times larger because pulse temperature is significantly smaller deriving sun temperature of 10000 K confirming the earlier finding:

$$300\ K/1.0328^2 = 286.2\ K \rightarrow x36/(1+1/81) = 9999.81 \approx 10000\ K$$

Sun gravitation pulses will have both infrared proton radiation and violet electron radiation, while earth pulses generated by earth will express only infrared radiation. In their interaction with protons electron radiation are scattered in all directions with their mementa being diffused so that electron impact becomes negligible.

The gravitation pulses have $1/4$ probability of radiation in required specific direction in their plane of rotation, generating for two independent pulses the wave function format displaying probability $1/4$ x $1/4 = 1/16$. In three planes of space the probability of the event is increased to $3/16$. For earth gravitation pulse each step increases velocity by $\sqrt{2}$ so that four steps are needed establishing probability of the obtaining velocity of light event at:

$(3/16)^4 = 0.001234961$. Its invert at 809.742 is the rate of random attempts required to generate velocity of light. Above figure represents interaction between the sun and the earth taking place in macroscopic dimension only in one plane deformed only by the magnetic attraction of sun pulses by earth. The probability per kilogram of sun mass is:

$$(3/16)^4/2x10^{30} \equiv 6.17981x10^{-34}\ J \equiv (2/3xh_s/10)/1.11111 + 18x10^{-36}$$
$$\equiv (2/3xh_s/10)/1.11111 + 18hs_{electr}$$

Above value appears very reasonable, be cause $h_s/10 \equiv h/2\pi$ with uncertainty $18h_{selectr}$ expressing $(16+2)h_{selectr}$ true new electron parameters.

Up to now space quantization was based on findings of Stern and Garlic of 1921. Above calculation establishes that quantum values and space quantization is the consequence of the structure of the Universe space having independent 6 directions and hence generating probabilities of events occurrence. Probabilistic presentation has its limitations, because it hides several structure events observed in other theories. It is however very fundamental.

The author contends that theories that do not recognize relativistic aspects of physics that do not recognize the absence of dimensions of length and of time in the kinetic formulations of radiation and of the most important factor the probability of the occurrence of the reference event are incomplete. They describe valid details of physics and gravitation without comprehending its fundamental bases.

Personal end note

The Universe close to us is run like clockwork obeying exactly the laws of nature and suggesting presence of a Superior Intelligence of the Creator. It is in very long balance and not in chaos. The transformation of single pulses to the wonderful world around us is a miracle and has zero probability of happening on its own. It needs a Creator. Chaos is limited to uncertainties, which express the huge number of the degrees of freedom. The creation of the Universe is attributed by the author to God, also because of the extreme beauty of the World and the presence of many undiscussed features of life that cannot be explained by probability. The living world is built on intent and purpose. A child needs intent, parents and teachers to become a whole human. Human life is improbability surviving only on its past and reproduction. This is the topic for philosophers.

Formation of velocity of light from local velocities

Equivalent electron is an oscillator formed by (proton+electron) pair of mass 1.001×10^{-30} kg oscillating at infrared frequency. All orientations are attributed with equal probability of equal radiations, lately confirmed by CERN experiments. With one pulse directed at $+\alpha$ to axis Y and another at $-\alpha$ to the orthogonal axis X components of both pulses can be fully represented solely with orientations Y and X, because residuals cancel. Hence radiation in space can be fully expressed by equal radiations in six orientations of space. The creation of source-receiver axis limits the quantum orientations to 36 positions, 12 per plane with 5 or 10, if included are internal reflections converting the wave to the pulse. Two radiations displayed by 90° are needed to increase velocity from $\frac{1}{4} c_s$ to $\sqrt{2} \times \frac{1}{4} c_s$ and four to get c_s. For sun, radiation move along cubic orientations increasing velocity by $\sqrt{3} \times 1.1547005$ and require one repetition to obtain c_s. The basic step of force formation takes place between two planes of rotation of two Bohr structures displaced by 90° with probability of correct orientation in each plane being 1/6, the total probability 1/6×1/6 being increased by 3 events being possible in three planes of space and decremented by two orientations not reacting with the receiver. Thus the total number of quantum events or quantum rate per unit event is:

$$1/(3 \times 1/6 \times 1/6) - 2 = 10 \qquad (26.77)$$

Four steps, 4×10, over relativity correction 1.333 . . . derives 30 eV (1 eV at 300 K) expressing the total energy of Bohr atom. This is attributed with a new name **the quantum rate per event**. In 1D the ionization energy of Bohr structure formed by divalent impurities in NaCl crystal or lattice energy of NaCl crystal amount to 10 J or 10 eV/pair, which corrected for relativity derives $10/(1.11...^2 \times 1.0328 \approx 8$ eV measured by the author and quoted widely, Dekker [55, Table5.2]. Eight eV converted to atomic level over temperature squared because two atoms take place in the event derives 1D quantum:

$$h_s/10 = 8 \times 10^{-30}/(300 \text{ K})^2 \times (1+1/8) = 1.0 \times 10^{-34} \text{ J-s}$$

The classical SI value of $h_s/10$ is $h/4\pi$. To compare them 12.56 has to be decreased by two not reacting quanta with the receiver orientations and divided by relativity correction $1.1111^{1/2}$ deriving 10.018. Furthermore h_s comprises reflected energy converting waves to pulses so that $h = 1.51 h_s/2$ and $h/h_s = 0.755$ increased by $h/h_s = 1.1111^2 \times 1.1547^{1/2} = 1.0016$. The value 0.0016 expresses relativity factor over 20: $0.0328 \times 300/6000$, because the variance of 0.0004 from 0.0016 expresses $\frac{1}{2} k_s$.

A comparison between the classic SI μ_o and the model SI μ_{os} displays:

$$\mu_o = (1.0328^{4/5} + 1/5) \mu_{os}$$

indicating that μ_o refers to the radiation at the source including a component that is not reacting with the receiver. This is expressed in the minimum 1D model SI energy:

$$(1 \times 10^{-14})^{1/2} = 1/10^7 = 1/\upsilon_e = 10/(1/3 \ c_s) = \mu_{os}/10$$

And it's other derivatives

$$(1 \times 10^{-14})^2 = 1 \times 10^{-28} = \text{muo}10_s$$

$$\mu_{os}^5 \times 1/1000 = 1 \times 10^{-27}/ \ 1.111...^{1/2} = m_{ps}$$

$$(1 \times 10^{-14}) = (\mu_{os}/10)^2 = 1 \times 10^{-27}/1.111...^{1/2} = (\mu_{os}/10)^2$$

Triple energy $3.000\ 000\ 00 \times 10^{-14} = (1 \times 10^{-27})^{1/2}/1.111...^{1/2} = (m_{ps}^{1/2})/1.111...^{1/2}$

$$1/c_s^2 \times 1.111... = \mu_{os} \varepsilon_{os} \qquad (28.78)$$

$$(m_{ps}^{1/6})/1.111...^{1/2} = 30 \ \mu_{os}$$

All classical and model SI units express elementary product, sum or resultant event probability formation. SI units are not important at this level because kinetics has no units of distance or time. Events express intensity and frequency and if they are per atomic event only frequency.

26. Author's background

Author's started his electrical engineering studies in 1942 at the University of At. Andrews, and so his experience covered period of 71 years. Autor's father, col. Zygmunt ZOLEDZIOWSKI M.D., was an ophthalmologist in charge of a university hospital department and he took good care of author's education. The Great War II took us to Paris then to England, where the author passed his Scottish University entrance examination studying at the Edinburgh Academy. James C. Maxwell studied earlier at this school.

Three years of military service in Signals as radio operator in the First Polish Armored Division, the invasion of France, Belgium and Holland were followed by graduation from the Officers Signals School in Scotland. Then three years course at the College of Advanced Technology in Dundee, while being an external student of London University. College Diploma with commendation was rewarded in 1948. The author returned to Warsaw, his place of birth, to complete in 1951 M.Sc. Degree in Technology specializing in the High Voltage Technology. His M.Sc.Thesis dealt with measurements of partial discharges in dielectrics, published much later as two chapters in the book of Engineering Problems in Nuclear Research (CIBA). This study disclosed the importance of resonating frequencies.

In the period 1949-57 the author worked in the Central Institute of Electro technology, High Voltage Department in Warsaw being responsible for the cable section and he was part-time senior assistant at the Warsaw Technical University. The first job was the assembly of the 2.8 MV impulse generator, at that time the highest voltage source in the Central Europe. He designed doubling voltage of a transmission line based on measured corona losses. Then he took over the section of prophylactic testing of the electrical insulation in power systems. A mobile laboratory was built for tests on site in power systems. He took part in the manufacture of Schering Bridges, insulation testers, capacitive ratio moisture indicators for oil impregnated insulation and ionization detectors. He organized specialist courses and he was laureate of the 3rd Class Polish State Medal for the technical progress heading a group of engineers. His last post was the Scientist in charge of the Laboratory Equipment of the High Voltage Institute including the workshop with some 30 personnel.

He left Poland, because his wife refused to co-operate unethically with communist officials creating a personally dangerous situation in the Stalin period. The switchgear firm A. Reyrolle & Co in U.K. needed a specialist with electrical field experience. He was again working with minute electrical forces in oil impregnated insulation, with forces on the surface of insulators magnified by the moisture producing internal flashover and with field problems related to switchgear. Flashover of external polluted insulation was another topic he researched for many years.

In 1960 he was appointed a lecturer and then a senior lecturer at the College of Advanced Technology, which soon afterwards became the University of Salford. He collaborated also with Manchester University on the effects of pulsed, direct and alternating electrical current on respiration and on heart by electrically inducing muscular contraction, heart fibrillation and defibrillation. The three room laboratories, which he supervised, had beside several high voltage sources (including a bank of 20 μF 100 kV capacitors), spectrometers, pulse height analyzers, image intensifiers, microscopes, high speed oscilloscopes and ancillary equipment. His responsibility was to plan and provide research for visitors from abroad, for M.Sc. projects and annually for 2-3 final year students' projects. He supervised two Ph.D. students, while his lectures mainly dealt with electrical power systems. Electrical pulses in solids, in liquids and in gases were studied in detail. Sparks up to 6 m long were generated to model flashover of polluted insulation.

He is the author and co-author of four books, and of chapters in a book and of some 70 publications and contributions in Scientific Journals and at International Conferences. He had the privilege to be the co-chairman of the first International Conference on the flashover of polluted insulation at Madras in India. His name is in Dictionaries of International Biography 1976 and 1979 and in others. He was member of the New York Academy of Science. His research was facilitated by obtaining a long term contract from the Atomic Energy Authority providing a technician to carry experiments to study aging of epoxy resins subjected to high voltage impulses from a huge bank of 20 μF 100 kV capacitors. Relaxation characteristics were studied to assess absorbed charges in the solids. The logarithms of the relaxation currents displayed straight sections displaying time constants decreasing by a factor of about two. He was fascinated by this experimental result conflicting with the established theory of distributed relaxation times. He studied the topic intermittently for the next three decades. He was directed to concentrate his research on the forces in crystals and he studied relaxation and high field conduction together with his students and visitors. New theories are more likely to be studied in detail, if presented by a scientist with known background. Search procedure on Google with the name "S. Zoledziowski" following one of the topics of authors research or co-authors: citations, high voltage, epoxy resin, NaCl crystal, spark, oil, aging, dielectric constant, local field, gravitation, quantum, flashover, video, electric, probability, trees, impulse, insulation, power arc, magnetic, digital, citations, world scientist, paper, headgear, death, electric shock, rabbit, dog, fibrillation, Golinski, Calderwood, Lee, Sierota, Zdanowicz, Babula, Shibya, Slowikowski and others display over thousand citations repeating some of the original papers. This is an achievement since the author been officially retired for 30 years. The readers can check themselves the good standing of the author as a researcher.

After author's obligatory retirement he moved to Canada in 1983. He continued private unsupported research, while serving as volunteer in several Canadian and Polish Charitable Organizations and being recognized or mentioned on WEB. The University of Salford allowed the author to use the title of the Honorary Senior Lecturer. On 24 March/2012 he received Gold Army Medal awarded to foreigners by the Minister of Defense of Poland for his contribution to history of the Polish Army on the international level. Being a widower he filled his time in pursuing private research at home using international physical constants as reliable experimental data, the Internet and WEB pages for his sources.

The model presented is not based on exceptional ability or brilliance of the author, but on eleven years of hard unsupported work initially on unspecific topic, on a long life of research experience with minute electrical forces, an extensive experience of model building and on the author's stubbornness and curiosity. The presented concepts do not negate the classical SI system. They expand them. The age and health puts time pressure on the author to publish, before all possible assessments, corrections and editing have been made. Any short comments are welcome from the readers and may be included in the next edition, with authors being promised full acknowledgement and E-mail answer.